Sustainable Textiles: Production, Processing, Manufacturing & Chemistry

Series Editor

Subramanian Senthilkannan Muthu, Head of Sustainability, SgT and API, Kowloon, Hong Kong

More information about this series at http://www.springer.com/series/16490

Subramanian Senthilkannan Muthu
Miguel Angel Gardetti
Editors

Sustainability in the Textile and Apparel Industries

Sustainable Textiles, Clothing Design and Repurposing

Springer

Editors
Subramanian Senthilkannan Muthu
Head of Sustainability
SgT and API
Kowloon, Hong Kong

Miguel Angel Gardetti
Centro de Estudios para el Lujo Sustentable
Buenos Aires, Argentina

ISSN 2662-7108 ISSN 2662-7116 (electronic)
Sustainable Textiles: Production, Processing, Manufacturing & Chemistry
ISBN 978-3-030-37928-5 ISBN 978-3-030-37929-2 (eBook)
https://doi.org/10.1007/978-3-030-37929-2

This Springer imprint is published by the registered company Springer Nature Switzerland AG
The registered company address is: Gewerbestrasse 11, 6330 Cham, Switzerland

Preface

Eighty percent of the impacts are decided at the design stage, and hence it is crucial to implement sustainability aspects and strategies in the design stage. The crux of the book is to deal around the sustainability aspects in textile design and clothing design. This volume addresses many important topics such as innovative sustainable apparel design: application of CAD and upcycling process, repurposing of fashion, etc. It begins with the work titled *Sustainable Textile Designs Made from Renewable Biodegradable Sustainable Natural Abaca Fibers* developed by Feristah Unal, Ozan Avinc, and Arzu Yavas; it gives detailed information regarding abaca fiber, its production, its chemical structure, its physical and mechanical properties, and also sustainable textile designs made from sustainable abaca fiber.

The following chapter, *Analysis of Zero Waste Patternmaking Approaches for Application to Apparel*, written by Ellen McKinney, Sunhyung Cho, Ling Zhang, Rachel Eike, and Eulanda Sanders, conducts a systematic review on zero waste patternmaking and its application to the apparel industry.

Then, Mercy Rugedhla, Disele P. L. P, Moalosi R., and Fidzani L.C. develop the chapter titled *Factors that Affect Sustainability in the Textile Design Industry in Kadoma, Zimbabwe.* They explore the need to develop sustainable textiles and materials in Kadoma, Zimbabwe, and outline the factors that affect sustainability in textile designing companies in Kadoma, explaining how the textile industry can gain sustainability through the use of sustainable textile production and people-centered designing methods.

Subsequently, in chapter *Contributions to Sustainable Textile Design with Natural Raffia Palm Fibers,* by Feristah Unal, Arzu Yavas, and Ozan Avinc, the purpose of writing is to detail the structure, properties, production methods, and end use applications and designs of the raffia palm fibers.

Moving on to the next chapter, *Innovative Sustainable Apparel Design: Application of CAD and Redesign Process,* Chanmi Hwang and Ling Zhang present insights for designers, researchers, and educators seeking innovative ways to practice redesign activities within sustainable apparel design methods at micro- and macro-levels of the apparel industry.

The following chapter entitled, *Bacteria Working to Create Sustainable Textile Materials and Textile Colorants Leading to Sustainable Textile Design*, by Fatma Filiz Yıldırım, Arzu Yavas, and Ozan Avinc, explores in detail the possibilities and use of sustainable bacterial cellulose for textile substrates and sustainable bacterial pigments as sustainable textile colorants.

Later, R. Rathinamoorthy, in the chapter entitled *Sustainable Clothing Designs for Fashion: Design Strategies and Its Implementation Possibilities*, analyzes the importance of implementing the sustainable strategies in design stage along with the role designers. It also discusses applications of design for disassembly (DfD), design for zero waste, design for longevity, design for co-design, design for end of life (EoL), etc. in the fashion industry.

In the following chapter, *Contribution of UV Technology to Sustainable Textile Production and Design,* Feristah Unal, Ozan Avinc, Arzu Yavas, Hüseyin Aksel Eren, and Semiha Eren explore ultraviolet (UV) and UV technology, usages, contributions, and production efficiency advantages to sustainable textile production and design (e.g., the use of UV technology in decolorization and purification of textile wastewaters, pretreatment and surface modification processes prior to coloration processes, UV curing process, and pilling problem prevention).

In the next chapter, *Repurposing Design Process* by Rachel Eike, the term "repurposing" is used to describe the process that utilizes discarded textiles to create new fashion (textile-based) products. It provides and analyzes the four repurposing levels through case study application to detail the creative design.

Dr. Sheetal Jain in his chapter, *Doodlage: Reinventing Fashion via Sustainable Design*, explores the emergence of an alternative notion to luxury, away from the ubiquitous luxury conglomerates and their dominant luxury brands.

Then, *Sustainability in Textile Design with Laser Technology*, developed by Feristah Unal, Arzu Yavas, and Ozan Avinc, analyzes the contribution of laser technology, as a dry and clean method, to sustainable textile production and design.

Finally, Veena Rao, Rajesh Kumar, Aysha Shaima, and Venkatachalam A, in their chapter *University Intervention in Inculcating Design Practices for Sustainable Fashion*, explore the model and role that involves stakeholders and institutions of higher education on ecological sustainability and sustainable design practices among the budding designers.

Kowloon, Hong Kong — Subramanian Senthilkannan Muthu
Buenos Aires, Argentina — Miguel Angel Gardetti

Contents

Sustainable Textile Designs Made from Renewable Biodegradable Sustainable Natural Abaca Fibers

Feristah Unal, Ozan Avinc, and Arzu Yavas

Abstract Nowadays, interest in sustainable and renewable-origin textile fiber usage on textile design and fashion is increasing with the increasing environmental problem. Therefore, sustainable textile designs made from renewable biodegradable, sustainable, and natural fiber alternatives has become a recent trend. The textile industry is now turning its face to biodegradable and environmentally friendly textile resources for their designs to leave a livable world for future generations. In the recent years, abaca fiber which is a renewable, biodegradable, sustainable, eco-friendly, and natural fiber alternative for the textile designs comes into prominence with their ecological properties and superior daily usage performance. Indeed, abaca fibers are biodegradable and renewable natural fibers and are listed among the sustainable textile resources. The abaca plant endemic in Asia is usually grown in the Philippines, Ecuador, Borneo, and Sumatra. Abaca fibers display a wide range of different designed applications owing to their promising properties. In many years, abaca fibers have been used for ship navy, rope, fishing net, etc. Also, the use of abaca fibers as paper is quite common. In the recent years, natural sustainable textile fibers have been preferred not only in apparel sector but also in technical textile applications such as composite material designs and products. For instance, many different designed products such as garments, clothing, shawls, curtains, furnitures, carpets, mats, hats, coasters, pillows, packaging materials, handicraft products, decoration materials, decorative lamps and lampshades, flower-wreath ornaments, multipurpose baskets, tabletop accessories, sports equipments, ropes, fishing nets, bags, papers, money papers, cigarette filter papers, wrapping papers, tea bags, cardboards, chipboards, liquid filtration filters, and various industrial textile materials such as medical articles and composites for different purposes can be made from abaca fibers. Indeed, nowadays, abaca fibers can be utilized in the place of glass fibers in automotive industry technical textiles and composite applications. In this chapter, information regarding abaca fiber, its production, its chemical structure, and its physical, chemical, and mechanical properties and sustainable textile designs made from sustainable abaca fibers, is given in detail.

F. Unal · O. Avinc (✉) · A. Yavas
Textile Engineering Department, Engineering Faculty, Pamukkale University, Denizli, Turkey
e-mail: oavinc@pau.edu.tr

S. S. Muthu, M. A. Gardetti (eds.), *Sustainability in the Textile and Apparel Industries*, Sustainable Textiles: Production, Processing, Manufacturing & Chemistry, https://doi.org/10.1007/978-3-030-37929-2_1

Keywords Abaca fiber · Manila hemp · Design · Sustainable · Biodegradable · Renewable · Sustainability · Natural fiber · Abaca

Introduction

It is known that fibers are first used to meet the dressing requirements of people and their use area is increasing day by day (Bellmann et al. 2005; Mert and Çopur 2019). However, environmental pollution has also started to increase with the increase in the utilization area of textile fibers. Therefore, production without consuming natural resources and damaging the environment has become important all over the world. This means sustainability. In other words, sustainability means ensuring the continuity of diversity and productivity with the available resources and not to jeopardize the quality of life of future generations. Interests in sustainable, recovery, biodegradable, and environmentally friendly fibers are also increasing with the increasing awareness of production and consumption in the world. Therefore, nowadays, interest in sustainable and renewable-origin textile fiber usage on textile design and fashion is increasing with the widely known increasing environmental problems. Consequently, sustainable textile designs made from renewable biodegradable, sustainable, and natural fiber alternatives have become a recent trend. The textile sector is now turning its face to biodegradable and environmentally friendly textile resources for their designs to leave a livable world for future generations.

The most important sustainable alternatives are plant-based textile fibers which belong to the natural fiber group (Korkmaz 2015; Doğan 2011; Rahman et al. 2009; Santoso et al. 2017). More than 1000 fiber plants in the world are used today for different purposes. Many of these plant fibers still meet local needs, even if they are not economically important (Bellmann et al. 2005; Mert and Çopur 2019). In the recent years, abaca fiber which is a renewable, biodegradable, sustainable, eco-friendly, and natural fiber alternative for the textile designs comes into prominence with their ecological properties and superior daily usage performance. The abaca plant, which has the appearance of a palm tree, is also known as a fiber plant (Baser 1992). These plant fibers, also known as Manila hemp, are also known as cord fibers due to their common usage for cords (Saleeby 1915; Jarman 1977). Indeed, many years, abaca fiber has been used for ship navy, rope, fishing net, etc. Abaca natural fibers are grouped in the leaf fibers section (Vijayalakshmi et al. 2014; Future Fibres 2019; Mwaikambo 2006; Del Río and Gutiérrez 2006; Lewin 1998; Paridah et al. 2015; Bledzki et al. 2007). Abaca fibers are obtained from leaf sheaths which form the false stem and stem of the abaca plant together with the central core (Saleeby 1915; Jarman 1977; Saragih et al. 2018). Abaca plant, which attracts attention with its resemblance to banana plant, is a nonedible banana species (Saleeby 1915; Jarman 1977; Blackburn 2005; Mert 2009; Saragih et al. 2018; Kumar et al. 2017).

However, it is known that the yield of the fibers obtained from the edible banana plant is less than half the yield of the fibers obtained from the abaca plant (Saleeby 1915; Jarman 1977; Saragih et al. 2018).

There are more than 400 varieties of abaca plant, which are cultivated in humid areas such as the Philippines, and only 20 of them are of commercial importance (Baser 1992; The Good News Pilipinas Team 2016; Saragih et al. 2018). Abaca fiber production in the world is 100–140 thousand tons, and 90–94% of the fiber production is obtained from the Philippines (Baser 1992; Lewin 1998; Yazıcıoğlu 1999). Abaca fiber is also produced in Borneo and Sumatra (Lewin 1998). The majority of abaca fibers produced in the Philippines remain in the domestic market, while raw fiber exports are carried out by Ecuador (Amatta 2017; FAO Statistics 2016; Chang 2013). Abaca fiber and manufacturing industry exports declined due to the decrease in global demand between 2011 and 2013 and recovered in 2014. Generally, abaca fiber is exported to the European Union, Japan, and China (Amatta 2017; FAO Statistics 2016; Chang 2013). Abaca fibers displaying high strength and extremely good fiber length (varying between 2 and 3 m) are now used as industrial raw materials in the automotive industry (Wühhlbauer and Hutter 2002; Youtube 2016; Saleeby 1915; Suantara and Oktaviani 2015; Onggo and Judawisastra 2018). In the recent years, natural fibers have been preferred as industrial raw materials in the automotive industry. Because of their ecological advantages, in composite structures, natural fibers are now used in the place of synthetic fibers. This has made the use of natural fibers such as abaca fibers in different industrial areas widespread (Wühhlbauer and Hutter 2002; Youtube 2016; Suantara and Oktaviani 2015; Onggo and Judawisastra 2018). In this sense, the use of abaca fiber-reinforced composite structures is also becoming widespread. Composite structures reinforced with abaca fibers have been used for floor protection for passengers and patented by Daimler-Chrysler researchers (Wühhlbauer and Hutter 2002; Agung et al. 2011; Hadi et al. 2011; Çavdar and Boran 2016). The fact that renewable abaca fibers are long fibers and have a porous structure with high strength and thinness shows that they can be used for also paper production (Del Rio et al. 2006; Blackburn 2005). Abaca fibers utilized in the production of special papers such as tea and coffee bags, money papers, and cigarette filter papers generally have a pale white color (Future Fibres 2019; Del Rio et al. 2006; Blackburn 2005; Kaya and Yazicioğlu 1992). Ropes, rugs, furnitures, and carpets are also made from abaca fibers (Youtube 2016; Shibata et al. 2002; Thakur and Thakur 2014; Suantara and Oktaviani 2015). It is also widely utilized in the production of handicrafts. Furthermore, abaca fibers are used in the manufacture of marine ropes and fishing industry because of its resistance to moisture and saltwater (Baser 1992; Agung et al. 2011; Del Rio et al. 2006; Rahman et al. 2009; The Good News Pilipinas Team 2016; Blackburn 2005; Kaya and Yazicioğlu 1992; Mert 2009; Suantara and Oktaviani 2015; Kumar et al. 2017). Another area of application is the use of abaca fiber wastes as organic fertilizer (Future Fibres 2019).

Although the abaca fibers display many different usage applications, its use in the industrial area is still not sufficient because of the lack of proper and sufficient equipment and machinery for the production and processing of abaca fibers

(Wühhlbauer and Hutter 2002). Considering this, the production, processing, and production conditions of abaca fibers need to be improved. In this sense, many different projects, collaborations, and feasibility studies were initiated and carried out. For instance, technical cooperations are carried out with the experienced companies of the automotive industry. Moreover, projects are supported by the German Agency of GTZ, as well as scientific institutions and local experts such as the National Abaca Research Center at Leyte State University. Environmental education, sustainable habitat conservation, and ecological assessment of the production and processing of abaca (Manila hemp) are also included in the project by the European Nature Heritage Fund (EURONATURE) (Wühhlbauer and Hutter 2002).

In the recent years, natural sustainable textile fibers have been preferred not only in apparel sector but also in technical textile applications such as composite material designs and products. In this chapter, information regarding abaca fiber, its production, its chemical structure, and its physical, chemical, and mechanical properties and sustainable textile designs made from sustainable abaca fibers, is given in detail. Firstly, brief information about abaca plant, its history, and cultivation is given. Then, the methods of obtaining abaca leaf fibers were classified. And then, chemical structure and physical, chemical, mechanical properties of abaca leaf fibers are explained in detail. Finally, the contribution of sustainable abaca fibers to textile design and textile designs made from sustainable abaca fibers and different application areas of abaca fibers are comprehensively reviewed.

Abaca Plant

Abaca is an Asian plant (Musaceae 2019; Saragih et al. 2018). It is generally cultivated in humid tropical regions such as Ecuador, the Philippines, and eastern Indonesia. It is also cultivated in Borneo and Sumatra. Abaca is also known as Manila hemp since this plant has been grown in Manila which is the capital of the Philippines (Vijayalakshmi et al. 2014; Future Fibres 2019; Mwaikambo 2006; Lewin 1998; Paridah et al. 2015; Bledzki et al. 2007; FAO Statistics 2016; Wühhlbauer and Hutter 2002; Agung et al. 2011; Hadi et al. 2011; Del Rio et al. 2006; Cai et al. 2015; Spencer 1953; Wester 1927; Bledzki et al. 2009; Berkley et al. 1949; Saleeby 1915; Liu et al. 2012; Parrott-Sheffer 2019; The Good News Pilipinas Team 2016; Blackburn 2005; Kaya and Yazicioğlu 1992; Mert 2009; Yazıcıoğlu 1999; Suantara and Oktaviani 2015; Saragih et al. 2018. Abaca is a flowering plant belonging to the family Musaceae (Musaceae 2019; Saragih et al. 2018). The scientific name of the abaca plant is *Musa textilis* (Fig. 1) (Vijayalakshmi et al. 2014; Future Fibres 2019; Mwaikambo 2006; Lewin 1998; Paridah et al. 2015; Bledzki et al. 2007; FAO Statistics 2016; Wühhlbauer and Hutter 2002; Agung et al. 2011; Hadi et al. 2011; Del Rio et al. 2006; Cai et al. 2015; Spencer 1953; Wester 1927; Bledzki et al. 2009; Berkley et al. 1949; Liu et al. 2012; Parrott-Sheffer 2019; The Good News Pilipinas Team 2016; Blackburn 2005; Kaya and Yazicioğlu 1992; Mert 2009; Yazıcıoğlu 1999; Suantara and Oktaviani 2015; Saragih et al. 2018). Abaca

Fig. 1 Abaca plant appearance (Wikimedia Commons 2018a) and its fruit (Wikimedia Commons 2018b)

plant is formed by hybridization of 30–50 species, and in general its structure is similar to banana varieties (Vijayalakshmi et al. 2014; Future Fibres 2019; Mwaikambo 2006; Lewin 1998; Paridah et al. 2015; Bledzki et al. 2007; FAO Statistics 2016; Wühhlbauer and Hutter 2002; Agung et al. 2011; Hadi et al. 2011; Del Rio et al. 2006; Cai et al. 2015; Spencer 1953; Wester 1927; Bledzki et al. 2009; Berkley et al. 1949; Liu et al. 2012; Blackburn 2005). Indeed, it is also similar to banana plant. Abaca is known as nonedible banana plant (Saleeby 1915; Jarman 1977). It is thinner than banana plant and its leaves are smaller (Vijayalakshmi et al. 2014; Spencer 1953; Saleeby 1915).

The plant age of the abaca plant (Manila hemp), which is in the perennial herbaceous plants group, ranges from 8 to 20 years on average (Baser 1992; Lewin 1998). Plant height can grow up to 8–22 m (Agung et al. 2011; Spencer 1953). Abaca plant consists of 12–30 stalks spreading from the same stem (Lewin 1998; Category Archives 2019; Parrott-Sheffer 2019). The stem of the plant is formed by overlapping more than one leaf (Spencer 1953; Parrott-Sheffer 2019). There is 90% water and 2–5% fiber in the stems of the plant (Lewin 1998; Spencer 1953; Yazıcıoğlu 1999). The abaca plant grows from the central core and consists of tightly wrapped crescent-shaped sheaths (Lewin 1998; Spencer 1953). The diameter of the stem decreases as we move toward the upper parts of the plant (Lewin 1998; Spencer 1953). The diameter of the stem measured in the upper parts of the abaca plant, which is known to have a large stem, is between 4 and 6 in (Spencer 1953). Plants have irregular line crossings (Spencer 1953). The abaca plant has a mucronate leaf type. Mucronate shows that the leaf ends of the plant have a hard spiny structure.

The leaves of abaca plant are usually overlapping and drooping. Leaf length varies from 3 to 6 m (Spencer 1953). Abaca plant has a sharp and thin leaf type and its leaf width is 30 cm (Vijayalakshmi et al. 2014; Lewin 1998; Parrott-Sheffer 2019). Generally, the abaca plant is known to have a narrower leaf width than the banana plant (Spencer 1953). The fruit of the abaca plant is not edible because it contains many black seeds (Spencer 1953; Wester 1927; Parrott-Sheffer 2019). If the abaca plant is not cut for fiber production after its ripening, its roots are broken down and the plant dies (Spencer 1953; Wester 1927). According to FAO (Food and Agriculture Organization) data, in 2014, abaca production was 8.4 thousand tons only in Ecuador, and 66.0 thousand tons abaca fiber production was recorded in the Philippines. And the abaca plant production in other countries was 2.3 thousand tons (FAO Statistics 2016). While the entire abaca plant produced in Ecuador is used as fiber, only 10 thousand tons of abaca plant grown in the Philippines is used for fiber because the majority of which is reserved for paper pulp (FAO Statistics 2016). Most of the produced abaca fiber is exported especially to Europe, Japan, and the USA (Amatta 2017; FAO Statistics 2016).

History of Abaca Plant

Abaca plant (*Musa textilis*), which is commercially known as Manila hemp and native to the Philippine islands, is a species of tree-like plant (Berkley et al. 1949; Saleeby 1915; Suantara and Oktaviani 2015). The abaca plant, which has been grown up as a commercial product in the Philippines since the sixteenth century, began to be grown in Manila in the 1500s (Amatta 2017; Youtube 2016; Berkley et al. 1949). It is known as the most grown fiber in the early twentieth century (Saragih et al. 2018). It has been used as a rope in ship navy in the 1800s since it has a very good strength and is highly resistant to saltwater, and it is still being used as rope in ship navy today (Youtube 2016; Berkley et al. 1949; Saleeby 1915). In the early 1880s, after being aware of its commercial usefulness, it began to be called "Manila hemp" by the British and Americans, but Manila hemp name is not the correct word for abaca plant because abaca is completely different from the hemp plant (Lewin 1998; Berkley et al. 1949; Parrott-Sheffer 2019; Kaya and Yazicioğlu 1992). In the 1920s, the abaca plant was cultivated on 485,000 ha of land, and a vast majority of 177,312 metric tons of abaca, worth $29,950,000, were exported to the USA (Wester 1927). In 1925, six types of abaca plants were sent for planting from the Philippines to Panama by the US Department of Agriculture (Berkley et al. 1949; Parrott-Sheffer 2019). These plants were grown by United Fruit Co. in an area in the Changuinola Valley near Almirante in the Republic of Panama, and the cultivation of the original experimental plants was extended to approximately 2000 ha until 1941. At the end of 1941, plans were made for further expansion of abaca cultivation, and in 1942, abaca planting was started (Berkley et al. 1949). In the early 1942s, due to insufficient abaca fibers produced in the Philippines, abaca production started in the western hemisphere (Berkley et al. 1949). In accordance with the contract with the US government, all the operations from growing the abaca plant to fiber production were carried out by United Fruit Co. At the end of the project, abaca farming was expanded to approximately 28,000 ha in Panama, Costa Rica, Honduras, and Guatemala (Berkley et al. 1949). It is known that abaca fiber grown in Central America is different from abaca fiber grown in the Philippines due to the number of factors. Abaca fiber production has two varieties, Bungulanon and Maguindanao in Central America, while many varieties are grown in the Philippines (Berkley et al. 1949). Today the Philippines is still the world's largest abaca producer (Parrott-Sheffer 2019).

Cultivation and Agriculture of Abaca Plant

The main origins of abaca plant cultivation are Mindanao and Visayas and Bicol regions in the southern part of the Philippines, where rain forests and very humid atmospheric conditions exist (Baser 1992; Berkley et al. 1949; Blackburn 2005) and abaca plants grow well on very rich, loose, loamy soils with good drainage (Parrott-Sheffer 2019). The best regions for growing abaca plants are those with humid tropical weather and temperatures between 22 and 28 °C. In addition, rainfall from 1800 to 2500 mm per year is required in areas where abaca plant is cultivated. Besides, the optimum height of the area should be between 100 and 140 m above sea level (Blackburn 2005). Rain and sunlight are known as the main factors in abaca production. When more sunlight is combined with a lack of rain, the development of the abaca plant is adversely affected (Blackburn 2005). Abaca plant grows better in shaded areas and fiber yield is better (Kaya and Yazicioğlu 1992). The abaca plant can commercially produce fibers for 15–20 years in a well-grown environment (Blackburn 2005; Mert 2009).

Abaca cultivation is generally carried out at the beginning of the rainy season, i.e., winter season (Parrott-Sheffer 2019; Blackburn 2005). The average plant age ranges from 8 to 20 years (Baser 1992). The harvesting of the abaca plant takes place when the plant begins to bloom. It is usually carried out between 18 and 24 months from the first shoots (Baser 1992; Category Archives 2019; Parrott-Sheffer 2019; Blackburn 2005; Kaya and Yazicioğlu 1992; Mert 2009). It is known that the fibers have high strength during this period (Kaya and Yazicioğlu 1992). The harvesting period of the abaca plant, which is harvested 2 or 4 times a year, is carried out at 3–4-month intervals (Youtube 2016; Category Archives 2019; Parrott-Sheffer 2019; Blackburn 2005; Kaya and Yazicioğlu 1992). The harvesting process can be carried out in two ways as amputate the apex of the plant and overturning. The cut leaves are put in the bulk and prepared for the next step (Category Archives 2019; Kaya and Yazicioğlu 1992).

In the next step, the abaca fibers are separated from the leaves by various fiber extraction methods. Very bright and robust fibers are obtained. Color of the abaca fibers can be between white and brown (Baser 1992). Important factors in abaca plant cultivation are disease resistance of the plant, yield, and vegetation of the grown region (Wühhlbauer and Hutter 2002; Saleeby 1915). These factors affect the fiber yield and fiber production cost significantly. Since the strength of the fiber depends on the maturity stage of the plant, it is very important to harvest at the optimum time (Wühhlbauer and Hutter 2002; Saleeby 1915). When the cultivation areas, production yield, and consumption status of abaca plant are examined, it is stated that there was 252,577 ha cultivation area in the 1980s, but day-to-day cultivation area decreased to 147,920 ha. However, there was no significant variability in fiber yields, indicating an increase in average fiber yields (Mert and Çopur 2019). It is known that the abaca industry in the Philippines has been steadily going on for the last 10 years (Amatta 2017). Nevertheless, the limited processing activities of most abaca plantations and the lack of adequate technical knowledge in the proper extraction and processing of abaca fibers should be improved in order to increase the abaca fiber productivity (Amatta 2017; Saleeby 1915).

According to the United Nations Food and Agriculture Organization, abaca growing is reported to be as an important option for erosion control and biodiversity rehabilitation. It was recorded that the erosion and sedimentation problems in the coastal areas were minimized when planting with abaca (Sustainable fibres: what is abaca? 2016).

Abaca Leaf Fiber Production Methods

Abaca fibers, also known as Manila hemp, are stiff fibers that are completely different from the real hemp (Vijayalakshmi et al. 2014; Mwaikambo 2006; Lewin 1998; Parrott-Sheffer 2019; Kaya and Yazicioğlu 1992). Sustainable, biodegradable, renewable, and natural abaca fibers are included in the group of plant fibers, and abaca fiber is a leaf fiber (Vijayalakshmi et al. 2014; Future Fibres 2019; Mwaikambo 2006; Del Río and Gutiérrez 2006; Lewin 1998; Paridah et al. 2015; Bledzki et al. 2007; Shibata et al. 2002; Rahman et al. 2009). Abaca fibers are obtained from the leaf sheaths of the abaca plant. Sheaths per root are divided into four groups. These are known as outer sheaths, sheaths adjacent to the outer sheath, middle sheaths, and inner sheaths (Lewin 1998). There are three different layers in each leaf sheath (Lewin 1998; Mert 2009; Yazıcıoğlu 1999). The outer layer comprises stacks of fibers disseminated throughout the soft tissue matrix, including the epidermis. The middle layer is composed of fibrovascular tissues carrying water. The inner layer contains soft cellular tissues (Lewin 1998; Yazıcıoğlu 1999). Fibers with different degrees and qualities can be produced from different fiber sheaths within a single root (Lewin 1998; Mert 2009; Yazıcıoğlu 1999). The strongest, most resistant fibers are obtained from the outer sheaths of the abaca plant (Parrott-Sheffer 2019; Mert 2009; Yazıcıoğlu 1999). The abaca plant harvesting in order to obtain abaca fibers and then abaca yarns are shown in Fig. 2.

The trunks of the abaca plant are cut at an angle and the leaves are removed and different fiber production methods are available today (Lewin 1998; Mert 2009; Yazıcıoğlu 1999). Substances such as hemicellulose, lignin, and pectin on the abaca

Fig. 2 Harvesting of the abaca plant and separation into sheaths (Wikimedia Commons 2018c), manual separation of fibers (Wikimedia Commons 2018d), drying (Wikimedia Commons 2018e), and abaca fiber (Wikimedia Commons 2019), respectively

fiber are removed with these applied different methods (Baser 1992; Mwaikambo 2006). Thus, a smooth, fairly strong, and elastic fiber with a uniform diameter can be obtained (Mwaikambo 2006; Agung et al. 2011; Spencer 1953). Abaca leaf fibers become a superior fiber with high tensile and folding strength, floatage, high porosity, resistance to saltwater, and fiber length up to 3 m (Vijayalakshmi et al. 2014; Future Fibres 2019; Wühhlbauer and Hutter 2002; Del Rio et al. 2006; Bledzki et al. 2009; Saleeby 1915; Parrott-Sheffer 2019). The best abaca fiber is described as a thin, bright, light beige color and a very strong fiber (Vijayalakshmi et al. 2014; Future Fibres 2019; Saleeby 1915). Moreover, abaca fiber is known as the most durable fiber among natural fibers (Vijayalakshmi et al. 2014; Future Fibres 2019; Saleeby 1915; The Good News Pilipinas Team 2016; Mert 2009).

Abaca Leaf Fiber Production with Retting Process

The woody parts of the plant must be cleaned to obtain fiber from the abaca plant. The oldest abaca leaf fiber production method is the retting method. The retting process makes it easier to separate the fiber bundles from the woody parts of the plant. Moreover abaca fiber production with retting process is the whole of the chemical and biological processes applied to remove the noncellulosic materials from the fiber (Mwaikambo 2006; Farnfield and Alvey 1975; Paridah et al. 2011). In this method, the harvested leaves are kept in an area or under water for 2–3 weeks. In this way, pectinase substances that bind the fiber with other plant tissues are softened, and pectinase substances are removed by the action of microorganisms (Paridah et al. 2011). There are two main conventional retting types: dew retting and water retting.

In the dew retting method, plant stems are left to rot in the field. In this method, the fibers are separated from the stems without any deterioration in the quality of the fibers (Mwaikambo 2006; Paridah et al. 2011; Van Sumare 1992; Tarakçıoğlu 1978). In this process, which usually is made with native soil fungi, foreign substances such as pectin and hemicellulose on the fiber are removed by free *polygalacturanase* (PGase) and *xylanase* fungi (Paridah et al. 2011). Although dew retting is the most popular method in Europe, geographic location is very important for the implementation of this method. The fibers produced by the dew retting method are generally coarser and of lower quality than those produced by the water retting technique (Mwaikambo 2006; Paridah et al. 2011; Van Sumare 1992; Tarakçıoğlu 1978).

In the water retting process, the leaves of the abaca plant need to be wetted. For this reason, it is considered to be appropriate to apply this method on the riverside (Mwaikambo 2006; Paridah et al. 2011). This process is carried out in an aqueous environment, and anaerobic, pectinolytic bacteria provide the disintegration of pectic substances and the release of the abaca fibers (Paridah et al. 2011). This is an expensive method because too much clean water is required to achieve this process, but high-quality fibers are produced. The fermentation wastes released to the environment after this process are quite high (Mwaikambo 2006; Paridah et al. 2011; Tarakçıoğlu 1978).

The treatment time changes from 14 to 28 days in both methods (Paridah et al. 2011; Tarakçıoğlu 1978). There are also chemical and enzyme retting methods applied to abaca fibers. These methods require more dominance and control than dew and enzyme retting methods (Paridah et al. 2011). Hydrochloric acid (HCl) and alkali are applied with normal pressure or high pressure in the retting process with chemicals. It is neutralized by washing after the treatment (Tarakçıoğlu 1978). It was reported in the literature that the tensile strength of fibers obtained from the retting processes using sodium hydroxide and sodium benzoate is lower than this of fibers obtained from the wetting method in water. According to the studies in the literature, it is known that enzymatic methods are relatively better than chemical methods and fiber quality leads to better and less environmental mass. In addition, chemical retting processes that require very high energy tend to damage the fibers (Paridah et al. 2011). The method of retting with enzymes is mainly carried out by pectic enzymes produced by bacteria. Bacteria multiply and produce extracellular pectinases during retting processes. Thus, the pectin in the fiber structure is dissolved, and abaca fibers are separated from the cortex surrounding the fibers. Today, with the development of biotechnology tools, such enzymes could be produced commercially, making the enzymatic retting processes more popular for the manufacturing of long abaca fibers (Paridah et al. 2011).

Fig. 3 Manually scraping process the abaca fibers (Wikimedia Commons 2018f)

Abaca Leaf Fiber Production with Mechanical Processes

In order to obtain fibers from the leaf sheaths which are separated after the harvesting of the abaca plant, a cleaning process called scraping is performed. This cleaning process carried out by scratching of raw material can be done manually or with the help of machines (Category Archives 2019; Kaya and Yazicioğlu 1992; Yazıcıoğlu 1999; Suantara and Oktaviani 2015). The most commonly used scraping methods in the Philippines are known as manual scraping and spindle scraping. The manual scraping method is a simple but laborious method performed entirely manually. Abaca leaf sheaths which are cut into scraps, 50–80 mm wide, 0.6–2 m long, are clamped between a wooden block and a serrated knife, and plant leaves are scraped off by dragging (Fig. 3) (Vijayalakshmi et al. 2014; Lewin 1998; Category Archives 2019; Saleeby 1915; Parrott-Sheffer 2019; Blackburn 2005). Thus, the abaca fibers are scraped from the epidermis and parenchyma (Mert 2009).

Low scraping density reduces the fiber quality. This method, which requires intensive effort, has been replaced by scraping mechanisms developed to increase productivity by minimizing effort (Lewin 1998; Kaya and Yazicioğlu 1992). Another method used in the Philippines is spindle scraping method. In this method, the abaca fibers are obtained by winding them on a conical shaft actuated by an electric motor. The obtained abaca fibers are then naturally dried under the sun (Category Archives 2019).

The abaca fibers obtained by this method are whiter and brighter than the abaca fibers obtained by the manually scraping method (Category Archives 2019). In order to achieve the desired quality of the abaca fiber, machines have been designed in such a way that the density of the blade teeth is higher and the applied pressure is increased. This method, which is performed entirely by machines, is called decortication. This method is used in part of Central America, Indonesia, and the Philippines to separate abaca fiber bundles from the leaves and bottom of fiber plants (Lewin 1998; Liu et al. 2012; Peralta 1996; Yazıcıoğlu 1999). The fibers obtained by this method are called *Deco* (Mwaikambo 2006; Lewin 1998; Yayock et al. 1988). The machine used to implement this method is known as a decorticator machine (Mwaikambo 2006; Yayock et al. 1988). The decorticator machine consists of a 30 m long conveyor belt, a crusher press, a pair of crusher rollers, a rope belt, and a decorticator. The leaves are conveyed to the crushing press by the conveyor belt. It then passes between the crushing rollers. With the set of swivel wheels with blunt blades, the crushed leaves are conveyed to the decorticator by a rope conveyor. First, half of the leaf is subjected to decortication and then the other half is treated with the same process. Thus, the woody structure on the abaca fibers is removed. The leaves which exposed decortication pass to the scrubbing machine through which the fibers are unwound along another pair of pressing rollers, where the dirty, mixed ends are cut with scissors. Properly cleaned abaca fibers are obtained with these machines, which provide an intact blade system and a pressure of 20,000 kN/m^2. 16,000–35,000 kg per hour abaca leaves can be processed by decortication method (Vijayalakshmi et al. 2014; Mwaikambo 2006; Lewin 1998; Yayock et al. 1988; Peralta 1996).

Fiber yield is known to be approximately 2–3% of the abaca plant weight. Between 200 and 800 kg of abaca fiber is produced from abaca plant grown per hectare of land (Vijayalakshmi et al. 2014; Lewin 1998; Peralta 1996; Yazıcıoğlu 1999). Another machine developed for this purpose is known as raspador (Mert 2009). The raspador machine, which was used before decorticator machines, is still known to be used in some areas today. However, the decorticator machine is more widely used in the fiber production because it is much faster and automatic than the raspador machine (Mert 2009). The amount of fiber obtained from an abaca leaf may vary according to the manufacturing method. The abaca fiber is obtained up to 1.2–1.5% of a leaf according to the manual scraping method, 1.5–2.5% with semi-mechanical tools or machines, and 3–4% of the scraping method with decortication machine (Mert 2009; Yazıcıoğlu 1999). After washing, the fibers, which were scraped by these mechanical methods, are dried in semi-shaded areas and under the sun or by a hot air dryer at 100 °C (Mwaikambo 2006; Lewin 1998; Category Archives 2019; Saleeby 1915; Yayock et al. 1988; Peralta 1996; Kaya and Yazicioğlu 1992; Mert 2009). If the fibers are not well cleaned, the fiber color turns brown when dried in the sun. Furthermore, it is known that there is no reduction in the strength of the dried fibers immediately after scraping, and the fibers even have a brighter appearance (Kaya and Yazicioğlu 1992; Mert 2009). The dried fibers are combined and classified in various degrees according to their color, fineness, length, strength, cleanliness, and processing method (Mwaikambo 2006; Lewin 1998; Category Archives 2019; Saleeby 1915; Yayock et al. 1988; Peralta 1996; Kaya and Yazicioğlu 1992; Mert 2009). The official standard qualities of the fibers depend on the extraction form, manual scraping, spindle scraping, and decortication (Category

Archives 2019). And then abaca fibers are baled according to their qualities and classification types.

Gum Removal Process from Abaca Leaf Fibers

The fiber extraction process, which starts with abaca plant harvesting, continues with the degumming process after the retting process. Degumming process, which can be applied to similar leaf fibers, is carried out by chemical and biological processes applied on abaca fibers. In this way, the adhesive materials holding the fibers together are removed from the abaca fiber and the abaca fiber bundles are separated. Generally, the abaca fibers are treated with hot soap or alkali solution (Cai et al. 2015; Boopathi et al. 2012; Suantara and Oktaviani 2015). Internal microstructure and tensile properties of abaca fibers, which were treated with alkaline, were investigated. Structure of abaca fiber contains hemicelluloses, lignin, and binders such as pectin. In the studies performed, it was observed that the binding agents in the structure of abaca fiber, which was treated with 10% NaOH solution for 30 min, were separated from the fiber structure. The alkali-treated abaca fibers are washed with water until pH 7 (Cai et al. 2015). Alkali (i.e., mercerization) process is also applied for surface modification of abaca fibers because it is popular and cost-effective. It was observed that the surface areas of the alkali-treated fibers enhanced when compared to those of untreated fibers. It is also known that abaca fibers used in composite structures tend to form a better interface with other composite matrices (Cai et al. 2015; Boopathi et al. 2012; Gassan and Bledzki 1999; Sun et al. 1998; Suantara and Oktaviani 2015).

The lumen size of abaca fiber decreases after alkali treatment. At the same time, the cell wall swollens with increasing alkali concentration. The lumen structure at the center of the alkali untreated abaca fiber is clearly visible under the SEM microscope but the lumen structure of the abaca fiber treated with high concentration of NaOH is no longer visible (Cai et al. 2015; John and Anandjiwala 2008).

It was observed that the binder agents are removed from the abaca fiber bundles after the alkali treatment. Untreated abaca fibers are twistless. Fiber bundles were observed to be significantly twisted after 5% NaOH treatment. Binding substances such as pectin, lignin, and hemicellulose were removed from the abaca fiber during the alkali treatment, and the abaca fiber bundle was separated into basic fibers. After 10% NaOH treatment, the longitudinal twisted of the fiber bundles can be observed (Cai et al. 2015; Sun et al. 1998). The factors affecting the bending behavior of abaca fibers are increased microfibril angle and longitudinally narrowing structure of microfibrils. This theory was proved by Nakano et al. The longitudinal bending of the microfibrils is associated with an end-to-end reduction of chain segments in the amorphous region of cellulose (Cai et al. 2015).

Chemical Structure of Abaca Fiber

Abaca leaf fibers, environmentally friendly fibers and belonging to plant fiber family, consist of long thin cells that form part of the support structure of the leaf (Future Fibres 2019). It usually has a lignocellulosic structure (Punyamurthy et al. 2012). Abaca fibers contain impurities such as hemicellulose, lignin, and pectin (Punyamurthy et al. 2012; Yazıcıoğlu 1999; Onggo and Judawisastra 2018). The celluloses in the structure of abaca fibers are held together by amorphous hemicelluloses. The fibers within the abaca plant are held by lignin, commonly known as plant cell adhesive. Such materials in the structure of abaca fibers prevent interfacing adhesion of the abaca fibers during their use in polymer matrices and composite structures (Punyamurthy et al. 2012). Cellulose is an essential component with a strong effect on the plant fibers (Bledzki et al. 2009; Punyamurthy et al. 2012; Onggo and Judawisastra 2018). Its strong effect on fibers is associated with the covalent bonding in the structure (Onggo and Judawisastra 2018). Biocompatibility and biodegradability are the most important properties of cellulose (Rahman et al. 2009; Saragih et al. 2018). Essentially cellulose is defined as a non-branched macromolecule containing chains of varying length. Generally, the most natural cellulosic fibers have the general formula $C_6H_{10}O_5$, which consists of carbon (C), hydrogen (H), and oxygen (O). Also, those contain 60–70% cellulose (Thakur and Thakur 2014). Hemicelluloses consist of 5- and 6-ring carbon rings. Hemicelluloses contain a group of polysaccharides (excluding pectin) having a branched structure. These consist of mixtures of molecular weight polysaccharides smaller than cellulose. The hemicelluloses remains together with the celluloses, after the lignin has been separated from the fiber structure (Thakur and Thakur 2014).

Among all the components of the cell wall, only lignin is composed of excessively branched polymers. Lignin, which has a very complex structure, consists of phenylpropane units organized in a three-dimensional structure. The different structures in the lignin are linked to each other by carbon-carbon and ether bonds as opposed to linear or branched chains, such as carbohydrates. The lignin in the fiber structure plays an important role in protecting cellulose and hemicellulose from harsh environmental conditions such as water. All natural cellulosic fibers comprise different amounts of cellulose, hemicellulose, and lignin (Thakur and Thakur 2014). The lignin content of abaca fibers is also very high (Future Fibres 2019). According to some sources in the literature, the content of lignin in abaca fibers is around 15% (Future Fibres 2019), while in some sources it was stated that 13.2% of total abaca fiber is composed of lignin (Del Río and Gutiérrez 2006; Thakur and Thakur 2014). The fiber contents of abaca fibers obtained from different sources are given in Table 1 (Baser 1992; Vijayalakshmi et al. 2014; Mwaikambo 2006; Lewin 1998; Thakur and Thakur 2014; Punyamurthy et al. 2012; Riccio and Orchard 1999; Idicula et al. 2006; Han and Rowell 1997; Haque et al. 2010; Blackburn 2005; Kalia and Avérous 2011).

Table 1 Chemical content of Abaca fibers (Baser 1992; Vijayalakshmi et al. 2014; Mwaikambo 2006; Lewin 1998; Thakur and Thakur 2014; Punyamurthy et al. 2012; Riccio and Orchard 1999; Idicula et al. 2006; Han and Rowell 1997; Haque et al. 2010; Blackburn 2005; Kalia and Avérous 2011; Yazıcıoğlu 1999; Onggo and Judawisastra 2018; Saragih et al. 2018)

Cellulose content (%)	Hemicellulose (%)	Lignin (%)	Pectin (%)	Ash (%)	Moisture content (%)	References
61–64	21	12	0.8	–	14	Mwaikambo (2006), Riccio and Orchard (1999), Kalia and Avérous (2011)
63–64	10	5	5	–	10	Baser (1992)
63.2–70.1	19.6–21.8	5.1–5.7	0.5–0.6	–	10	Punyamurthy et al. (2012), Idicula et al. (2006)
61.50	14.90	9.70	–	4.80	–	Lewin (1998), Punyamurthy et al. (2012), Han and Rowell (1997)
56–63	–	12–13.2	–	–	5–10	Thakur and Thakur (2014), Punyamurthy et al. (2012), Taj et al. (2007)
68.5	19	12–13.2	–	4.8	10–11	Vijayalakshmi et al. (2014), Thakur and Thakur (2014), Haque et al. (2010)
76.6	14.6	8.4	0.3	–	–	Blackburn (2005)
63.20	10.60	5.10	0.50	–	10	Yazıcıoğlu (1999)
60	–	12–13	1	–	–	Onggo and Judawisastra (2018)
56–68	19–25	13.6	0.5–1	–	1.4	Saragih et al. (2018)
55–64	18–23	5–18	1	–	–	Subyakto and Gopar (2018)

The chemical structures and components of abaca fibers have been compiled from different studies in the literature (Tables 1 and 2). According to Table 1, cellulose, hemicellulose, lignin, and pectin contents of abaca fibers vary between 55 and 76.6%, between 10 and 25%, between 5 and 18%, and between 0.3 and 5%, respectively (Table 1). At the same time, it was stated in Tables 1 and 2 that the moisture content of abaca fiber is at most 14% as a result of different research sources (Baser 1992; Vijayalakshmi et al. 2014; Mwaikambo 2006; Lewin 1998; Thakur and Thakur 2014; Rahman et al. 2009; Punyamurthy et al. 2012; Riccio and Orchard 1999; Idicula et al. 2006; Han and Rowell 1997; Haque et al. 2010; Blackburn 2005; Kalia and Avérous 2011; Yazıcıoğlu 1999; Onggo and Judawisastra 2018; Saragih et al. 2018).

Table 2 Comparison of chemical properties of abaca fibers with other fibers (Vijayalakshmi et al. 2014; Lewin 1998; Taj et al. 2007; Blackburn 2005; Kalia and Avérous 2011)

Fiber type	Cellulose (%)	Hemicellulose (%)	Lignin (%)	Pectin (%)	Ash (%)	Moisture content (%)	References
Abaca	63.2–70.1	19.6–21.8	5.1–5.7	0.5–0.6	4.80	14	Baser (1992), Vijayalakshmi et al. (2014), Mwaikambo (2006), Lewin (1998), Punyamurthy et al. (2012), Riccio and Orchard (1999), Idicula et al. (2006), Han and Rowell (1997), Taj et al. (2007), Haque et al. (2010), Blackburn (2005), Kalia and Avérous (2011)
Banana	60–67.6	6–19	5–10	3–5	–	10–11	Lewin (1998), Taj et al. (2007), Abaca Fiber Production Process in Antique (2017)
Cotton	82–96	2–6	0.5–1.5	5–7	1–1.8	6–8.5	Vijayalakshmi et al. (2014), Lewin (1998), Taj et al. (2007), Kalia and Avérous (2011)
Bamboo	26–43	30	21–31	–	–	–	Lewin (1998), Taj et al. (2007), Kalia and Avérous (2011)
Linen	71–82	18.6–20.6	2.2–4	2.3	3.4	7.7	Vijayalakshmi et al. (2014), Lewin (1998), Taj et al. (2007), Blackburn (2005), Kalia and Avérous (2011)
Hemp	70.2–88	17.9–22.4	3.7–6.8	0.9	3.9	1.8	Vijayalakshmi et al. (2014), Lewin (1998), Taj et al. (2007), Blackburn (2005), Kalia and Avérous (2011)
Jute	51–84	13.6–20.4	12–13	0.2	1.0	1.5	Vijayalakshmi et al. (2014), Lewin (1998), Taj et al. (2007), Blackburn (2005), Kalia and Avérous (2011)

(continued)

Table 2 (continued)

Fiber type	Cellulose (%)	Hemicellulose (%)	Lignin (%)	Pectin (%)	Ash (%)	Moisture content (%)	References
Ramie	68.6–80	13.1–16.7	0.6–0.7	1.9	–	–	Lewin (1998), Taj et al. (2007), Blackburn (2005), Kalia and Avérous (2011)
Sisal	60–78	10–18.1	5.9–11	10	1.0	4.0	Vijayalakshmi et al. (2014), Lewin (1998), Taj et al. (2007), Kalia and Avérous (2011)
Kenaf	31–57	21.5	15–19	2	–	–	Lewin (1998), Taj et al. (2007), Kalia and Avérous (2011)
Pineapple	70–82	16–22.2	5–13	1–3	–	–	Lewin (1998), Taj et al. (2007), Kalia and Avérous (2011)
Coconut	36–46	0.15–0.3	41–45	3–4	–	–	Lewin (1998), Taj et al. (2007), Kalia and Avérous (2011)

Physical, Chemical, and Mechanical Properties of Abaca Fiber

When the cross-sections of the abaca leaf fibers are examined, it is seen that there are large oval or circular lumens, consisting of polygonal-shaped bundles with very thin walls. When viewed longitudinally, it is observed that there is a narrow and narrowing fiber structure which has a smooth and bright appearance (Lewin 1998; Cai et al. 2015; Kaya and Yazicioğlu 1992; Onggo and Judawisastra 2018). The abaca fiber bundles are surrounded by a layer of stem cells filled with silica (Lewin 1998). This layer outside of the fiber bundles is characteristic for abaca fibers (Kaya and Yazicioğlu 1992). The well-dispersed X-ray diffraction of abaca fibers indicates that the helical fibers are dense and dominant. Therefore, the spiral angles of the abaca fibers are indicated as 22.5° (Lewin 1998; Chakravarty and Hearle 1967). The lumen in the natural fibers, such as abaca fibers, contributes to high sound absorption performance and plays a greater role on transverse thermal conductivity than the crystal structure and chemical compounds of the fiber (Cai et al. 2015; Li et al. 2008; Liu et al. 2012). In this sense, the fiber lumen in abaca fibers is quite large due to the cell channel (Mwaikambo 2006; Yazıcıoğlu 1999). The cell wall of abaca fibers is narrow (Yazıcıoğlu 1999). Furthermore, the lumen in abaca fiber is in the center of the fibers and has a hollow structure that affects the properties of the fiber (Cai et al. 2015; Li et al. 2008; Liu et al. 2012).

Table 3 Physical and mechanical properties of abaca fibers (Vijayalakshmi et al. 2014; Mwaikambo 2006; Bolton 1994; Shibata et al. 2003; Ramnath et al. 2014; Onggo and Judawisastra 2018)

Density (g/m^3)	Diameter of fiber (μm)	Tensile strength (MPa)	Initial module (GPa)	Elongation at break (%)	References
1.50	17.0–21.4	12	41	3.4	Mwaikambo (2006), Bolton (1994)
–	19.2	813	33.6	–	Shibata et al. (2003)
1.50	15.0–26.0	980	41	1.1	Vijayalakshmi et al. (2014)
1.50	10–30	430–813	31.1–33.6	2.9	Ramnath et al. (2014)
–	10	381	–	1.1	Onggo and Judawisastra (2018)

When the chemical properties of abaca fibers were examined, it was observed that they could not be hydrolyzed by acids due to their high lignin content. However, when treated with hot alkali, it dissolves, easily oxidizes, and easily condenses with phenol (Vijayalakshmi et al. 2014). When the physical and mechanical properties of abaca fibers are examined, its density is 1.50 g/m^3 (Vijayalakshmi et al. 2014; Mwaikambo 2006; Bolton 1994; Shibata et al. 2003; Ramnath et al. 2014), and fiber diameter varies between 10 and 30 μm (Vijayalakshmi et al. 2014; Mwaikambo 2006; Bolton 1994; Shibata et al. 2003; Ramnath et al. 2014; Onggo and Judawisastra 2018). It has been proved as a result of the experimental studies that the breaking strength of abaca fibers is 980 MPa and initial modules of the abaca fiber vary between 31,1 and 41 GPa (Vijayalakshmi et al. 2014; Mwaikambo 2006; Bolton 1994; Shibata et al. 2003; Ramnath et al. 2014). Table 3 shows the physical and mechanical properties of abaca fibers (Vijayalakshmi et al. 2014; Mwaikambo 2006; Bolton 1994; Shibata et al. 2003; Ramnath et al. 2014).

Abaca fiber is known to be one of the most powerful natural fibers (Blackburn 2005; Saragih et al. 2018). When examined in terms of fiber properties, the properties of abaca fibers are similar to the properties of sisal fibers (Table 4) (Mwaikambo 2006; Quinion 1996), but it was reported that the strength of abaca fibers may be up to three times stronger than that of sisal fibers (Blackburn 2005). It was earlier reported that abaca (Manila hemp) fiber has high tensile strength (381–980 MPa) and young modulus (31,1–41 GPa) (Table 4). It has been shown that the abaca fibers are better than other strong fibers such as sisal fibers, which have tensile strength of 511–635 MPa and young modulus between 9,4 and 22,0 GPa (Table 4).

Understanding the extremely good physical and chemical properties and fiber structure-function relationship of abaca fibers is critical for its effective use in different industrial applications (Bledzki et al. 2007; Cai et al. 2015; Bledzki and Gassan 1999). The strengths or weaknesses aspect of abaca fibers are evaluated depending on where they will be used. The strengths aspect of abaca fibers is its high water absorption capability. If abaca fibers are to be used in composite materials, this situation is considered as a weak feature since it will form a weak interface with the matrix material of the composites. Therefore, the use of abaca fibers in the

Table 4 Comparison of physical and mechanical properties of abaca fibers with other fibers (Shibata et al. 2002; Cai et al. 2015; Saxena et al. 2011; Bledzki et al. 2006; Ramnath et al. 2014; Subyakto and Gopar 2018)

Fiber types	Density (g/m^3)	Tensile strength (MPa)	Initial module (GPa)	Elongation at break (%)	Hygroscopic (%)
Abaca	1.50	381–980	31.1–41	2.9–3.4	14
Banana	1.35	540–550	20	5–6	10–11
Sisal	1.41	511–635	9.4–22.0	2.0–2.5	11
Linen	1.50	345–1035	27–39	2.7–3.2	8
Jute	1.3–1.46	393–800	13–26.5	1.5–1.8	12
Cotton	1.60	300–590	5.5–12.6	3–10	8
Bamboo	0.8	220–1000	22.8	1.3	–
Hemp	1.48	550–900	70	1.6	8–10
Ramie	1.5	220–938	44–128	2–3.8	12–17
Pineapple	1.07–1.52	170–1627	6.21–82.5	1.6	11.8
Glass	2.50–2.55	2000–3500	70.0–73	2.5–3	–
Aramid	1.38–1.47	3000–3150	63.0–67.0	3.3–3.7	–
Carbon	1.7–2.1	4000	230–240	1.4–1.8	–

industrial areas is limited to certain areas. Studies initiated and carried out to improve fibers are mainly focused on fiber-matrix interface adhesion and water absorption reduction (Cai et al. 2015; Bledzki et al. 2006).

Contribution of Sustainable Abaca Fibers to Textile Design

The contribution of sustainable abaca fibers to textile design is indeed promising. The main reason why abaca fibers are used in textile design studies is its sustainable natural character (Korkmaz 2015). The harmful chemicals used in the production stages of the textile and apparel sector, especially the high water consumption and related water pollution in dyeing-printing and finishing enterprises, the high energy consumption used in each production step, and the air pollution which is a result of all these, are the biggest obstacles to the sustainability of the textile sector. In addition, the implementations of sustainability are made possible by reduce, reuse, and recycle. These apply also to textile and fashion design applications (Korkmaz 2015). Therefore, the first important thing is the raw material selection (Korkmaz 2015; Doğan 2011). The raw materials for the textile industry are fibers. As abaca fibers are natural fibers, they are an ecologically important fiber in terms of being lost in nature after production and use (Korkmaz 2015; Doğan 2011). The use of pesticides during their production is carried out to a minimum, that is to say, organic production (Amatta 2017). It is also possible to color abaca fibers with natural dyes, and even natural dyes can be obtained from wastes, and sustainability can be taken one step further (White Champa 2019). Industrial natural (root) dyed abaca fibers can be obtained using inexpensive plant wastes such as rose, lavender oil flowers, turnip, and wine residues and even in any kind of mown grass waste. This reduces the chemical usage during the dyeing process (White Champa 2019). In addition, the use of rotting methods to obtain fibers and even the use of enzymes and bacteria also aim to minimize the use of chemicals (Paridah et al. 2011). However, synthetic fibers, which are highly preferred today, are petroleum based and require energy and chemicals based on nonrenewable resources during their production. Therefore, unfortunately, it is not possible to talk about environmentally friendly production in the production of synthetic fibers. In addition, synthetic fibers take longer years to disappear in nature than abaca fibers (Korkmaz 2015).

One cannot imagine that a stiff fiber from an abaca plant can be used as an alternative to traditional raw materials in avant-garde clothing and different fashion accessories (My Useful Tips 2012). But in some parts of the world, abaca fibers have been used for many various purposes even before the First World War and have been beneficial. Fashion designers and fashion entrepreneurs support and use abaca fibers. The National Association of Sustainable Fashion Designers is one of the major organizations called sustainable fashion that advocates the use and production of domestic materials, fashion, clothing, and accessories in its own initiative. Ditta Sandico-Ong, a Filipino fashion designer, is successful in handicraft and is an active supporter of a clean environment. Ditta Ong combines her artistic achievements with apparel designs and other fashion lines using abaca plant-borne fibers (My Useful Tips 2012). Designed by Jontie Martinez, part of the Maria Clara exhibition, this Filipino dress was on display at The Block at SM City North EDSA. It consists of two parts, kimono (the blouse) and the panuelo (the shawl). With today's technology, abaca fiber can be as delicate and smooth as silk, but the coarse texture of this fragment is remarkable (Manila Hemp 2009). Many different application areas and design types of abaca fibers are investigated below in more detail.

Textile Designs Made from Sustainable Abaca Fibers

In the recent years, abaca fiber which is a renewable, biodegradable, sustainable, eco-friendly, and natural fiber alternative for the textile designs comes into prominence with their ecological properties and superior daily usage performance. Indeed, abaca fibers display a wide range of different designed applications due to their promising properties (Figs. 4 and 5). Many different designed products such as garments, clothing, shawls, curtains, furnitures, carpets, mats, hats, coasters, pillows, packaging materials, handicraft products, decoration materials, decorative lamps and lampshades, flower-wreath ornaments, multipurpose baskets, tabletop accessories, sports equipments, ropes, fishing nets, bags, papers, money papers, cigarette filter papers, wrapping papers, tea bags, cardboard, chipboard, liquid filtration filters, and various industrial textile materials such as medical articles and composites for different purposes can be made from abaca fibers (Figs. 4 and 5) (Vijayalakshmi et al. 2014; Future Fibres 2019; Amatta 2017; Shibata et al. 2002; Abaka Weaving of Shelmed 2012; Fabric 2019; Our 7107 Islands 2019; Blackburn 2005; Onggo and Judawisastra 2018; Saragih et al. 2018).

Fig. 4 Abaca fiber and abaca fabrics made from abaca fiber

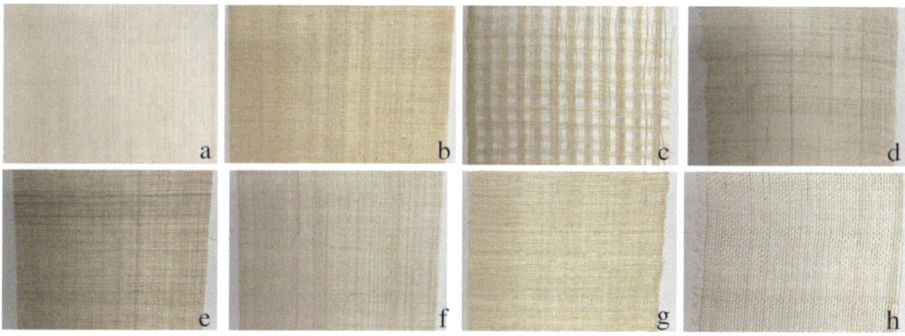

Fig. 5 (**a**) Woven fabric made from polyester and abaca fibers; (**b–h**) Different greige woven fabric examples made from 100% abaca fibers

There is a developing niche market for clothes, curtains, and furnitures obtained using abaca fibers. Fiber craft products such as carpets, mats, hats, coasters, pillows, and bags obtained using abaca fibers are in high demand (Figs. 6 and 7). The second foreign exchange gain after abaca raw fiber exports is the trade of abaca products (Vijayalakshmi et al. 2014). Jontie Martinez designed a dress made from abaca fiber which is a part of the Maria Clara exhibition. This design consists of a shawl and blouse (Manila Hemp 2009). Abaca fibers are also used in the production of traditional costumes in the Philippines, often referred to as Barong Tagalog, which are usually worn by men on special occasions such as weddings (Saragih et al. 2018; Barong Tagalog 2006; Ceylanoğlu 2013).

The fabric obtained by the weaving of abaca fibers by manual or mechanical methods is called Jusi (Barong Tagalog 2006; Ceylanoğlu 2013). According to Filipino culture, this special clothing, which is used especially for weddings, is not only produced from abaca fibers. It is known that this special garment can also be produced from banana and pineapple fibers (Barong Tagalog 2006; Ceylanoğlu 2013). Nowadays, these special clothes can also be produced from abaca/cotton fiber blended and abaca/silk/pineapple fiber blended woven fabrics (Barong Tagalog 2006; Ceylanoğlu 2013). Abaca fiber can be used even in high tensile components due to their extremely high mechanical strength and fiber length. Abaca fibers therefore offer great potential for different industrial applications (Future Fibres 2019;

Fig. 6 Bag made from greige abaca fibers

Fig. 7 Bell silhouette skirt made with weaving process using greige natural abaca fiber (Gündoğan et al. 2018)

Suantara and Oktaviani 2015). Abaca fibers are also used in manufacture of chipboard, which is also considered as composite material, that is obtained by bonding wood particles or cellulosic materials with adhesive materials (Suantara and Oktaviani 2015). Abaca fibers are not only used to produce fabrics but also are used extensively in papermaking (Mert and Çopur 2019; Lewin 1998; Odabaş Serin and Gümüşkaya 2006; Kaya and Yazicioglu 1992; Mert 2009, Suantara and Oktaviani 2015). Today, modern paper and cardboard production can be made from short and long fiber trees, fiber plants [short-staple annual plants such as straw, reed, and paddy; and long-staple fiber plants such as lignocellulosic fibers (linen, hemp, jute, ramie, etc.) and abaca fibers waste paper and waste cardboard (linen, hemp, jute, ramie, etc.) and abaca fibers (Mert and Çopur 2019; Vijayalakshmi et al. 2014; Odabaş Serin and Gümüşkaya 2006; Mandegani et al. 2016). The use of fiber plants such as abaca as an alternative raw material source to woody forest trees in papermaking has some advantages. These are shorter growth time of abaca plant, lower lignin content compared to woody forest trees of abaca plant, and thus the need for less energy and chemicals in the pulp production process (Mert and Çopur 2019; Odabaş Serin and Gümüşkaya 2006; Santoso et al. 2017). Abaca fibers are widely used in tea and coffee bags, case papers, cigarette filter papers, liquid filtration filters, high-quality writing papers, vacuum bags, and special papers such as money paper (Vijayalakshmi et al. 2014; Future Fibres 2019; Hadi et al. 2011; Alan and Tercan 2012; Blackburn 2005; Mert 2009; Mandegani et al. 2016; Santoso et al. 2017). For example, Japan's yen banknotes contain 30% abaca fiber (Vijayalakshmi et al. 2014; Future Fibres 2019).

Abaca fibers which were used in ship navy due to their resistance to saltwater in the ancient times are still used in rope and fishing nets (Vijayalakshmi et al. 2014; Future Fibres 2019; Blackburn 2005; Kaya and Yazicioğlu 1992; Mert 2009; Yazıcıoğlu 1999; Suantara and Oktaviani 2015; Mandegani et al. 2016; Kumar et al. 2017). One of the advantages of abaca fiber usage in industrial applications is its biodegradability (Rahman et al. 2009; Cai et al. 2015; Haque et al. 2010; Ramnath et al. 2014; Suantara and Oktaviani 2015). It is also preferred in the automotive and aircraft industries due to its light weight, low cost, and high specific rigidity (Rahman et al. 2009; Cai et al. 2015; Haque et al. 2010; Ramnath et al. 2014).

Today, abaca fibers are used as fillers for soft applications in the automotive industry. However, given the strong tensile strength of abaca fibers, it can also be used for stiffer applications for an outer semi-building component instead of glass fiber in the reinforced plastic components (Future Fibres 2019; Hintermann 2005). Low cost and good mechanical properties are the main reasons why glass fiber is widely used in composite materials (Onggo and Judawisastra 2018). However, there is an increasing interest in natural fibers with superior mechanical properties due to nonrenewable nature, high density, and high production energy of glass fibers (Onggo and Judawisastra 2018; Subyakto and Gopar 2018). Due to the increasing environmental awareness, researches and studies on the use of natural fiber-reinforced composites, biocomposites, and thermoplastic and thermoset composites have been conducted, and many examples of studies in this field can be found in the literature (Hadi et al. 2011; Cai et al. 2015; Hintermann 2005; Bledzki et al. 2006; Saragih et al. 2018; Subyakto and Gopar 2018).

Dr. Hitoshi Takagi from Tokushima University in Japan worked with the Korean Institute of Materials Science (KIMS) and produced the DOST-ITDI-sponsored *Tryk ni Juan* (traditional three-wheel vehicle) using abaca fibers. Dr. Byung-Sun Kim, the principal researcher of the Korean Institute of Materials Science (KIMS), has made a detailed examination of the various applications of natural fiber composites in objects around us (The Good News Pilipinas Team 2016; Villareal 2016). They achieved a more, triple, heat-resistant structure than even bamboo fibers by solidifying the lumen structure and hollow spirals of abaca fibers (The Good News Pilipinas Team 2016). Abaca fiber is known for their light material properties and low density, and low-cost natural fiber composites can be produced from abaca fibers. Abaca fibers have also been used as roofing materials for traditional tricycle vehicles in the Philippines (The Good News Pilipinas Team 2016; Villareal 2016; Dost-Stii 2016). Dr. Kim stated that unlike steel, abaca fibers reduce the thermal conductivity and thus keep the temperature of these small vehicles cooler, taking into account the tropical temperature and humidity of the country (The Good News Pilipinas Team 2016). In addition, the abaca fiber composites are biodegradable and resistant to rusting, suggesting that they are alternative materials to materials such as stainless steel and iron (Rahman et al. 2009; The Good News Pilipinas Team 2016; Villareal 2016).

In addition, the use of abaca fiber bags instead of plastic market bags in the Philippines eliminates the environmental problems in half. The project initiated for this is still ongoing (The Good News Pilipinas Team 2016). Abaca fiber-reinforced polymer composites were investigated by Shibata et al. Ballyzki et al. conducted a study on abaca fiber-reinforced thermoplastics, especially using polypropylene (Hadi et al. 2011). Abaca fibers are also used in the underfloor protection for passenger cars manufactured by Daimler-Chrysler (Bledzki et al. 2007; Wühhlbauer and Hutter 2002; Hadi et al. 2011; Cai et al. 2015; Hintermann 2005; Scherübl 2006; Bledzki et al. 2006; Kalia and Avérous 2011).

In 2004, Daimler-Chrysler AG (Stuttgart, Germany) used the fibers of the abaca plant instead of fiberglass in the spare tire protector on Mercedes-Benz A-series vehicles. This product was patented in 2002 by Daimler-Chrysler researchers as a blend of polypropylene (PP) thermoplastic and abaca fibers. The production process was initiated by Rieter Automotive. Natural fiber-reinforced biocomposite structures have already been used in interior panels and upholstery of automobiles, but the use for exterior facades of automobiles has not yet become widespread. Abaca is known as the first natural fiber to meet the strong quality requirements for components used in the exterior of automobiles. It has become a preferred fiber, especially because of its high resistance to effects (such as stone throwing, exposure to different factors, and moisture) (Future Fibres 2019; Bledzki et al. 2007; Wühhlbauer and Hutter 2002; Agung et al. 2011; Hadi et al. 2011; Çavdar and Boran 2016; Rahman et al. 2009; Hintermann 2005; Scherübl 2006; Knothe et al. 2000; Bledzki et al. 2006; Kalia and Avérous 2011; Onggo and Judawisastra 2018; Subyakto and Gopar 2018). Mercedes-Benz used a mixture of polypropylene thermoplastic and abaca fiber in automobile body parts, and it has provided the replacement of glass fibers with natural fibers. Thus, automotive parts have been reduced in weight and have developed lighter, environmentally friendly, and recyclable parts (Future Fibres 2019; Hintermann 2005; Scherübl 2006; Bledzki et al. 2006; Subyakto and Gopar 2018).

Studies on the production of natural fiber-reinforced polypropylene composite structures are still continuing. In the literature, studies related to coconut fiber- and abaca fiber-reinforced composite structures can be often encountered. Crude fibers are treated with benzene diazonium salt and comparisons are made with untreated fibers. Chemically treated fiber-reinforced composites form structures with better mechanical properties than untreated raw composites (Rahman et al. 2009; Haque et al. 2010).

Conclusions

Consumption of resources and waste production are also increasing with the acceleration of industrialization. For this reason, many countries have started to give importance to the concept of sustainability in order not to face a big disaster in the future and in order to take necessary precautions. Sustainability is one of the most important topics in the textile and fashion world. Therefore, the use of biodegradable and renewable textile fibers is increasing. Abaca fiber is a renewable, biodegradable, sustainable, eco-friendly, and natural fiber. Abaca leaf fibers are among the preferred natural fibers day by day, thanks to their superior daily usage performance and ecological characteristics. The largest share in the export of abaca, a type of banana plant, belongs to the Philippines. Abaca fibers display a wide range of different designed applications due to their promising properties. For instance, many different designed products such as garments, clothing, shawls, curtains, furnitures, carpets, mats, hats, coasters, pillows, packaging materials, handicraft products, decoration materials, decorative lamps and lampshades, flower-wreath ornaments, multipurpose baskets, tabletop accessories, sports equipments, ropes, fishing nets, bags, papers, money papers, cigarette filter papers, wrapping papers, tea bags, cardboards, chipboards, liquid filtration filters, and various industrial textile materials such as medical articles and composites for different purposes can be made from abaca fibers. For instance, nowadays, abaca fibers can be used in the place of glass fibers in automotive industry technical textiles and composite applications. Indeed, the usage and application areas of abaca fibers are being developed, and abaca fiber replaces synthetic fibers mostly used in composite structures. The search for different alternative natural fibers which can be provided with superior performance with appropriate performance characteristics according to the area of use is increasing day by day. As a result of the recent research studies, as aforementioned, glass fibers used in composite structures can now be replaced by abaca fibers. Studies on the development of different uses of abaca fibers are still continuing. Thus, in the long run, it is aimed to leave a cleaner and more sustainable world to future generations.

Our world is experiencing the highest period in the use of plastic and polyester materials. That is why the interest in ecosystem-friendly abaca fibers as well as many natural other fibers is increasing by large companies. In particular, the use of different designs by designers such as Ditta Sandico-Ong and Jontie Martinez signals that abaca fibers will be added to fashion and design in different dimensions. In addition, the use of abaca fibers in design studies not only for fashion purposes but also in technical areas increases the interest in biodegradable abaca fibers, and it is believed that this trend is expected to increase. The introduction of abaca fibers in the automobile industry also indicates that the demand for abaca fibers will increase after 2024. Therefore, it is envisaged that abaca fibers will be used for both technical and aesthetic purposes and their use for design purposes will become widespread.

References

Abaca Fiber Production Process in Antique (2017). https://www.youtube.com/watch?v=1CVCHAD6ujo&t=3s. Accessed 20 July 2019

Abaka Weaving of Shelmed (2012) Proudly Philippine made. http://philippine-made.blogspot.com.tr/2012/03/abaka-weaving-of-shelmed.html. Accessed 18 July 2019

Agung EH, Sapuan SM, Hamdan MM, Zaman HMDK, Mustofa U (2011) Study on abaca (Musa textilis Nee) fibre reinforced high impact polystyrene (HIPS) composites by thermogravimetric analysis (TGA). Int J Phys Sci 6(8):2100–2106. https://doi.org/10.5897/IJPS11.285

Alan G, Tercan M (2012) Sıvı Filtrasyonunda Kullanılan Tekstiller. Tekstil Teknolojileri Elektronik Dergisi 6(1):47–53

Amatta JD (2017) FMA5: abaca fiber production in antique. https://www.youtube.com/watch?v=BkLo-7sfXy0. Accessed 4 July 2019

Barong Tagalog (2006) From Wikipedia, the free encyclopedia. https://en.wikipedia.org/wiki/Barong_Tagalog. Accessed 24 July 2019

Baser I (1992) Elyaf Bilgisi. Marmara Üniversitesi Döner Sermaye İşletmesi Teknik eğitim Fakültesi Matbaa Birimi, İstanbul

Bellmann C, Caspari A, Albrecht V, Doan TL, Mäder E, Luxbacher T, Kohl R (2005) Electrokinetic properties of natural fibres. Colloids Surf A Physicochem Eng Aspects 267(1–3):19–23

Berkley EE, Hessler LE, Burneston EB, Chew CF (1949) A study of the quality of Abaca fiber. United States Department of Agriculture, Economic Research Service, Washington, DC

Blackburn RS (2005) Biodegradable and sustainable fibres, vol 47. New York, Taylor and Francis

Bledzki AK, Gassan J (1999) Composites reinforced with cellulose based fibres. Prog Polym Sci 24(2):221–274

Bledzki AK, Faruk O, Sperber VE (2006) Cars from bio-fibres. Macromol Mater Eng 291(5):449–457

Bledzki AK, Mamun AA, Faruk O (2007) Abaca fibre reinforced PP composites and comparison with jute and flax fibre PP composites. Express Polym Lett 1(11):755–762. https://doi.org/10.3144/expresspolymlett.2007.104

Bledzki AK, Jaszkiewicz A, Scherzer D (2009) Mechanical properties of PLA composites with man-made cellulose and abaca fibres. Compos Part A Appl Sci Manuf 40(4):404–412

Bolton AJ (1994) Natural fibers for plastic reinforcement. Mater Technol 9(1–2):12–20. https://doi.org/10.1080/10667857.1994.11785012

Boopathi L, Sampath PS, Mylsamy K (2012) Investigation of physical, chemical and mechanical properties of raw and alkali treated Borassus fruit fiber. Compos Part B Eng 43(8):3044–3052

Cai M, Takagi H, Nakagaito AN, Katoh M, Ueki T, Waterhouse GI, Li Y (2015) Influence of alkali treatment on internal microstructure and tensile properties of abaca fibers. Ind Crops Prod 65:27–35

Category Archives: Abaca (19) Eco Kailla. https://ecokaila.wordpress.com/category/abaca/. Accessed 7 July 2019
Çavdar AD, Boran S (2016) Doğal Liflerin Otomotiv Sanayinde Kullanımı. Kastamonu Univ J For Fac 16(1):253–263
Ceylanoğlu Y (2013) Barong Tagalog: Klasik Bir Filipinler Geleneği. Muzatiklarinindegerlendirilmesi, https://muzatiklarinindegerlendirilmesi.wordpress.com/2013/05/09/barong-tagalog-klasik-bir-filipinler-gelenegı/. Accessed 24 July 2019
Chakravarty AC, Hearle JWS (1967) 45—Observations on the tensile properties of ultimate cells of some plant fibres. J Text Inst 58(12):651–656
Chang K (2013) Policy developments affecting jute and hard fibres markets and their implications for production and trade. FAO Commodity and Trade Policy Research Working Paper, Rome. http://www.fao.org/3/i3573e/i3573e.pdf. Accessed 5 July 2019
Del Río JC, Gutiérrez A (2006) Chemical composition of abaca (Musa textilis) leaf fibers used for manufacturing of high quality paper pulps. J Agric Food Chem 54(13):4600–4610
Del Rio JC, Jiménez-Barbero J, Chávez MI, Politi M, Gutiérrez A (2006) Phenylphenalenone type compounds from the leaf fibers of abaca (Musa textilis). J Agric Food Chem 54(23):8744–8748
Doğan Z (2011) Tekstil Sektöründe Atık Ekolojisi Uygulamaları. Akdeniz Sanat Dergisi 4(8):24–26
Dost-Stii (2016) Tryk ni Juan. https://www.youtube.com/watch?v=oh-xD_6jKzI. Accessed 19 July 2019
FAO Statistics (2016) Jute, Kenaf, Sisal, Abaca, Coir and Allied Fibres. Food and Agriculture Organization of the United Nations, Rome. http://www.fao.org/3/a-j5903m.pdf. Accessed 5 July 2019
Fabric. Save on crafts. https://www.save-on-crafts.com/meshabaca.html. Accessed 18 July 2019
Farnfield CA, Alvey PJ (1975) Textile terms and definitions. The Textile Institute, Redwood Burn Limited Trowbridge and Esher, Manchester
Future Fibres (2019) Food and Agriculture Organization of the United Nations. http://www.fao.org/economic/futurefibres/fibres/abaca0/en/. Accessed 4 July
Gassan J, Bledzki AK (1999) Alkali treatment of jute fibers: relationship between structure and mechanical properties. J Appl Polym Sci 71(4):623–629
Gündoğan M, Kalaycı E, Avinc O, Yavas A (2018) Egzotik Doğal Çevre Dostu Bitki Liflerinden Etek Tasarımları. Alanya 4th International Symposium on Fine Arts, University of Alanya Alaaddin Keykubat, Alanya, 3-5 May 2018
Hadi AE, Salit MS, Ahmad MMHM, Dahlan KZHM, Usman M (2011) Physical properties of Abaca (Musa textilis nee) fibre reinforced high impact polystyrene (HIPS) composites. Pertanika J Sci Technol 19(2):349–363
Han JS, Rowell JS (1997) Chemical composition of fibres. In: Rowell RM, Young RA, Rowell JK (eds) Paper and composites from agro-based resources. Lewis Publisher, Boca Raton, pp 83–134
Haque M, Rahman R, Islam N, Huque M, Hasan M (2010) Mechanical properties of polypropylene composites reinforced with chemically treated coir and abaca fiber. J Reinforced Plast Compos 29(15):2253–2261. https://doi.org/10.1177/0731684409343324
Hintermann M (2005) Automotive exterior parts from natural fibers. RIKO-2005, Hannover, 10 Nov 2005
Idicula M, Boudenne A, Umadevi L, Ibos L, Candau Y, Thomas S (2006) Thermophysical properties of natural fibre reinforced polyester composites. Compos Sci Technol 66(15):2719–2725. https://doi.org/10.1016/j.compscitech.2006.03.007
Jarman CG (1977) Banana fiber: a review of its properties and small-scale extraction and processing. Trop Sci 19(4):173–185
John MJ, Anandjiwala RD (2008) Recent developments in chemical modification and characterization of natural fiber–reinforced composites. Polym Compos 29(2):187–207
Kalia S, Avérous L (2011) Biopolymers: biomedical and environmental applications, vol 70. Wiley, Chichester
Kaya F, Yazicioğlu Y (1992) Lif Teknolojisi. Seçkin Ofset Matbaacılık, Ankara

Knothe J, Rebstock K, Schlosser T (2000) Natural fibre reinforced plastics in automotive exterior applications. In: 3rd International wood and natural fibre composites symposium, Kassel, pp 1–12

Korkmaz M (2015) Ev Tekstilinde Çevre Dostu Ürünler. Ar&Ge Bülten 2015 Eylül-Ekim/ Sektörel. http://izto.org.tr/demo_betanix/uploads/cms/yonetim.ieu.edu.tr/6420_1480068108.pdf. Accessed 3 July 2019

Kumar AM, Parameshwaran R, Kumar PS, Pal SK, Prasath MM, Krishnaraj V, Rajasekar R (2017) Effects of abaca fiber reinforcement on the dynamic mechanical behavior of vinyl ester composites. Mater Test 59(6):555–562

Lewin M (1998) Handbook of fiber chemistry, revised and expanded. CRC Press, New York

Li Y, Hu C, Yu Y (2008) Interfacial studies of sisal fiber reinforced high density polyethylene (HDPE) composites. Compos Part A Appl Sci Manuf 39(4):570–578

Liu K, Takagi H, Osugi R, Yang Z (2012) Effect of physicochemical structure of natural fiber on transverse thermal conductivity of unidirectional abaca/bamboo fiber composites. Compos Part A Appl Sci Manuf 43(8):1234–1241

Mandegani GB, Sumarto H, Perdana A (2016) Kertas Seni Berbahan Limbah Pewarna Alam Rumput Laut Jenis Sargassum, Ulva Dan Pelepah Pisang Abaka. Dinamika Kerajinan dan Batik 33(1):33–44

Manila Hemp (2009) My Manila, http://manila-photos.blogspot.com.tr/2009/08/manila-hemp.html. Accessed 22 July 2019

Mert M (2009) Lif Bitkileri. Nobel Bilim ve Araştırma Merkezi, Ankara

Mert M, Çopur O. Lif Bitkileri Üretiminin Artırılması Olanakları. http://www.zmo.org.tr/resimler/ekler/f810ebd27f4dbcf_ek.pdf. Accessed 3 July 2019

Musaceae. Wikipedia Özgür Ansiklopedi. https://en.wikipedia.org/wiki/Musaceae. Accessed 16 July 2019

Mwaikambo L (2006) Review of the history, properties and application of plant fibres. Afr J Sci Technol 7(2):120–133

My Useful Tips (2012) Abaca/banana fibers: an exotic fabric for "sustainable" fashion. http://www.myusefultips.com/abacabanana-fibers-an-exotic-fabric-for-sustainable-fashion/#sthash.0tiFJZ4B.dpbs. Accessed 11 Oct 2019

Odabaş Serin Z, Gümüşkaya E (2006) Kağıt ve Karton Endüstrisinde Odun Dışı Lignoselülozik Liflerin Kullanımı. I. Uluslararası Odun Dışı Orman Ürünleri Sempozyumu, KTÜ, Trabzon, Kasım

Onggo H, Judawisastra H (2018) Identifikasi Morfologi dan Sifat Mekanik Serat Abaka (Musa textilis) Identification of morphology and mechanical properties of abaca fiber (Musa textilis). Jurnal Ilmu dan Teknologi Kayu Tropis 7(1):34–38

Different abaca products that we are using everyday, Our 7107 Islands. https://www.our7107islands.com/different-abaca-products-that-we-are-using-everyday/. Accessed 18 July 2019

Paridah MT, Basher AB, SaifulAzry S, Ahmed Z (2011) Retting process of some bast plant fibres and its effect on fibre quality: a review. BioResources 6(4):5260–5281

Paridah MT, Juliana AH, Zaidon A, Khalil HA (2015) Nonwood-based composites. Curr For Rep 1(4):221–238

Parrott-Sheffer C. Abaca plant. Encyclopædia Britannica. https://www.britannica.com/plant/abaca. Accessed 17 July 2019

Peralta AG (1996) Pulp produced from decorticated abaca fiber. Tappi J 79(3):263–266

Punyamurthy R, Sampathkumar D, Srinivasa CV, Bennehalli B (2012) Effect of alkali treatment on water absorption of single cellulosic abaca fiber. BioResources 7(3):3515–3524

Quinion M (1996) Fibres from the earth. World Wide Words. http://www.worldwidewords.org/articles/fibres.htm. Accessed 12 July 2019

Rahman MR, Huque MM, Islam MN, Hasan M (2009) Mechanical properties of polypropylene composites reinforced with chemically treated abaca. Compos Part A Appl Sci Manuf 40(4):511–517. https://doi.org/10.1016/j.compositesa.2009.01.013

Ramnath BV, Manickavasagam VM, Elanchezhian C, Krishna CV, Karthik S, Saravanan K (2014) Determination of mechanical properties of intra-layer abaca–jute–glass fiber reinforced composite. Mater Des 60:643–652. https://doi.org/10.1016/j.matdes.2014.03.061

Riccio FA, Orchard LP (1999) Nonwood fibre resources: availability, infrastructure, and feasibility. In: Fifth international conference on woodfibre-plastic composites, Forest Products Society, Wisconsin, May 1999
Saleeby MM (1915) Abacá (Manila Hemp) in the Philippines:(Musa Textilis), San Francisco
Santoso B, Mastur M, Kadarwati FT (2017) Abaka (Musa Textilis Nee) Sebagai Sumber Serat Alam, Penghasil Bahan Baku Pulp/Abaca (Musa textilis Nee) As the source of natural fiber, producing raw material for. Perspektif 15(1):01–10
Saragih SW, Lubis R, Wirjosentono B, Eddyanto (2018) Characteristic of abaca (Musa textilis) fiber from Aceh Timur as bioplastic. In AIP conference proceedings. AIP Publishing, 14 Dec 2018
Saxena M, Pappu A, Sharma A, Haque R, Wankhede S (2011) Composite materials from natural resources: recent trends and future potentials. In: Advances in composite materials-analysis of natural and man-made materials. InTechOpen. https://doi.org/10.5772/18264
Scherübl B (2006) An innovative composite solution in the new Mercedes A Class-A successful story about the natural fiber Abaca. In: 6th Global wood and natural fibre composites symposium, Kassel, pA-12
Shibata M, Takachiyo KI, Ozawa K, Yosomiya R, Takeishi H (2002) Biodegradable polyester composites reinforced with short abaca fiber. J Appl Polym Sci 85(1):129–138
Shibata M, Ozawa K, Teramoto N, Yosomiya R, Takeishi H (2003) Biocomposites made from short abaca fiber and biodegradable polyesters. Macromol Mater Eng 288(1):35–43
Spencer JE (1953) The abaca plant and its fiber, Manila hemp. Econ Bot *7*(3):195–213. https://doi.org/10.1007/BF02984947
Suantara D, Oktaviani E (2015) Pemanfaatan Serat Kelapa Dan Serat Abaka Sebagai Bahan Baku Papan Partikel. Arena Tekstil 30(1):37–44. https://doi.org/10.31266/at.v30i1.1947
Subyakto S, Gopar M (2018) Tinjauan Penelitian Terkini tentang Pemanfaatan Komposit Serat Alam untuk Komponen Otomotif Review on current research on utilization of natural fiber composites for automotive components. Jurnal Ilmu dan Teknologi Kayu Tropis 7(2):92–97
Sun R, Fang JM, Goodwin A, Lawther JM, Bolton AJ (1998) Isolation and characterization of polysaccharides from abaca fiber. J Agric Food Chem *46*(7):2817–2822
Sustainable fibres: what is abaca? (2016). https://www.trustedclothes.com/blog/2016/04/27/sustainable-fibres-what-is-abaca/#_edn14. Accessed 11 Oct 2019
Taj S, Munawar MA, Khan S (2007) Natural fiber-reinforced polymer composites. Proc Pakistan Acad Sci 44(2):129–144
Tarakçıoğlu I (1978) Tekstil Terbiyesi ve Makinaları. Ege Üniversitesi Matbaası, Cilt 1 İzmir
Thakur VK, Thakur MK (2014) Processing and characterization of natural cellulose fibers/thermoset polymer composites. Carbohydr Polym 109:102–117. https://doi.org/10.1016/j.carbpol.2014.03.039
The Good News Pilipinas Team (2016) Tryk Nı Juan uses roof made with strong fiber abaca. https://www.goodnewspilipinas.com/tryk-ni-juan-uses-roof-made-strong-fiber-abaca/. Accessed 17 July 2019
Van Sumare C (1992) Retting of flax with special reference to enzyme-retting. In: Sharma HS, Van Sumare C (eds) The biology and processing of flax. M Publications, Belfast, pp 157–198
Vijayalakshmi K, Neeraja CY, Kavitha A, Hayavadana J (2014) Abaca fibre. Trans Eng Sci 2(9):16–19
Villareal M (2016) DOST launches Tryk Ni Juan: roof made with abaca fibers. https://outoftown-blog.com/dost-launches-tryk-ni-juan-roof-made-with-abaca-fibers/. Accessed 19 July 2019
Wester PJ (1927) Introduction of Manila hemp. J Heredity 18(3):117–118
White Champa (2019) Abaca, from fiber to fabric. http://whitechampa.com/craft/abaca/. Accessed 11 Oct 2019
Wikimedia Commons (2018a). https://commons.wikimedia.org/wiki/File:Musa_textilis_40zz.jpg.
Wikimedia Commons (2018b). https://commons.wikimedia.org/wiki/File:Musa_textilis_29zz.jpg.
Wikimedia Commons (2018c). https://commons.wikimedia.org/wiki/File:Abaca_sachsen-leinen_03.jpg.

Wikimedia Commons (2018d). https://commons.wikimedia.org/wiki/File:Abaca_sachsen-leinen_04.jpg.

Wikimedia Commons (2018e). https://commons.wikimedia.org/wiki/File:Fibra_de_abaca.jpg.

Wikimedia Commons (2018f). https://commons.wikimedia.org/wiki/File:COLLECTIE_TROPENMUSEUM_Een_man_bewerkt_de_stammen_van_Manilla-hennep_TMnr_10011401.jpg.

Wikimedia Commons (2019). https://commons.wikimedia.org/wiki/File:Abaca_weaving_in_Banton.JPG.

Wühhlbauer W, Hutter CP (2002) Utilization of abaca fiber as industrial raw material. In: Workshop sustainable development, natural fibers for modern technology, subsistence and biodiversity improvements projects in the Philippines, 11–17 March 2002

Yayock JY, Lombin G, Owonubi JJ (1988) Crop science and production in warm climate. Macmillan, London

Yazıcıoğlu G (1999) Pamuk ve Diğer Bitkisel Lifler. Dokuz Eylül Üniversitesi Tekstil Mühendisliği Bölümü Mühendislik Fakültesi Basım Ünitesi, İzmir

Analysis of Zero Waste Patternmaking Approaches for Application to Apparel

Ellen McKinney, Sunhyung Cho, Ling Zhang, Rachel Eike, and Eulanda Sanders

Abstract The apparel industry is a major contributor to environmental problems from textile manufacturing through garment production and distribution to consumer discard—donation, landfill, reuse, or otherwise (Gam et al., Int J Clothing Sci Technol 21(4):166–179, 2009). In 2015, there were estimated 400 billion square meters of fabrics produced worldwide and 60 billion square meters wasted during apparel production. Vennström (2012) stated that in the United Kingdom, around 2.35 million tons of waste comes from the clothing and textile industry per year, which estimates about 40 kilograms (kg) per person each year. Of the 40 kg (88 lbs) of apparel waste each year, 74% ends up in landfills (Vennström, 2012). Even though the fashion industry applies a variety of technologies to minimize the fabric waste in pursuit of cost reductions, it is still far from eliminating the waste of fabric during the cutting process. On average, 15% of fabric is wasted during the cut-and-sew garment production process (Rissanen & McQuillan, 2016). Professionals and researchers in apparel design and product development discipline are facing a tremendous challenge of combining the innovative patternmaking methods, aesthetics of apparel design, and fabric waste reduction.

Zero waste patternmaking offers a solution by utilizing the entire yardage of fabric, leaving no scrap left after the garment completion (Carrico and Kim, Int J Fashion Des Technol Educ 7(1):58–64, 2014). Further, zero waste has been touted as a means to more creative apparel design outcomes (Townsend & Mills, 2013). Many approaches to zero waste patternmaking have been proposed through a range of sources—blogs, websites, books, and articles (e.g., Townsend & Mills, 2013; Carrico and Kim, Int J Fashion Des Technol Educ 7(1):58–64, 2014; Fletcher, 2013; Antanavičiūtė & Dobilaitė, 2015). Unfortunately, many of these approaches are tied to the particular fabric width and finished garment size they are presented in.

The aim of this chapter is to conduct a systematic review of these approaches. In so doing, we may understand the key principles and be able to apply them across a range of fabric widths and garment sizes. The main method will be to collect a

E. McKinney (✉) · S. Cho · L. Zhang · R. Eike · E. Sanders
Department of Apparel, Events, and Hospitality Management, Iowa State University, Ames, IA, USA
e-mail: emckinne@iastate.edu; sunhyung@iastate.edu; lzhang1@iastate.edu; rjeike@iastate.edu; sanderse@iastate.edu

S. S. Muthu, M. A. Gardetti (eds.), *Sustainability in the Textile and Apparel Industries*, Sustainable Textiles: Production, Processing, Manufacturing & Chemistry, https://doi.org/10.1007/978-3-030-37929-2_2

representative sample of zero waste patternmaking approaches and analyze them. Approaches will be analyzed through visually for their pattern shapes, garment components, and three-dimensional finished garment shapes. Selected approaches will also be drafted, cut, and sewn to further understand the outcomes. Research questions include the following: (1) What are the major pattern design principles? (2) What outcomes do these results in? (3) What challenges exist with these approaches? These findings will provide information that will help designers successfully apply zero waste patternmaking methods in apparel design. Further, the research findings will provide focus for needed areas of future research.

Keywords Zero waste · Patternmaking · Garment · Design · Apparel · Fabric · Production · Apparel industry

Introduction

In the apparel industry, both the production of textiles and the waste of textiles during apparel production negatively affect the environment. Production waste of textiles is often caused by lack of creativity in design and patternmaking approaches (Gwilt 2013; Rissanen 2016). However, there is a potential solution to overcoming these fabric waste challenges: zero waste pattern making, where fabric waste is prevented through creative application of patternmaking methods.

The goal of this chapter is to provide an in-depth understanding of existing zero waste patternmaking methods. In this study, the authors aim to identify the major types of approaches to zero waste patternmaking to develop a classification system (Reynolds 2007). By developing a classification system, approaches may be grouped according to their similarities. This will allow designers to understand and select from myriad publicly available approaches with greater ease. The authors also aim to assess the pros and cons inherent in different patternmaking approaches. By contributing this knowledge to the patternmaking field, the authors hope to enable more apparel producers to adopt these innovative techniques and reduce textile waste in the apparel production process.

The following research questions guided this study:

1. What are the major approaches to zero waste patternmaking?
2. How can these approaches be understood relative to currently or historically used methods of patternmaking?
3. What strategies are used to avoid the production of textile waste in the patternmaking and apparel design process?
4. What kind of garment outcomes are achieved with zero waste patternmaking?
5. What are the benefits of zero waste patternmaking in the apparel manufacturing process?
6. What are the challenges of zero waste patternmaking in the apparel manufacturing process?

By identifying major approaches and relating them to existing patternmaking concepts, the authors hope to make zero waste patternmaking techniques more easily grasped and utilized by apparel industry professionals. Through the explanation of zero waste patternmaking benefits, the authors are better able to encourage the use of these techniques. Identification of challenges may give practical direction for future apparel researchers. Improved understanding of zero waste patternmaking approaches for increased use within the apparel industry is the targeted outcome of this chapter.

This chapter begins by reviewing the related literature including the apparel industry's contribution to environmental problems, the impact of patternmaking on production waste, challenges faced in the apparel industry, and an overview of zero waste patternmaking. Next, we explain the two-part method that was used to shed light upon zero waste patternmaking principles, so designers and patternmakers may be able to apply the production method to their works. In the Findings, we situate zero waste design approaches in relationship to existing patternmaking frameworks and illuminate patternmaking strategies for implementing zero waste design and garment outcomes and challenges when using zero waste techniques. Finally, we discuss how professionals in the apparel industry and researchers may use these findings to achieve zero waste patternmaking. Implications are provided for workflow within the apparel industry, as well as benefits and challenges to be considered. Ultimately, the findings from this study may empower apparel researchers, designers, and patternmakers to reduce the negative impacts of the apparel industry on the environment.

Literature Review

Apparel Industry Contributions to Environmental Problems

The apparel industry is a major contributor to environmental problems from textile manufacturing through garment production and distribution to consumer discard—donation, landfill, reuse, or otherwise (Jin Gam et al. 2009). These environmental problems cause negative changes to limited global resources, affecting the health of human lives on our planet. The apparel production process includes five stages: pre-production, production, distribution, utilization, and end-of-life disposal (Lawless and Medvedev 2016). Textiles and trims selected in the pre-production stage are cut and sewn during the production stage. The cutting is based upon the design, created by the fashion designer, and the corresponding garment pattern, as executed by the patternmaker. It is estimated that up to 80% of a product's environmental and economic cost is determined in the design stage (Fletcher 1999; Lawless and Medvedev 2016). There are negative environmental impacts not only from the production of textiles but also from the waste of textiles, during apparel production.

Textile production. First, fibers are grown or manufactured using plants, animals, or chemicals. Next, these fibers are spun into yarns. Yarns are knitted, woven, or otherwise constructed to produce textiles or trims. Finally, the textile is dyed and/or printed. Additionally, many textiles go through a finishing process where chemicals are added to imbue special properties to the textile (e.g., wrinkle-free finish).

Textile production may yield a number of negative impacts related to chemicals used in the growing and production processes (Fletcher 2012; Quinn 2015). For example, improper handling of pesticides used to stop insects from damaging cotton plants results in poison exposure, causing numerous negative health effects and deaths (Leonie 2017; Fletcher 2012). Use and contamination of water through dyeing and finishing processes diminishes or pollutes water resources, negatively affecting the livelihoods and health of community members that depend on these water resources (Ataniyazova 2003).

Problems in apparel production. In medium to large apparel companies, the designer and patternmaker roles are separated for maximum efficiency and output (Fletcher 2012). The designer is commissioned to sketch numerous styles each season, the patternmaker then develops the pattern, and a first prototype is sewn and reviewed by the designer (Fletcher 2012). This cycle may continue through several iterations before approval of a garment for large-scale production.

During the cutting phase of a garment's production, an average of 15% of the textile fabric is not used and simply tossed as waste (Rissanen 2016). In 2012, approximately 15% of all textile fabric utilized in the apparel industry equaled about 60 billion square meters of textiles being thrown away (Rissanen 2016). Apparel companies have an economic interest in maximizing fabric-cutting efficiencies, and this role is assigned to the technical support team (e.g., patternmakers, marker markers, etc.), who carry the responsibility for controlling costs through design and materials choices (Fletcher 2012; Rissanen 2016).

Nonetheless, despite industry's interest in efficient textile production, current practices do not promote reducing fabric waste in the best possible way, a fact that is largely caused by the separation of design procedures from patternmaking and technical design procedures (Fletcher 2012). Further, the current apparel industry mindset elevates the ideas of the designer over those of the patternmaker, creating a situation where suggestions to change designs to reduce waste are rarely made by the technical support team, for fear of encroaching on the designer's creativity and vision (Almond 2010; Fletcher 2012). It is important to recognize that garments represent "a finished product (that) embodies all the time and effort of all the people who contribute to the fiber production, wasting a percentage of the fabric is also wasting the embodied effort of all individuals who had contributed to the fabric production" (Saeidi and Wimberley 2018, p 243).

Impact of Patternmaking on Production Waste

Textiles wasted in the cutting process is primarily due to a lack of ingenuity in approaches to design and patternmaking (Gwilt 2013; Rissanen 2016). The pattern embodies information about how the textile is cut into pieces that can be sewn together to create the designed garment. Thus, it is essential for designers and patternmakers to work alongside each other during the development process to consider efficient textile use and design aesthetic simultaneously. Improving patternmaking through this collaborative method is essential in the reduction of textile waste.

Challenges

Professionals in apparel design and product development disciplines are facing the tremendous challenge of combining efficient apparel production, aesthetics of apparel design, and fabric waste reduction. However, DeLong et al. (2013) found that designers were interested in improving sustainability practices but lacked practical knowledge about how to integrate sustainability into their design process. Tangible information that gives designers directed guidance as to how they may integrate sustainable practices into their works is greatly needed.

Zero Waste Patternmaking

Zero waste patternmaking can be a solution to overcoming the textile waste challenges of the apparel industry. Zero waste patternmaking is defined as a technique or method for creating patterns without any wasted textile fabric. This special method brings sustainability into the initial stage of apparel product development "by integrating pattern cutting into the design process" (Rissanen 2016, p 11).

Conventional patternmaking methods are based on the three principles: dart manipulation, added fullness, and contouring (Joseph-Armstrong 2010). Dart manipulation refers to moving a dart from its original location of the basic pattern sloper. Added fullness regards providing volume to the garment with the help of gathers, pleats, flares, or tucks. Contouring is to fit a pattern to the body more nearly than the sloper does. Zero waste design approaches may be different from or share the same principles of conventional patternmaking method (Saeidi and Wimberley 2018).

Zero waste patternmaking methods may be beneficial in the reduction of waste of textiles; however, certain challenges must be considered, to help designers and patternmakers fully utilize methods. One challenge is that the methods frequently require more time to implement (Saeidi and Wimberley 2018). Variations in textile widths limit the repeatability of zero waste design techniques (Saeidi and Wimberley 2018). Most textiles are fabricated in the range from 36 to 60 in (approx. 91–152 cm) in width (Hatch 1993), which can significantly affect the layout of pattern pieces and/or finished garment size. Yet another challenge involves creative thinking regarding creating closures and constructing seams (Carrico and Kim 2014). The aim of this chapter is to conduct a systematic review of these approaches. In so doing, designers and patternmakers may better understand and apply the key principles of zero waste patternmaking, and researchers may become aware of areas needing future research.

Method

A representative sample of zero waste patternmaking approaches was collected and analyzed, in two stages. The first stage analyzed publicly available examples. Exemplifiable approaches were derived from these examples, using content analysis. In the second stage, zero waste patternmaking approaches were tested and analyzed by students in a graduate-level apparel design class, to provide tacit knowledge and further discussions. Findings from the two stages were synthesized to illuminate key zero waste patternmaking principles, so designers and patternmakers may be able to apply the production method to their works.

Sample and Data Collection

Stage 1: Publicly available collected examples. Sixty-four zero waste design cases were collected by searching online sources with the keywords "zero waste fashion design," "zero waste design," "zero waste pattern," and "zero waste fashion." The data were collected from publicly available online sources such as independent designers' websites, retrievable exhibitions, or publications. A single case was defined as a zero waste design in which images of both the flat pattern and the constructed garment were available. For each case, data were entered pertaining to when the case was published or collected, retrievable links or published sources, names of designer and design, and any additional relevant descriptions. Twenty-six designers were represented in the data set. The goal of this stage was to research current approaches to zero waste design to understand the methods based on publicly available examples.

Stage 2: Class part. Eight students were enrolled in a graduate-level apparel design class at a major Midwestern university in fall 2018. All students enrolled in the course happened to be female. The focus of the course was on the research, analysis, and application of experimental patternmaking techniques to original garments. The course began with a discussion of theory in patternmaking (Pedersen 2007; Lindqvist 2015; McKinney et al. 2016; Bye 2010) and a review of the traditional patternmaking principles: dart manipulation, added fullness, and contouring (Joseph-Armstrong 2010). The course investigated several types of sustainability-related patternmaking. Only the work related to zero waste patternmaking is reported in this chapter.

Patternmaking technique study and application (PMTSA). Each student was required to complete a PMTSA on a zero waste patternmaking method of her choice. PMTSA is a standardized process for testing and analyzing patternmaking approaches (McKinney et al. 2016). The steps are as follows. Each student created her selected pattern (Table 1) in half scale, using a Wolf brand size 18 half-scale form. A half-scale muslin first sample was sewn of each completed pattern. Each student studied her pattern to discover what she could learn and apply to future work. Students analyzed the following aspects: (a) type of patternmaking, dart manipulation, added fullness, and/or contouring, (b) basic principles and rules of the pattern method, (c) constancy of these rules and principles, and (d) rules of traditional patternmaking that are "broken" in this pattern.

Each student then created her own half-scale pattern by applying the rules and principles of the studied pattern in a different location of the garment. A muslin first sample of the new design was sewn. Each student again was prompted to closely review her pattern to determine areas for improvement, reflect on what she has learned in this phase, and identify how she could adapt her methods in future patterns, with the addition of the question: Where did you move the pattern manipulation and how well did that work? Each student gave a 10-min presentation to the class, including a presentation of their patterns and first samples and a description of their analyses.

Table 1 Selected zero waste patterns completed by students

Pattern designer	Name of zero waste pattern	Student
Anastasia Nash	Seamless trench vest	Courtney Barbour
Anastasia Nash	Zero waste blouse with Japanese airs	Sophia Luu
Eqos on DeviantArt	Gored cloak	Stephanie K. Hubert
Halston	Envelope top	Bingyue Wei
Holly McQuillan	Zero waste t-shirt	Jessica Ciarla
Holly McQuillan	Wrap dress	Megan Romans
Unknown	Circular hooded cape	Sunhyung Cho
Yeohlee Teng	Coat	Minsu Kim

Analysis Method

Stage 1: Content Analysis

Content analysis (Krippendorff 2019) was used to analyze the collected data. The content analysis method was to systematically analyze emerged patterns in images and text data (Neuendorf 2002). First, at the initial stage of data analysis, a coding scheme of paradigms was developed based on the existing conceptual frameworks of patternmaking. New garment-style subcategories emerged as additional cases were analyzed (Creswell 2014). A final list of codes was developed: (a) clothing paradigm (rectangular cut, tailoring matrix) (Lindqvist 2015), (b) garment silhouette (boxy, semi-fit, contour-fit), (c) use of traditional patternmaking techniques (dart manipulation, added fullness, and contouring) (Joseph-Armstrong 2010), (d) garment features present (neck hole, armscyes, waist area, or traditional seam lines), and (e) type of zero waste designs (e.g., Jigsaw (Rissanen 2008) or Transformational Reconstruction (Saeidi and Wimberley 2018)). The images and text of each case were imported into qualitative data analysis software (QDAS) and analyzed. During the coding process, image observation memos were recorded that related to potential concerns with manufacturability, marketability, construction, or garment fit, which may impact commercial viability. For example, in the initial round of coding, researchers identified basic and informatic codes as garment silhouette or presented garment features. The researchers carefully analyzed both of the two images (flat patterns and constructed garment) of design to understand the integrated structure and method of the design. Lastly, additional codes were added regarding construction methods and clothing paradigms based on the previous studies and Lindqvist's theoretical framework (2015). Finally, a report of code co-occurrences was generated to understand themes and trends within the clothing paradigms.

Lindqvist's (2015) two paradigms of clothing. To categorize zero waste design approaches, Lindqvist's (2015) two paradigms of clothing was selected as a starting theoretical framework. First, the paradigm of rectangular cuts arises from the perspectives of stressing relation to the body in ancient drapes, a whole cloth piece on the straight grainline. Three types of clothes can feature such historical garments: (a) djellaba, a wrapping-style garment; (b) tunics, a poncho-like-style garment; and (c) kaftan, a mantel-like-style garment (Hamre et al. 1980). Those types of garments are wrapped, folded, or naturally manipulated on the actual wearer resting on the shoulders or the waist. Another is the paradigm of tailoring matrix established in the drafting and draping system, the current standard design method in the fashion industry and education. This paradigm can be considered a "traditional" approach from the western perspectives since the tape measure was introduced in the beginning of the nineteenth century (Hollander 1994). This technique usually helps to generate fitted or "draped" garments that relate directly to the human body shape in contrast to the ancient styles.

Stage 2: Patternmaking Technique Study and Application

Following the students' presentations, the class analyzed the findings from the presented patternmaking techniques. This included the answers to the questions posed in the PMTSA assignment, as well as the pattern and garment outcomes. Students discussed use of traditional patternmaking rules, as well as those patternmaking rules that were broken during their design process. Students discussed the patternmaking principles present in their selected patterns. An important part of the discussion was the outcomes of drafting the pattern and sewing it with whatever instructions were provided by the pattern's designer. Through this part, many challenges to mass-producing these zero waste patterns were noted. In each part of the discussion, students noted similarities and differences. For example, in the discussion of traditional patternmaking rules used, many students found similarities in the use of the principle of *added fullness* in their selected patterns. However, differences were seen, for example, in the way openings were created for necks and armscyes. Some occurred through the use of slits, while others were created in a more traditional manner, with leftover pieces used creatively elsewhere in the garment. Aspects common to most zero waste patternmaking approaches were noted. Connections were drawn between major approaches to zero waste patternmaking and the outcomes achieved. For example, use of the patternmaking principle of *added fullness* resulted in loose, boxy silhouettes. The aim was to give a codified framework to the students' newly gained knowledge.

Findings

Zero Waste Design Approaches and Relationship to Existing Patternmaking Frameworks

As a result of Stage 1, the authors classified zero waste design approaches based on the existing conceptual frameworks of patternmaking using Lindqvist's (2015) two paradigms of clothing as an organizational framework (see Table 2). This attempt helps to provide better understanding of how recent approaches to zero waste design have been embodied by the apparel design community. By identifying major approaches and relating them to existing patternmaking concepts, the authors hope to make zero waste patternmaking techniques more easily grasped and utilized by apparel designers and patternmakers.

Paradigm of rectangular cuts. In the paradigm of rectangular cuts, 31 cases were identified with 3 garment styles: (a) tunic, (b) djellaba or kaftan, and (c) sandwich. These garments all demonstrated boxy silhouettes and did not include any or minimal dart, shoulder seam, or traditional armscye seam. They also were minimally cut

Table 2 Analysis of zero waste fashion design approaches

Perspectives	Paradigm of rectangular cuts		Paradigm of tailoring matrix	
	Loose/moderate fit		Fitted/contoured	
Style	Nontraditional armhole style		Existing RTW style	
Patternmaking	• Squared patterns	• Tunic, djellaba, or kaftan or sandwich style	• Tessellation or Jigsaw method	• Transformational Reconstruction method (Saeidi and Wimberley 2018)
	• T-shaped patterns		• Developing existing patterns	
	• Wrapping around		• "Fixed area" and modify the patterns to save the rests (Rissanen 2008)	

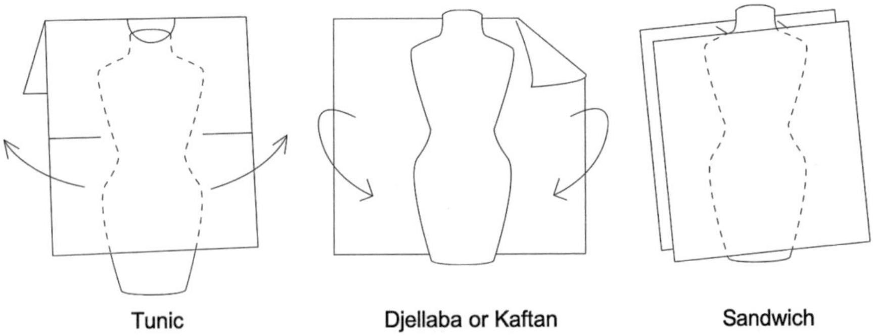

Fig. 1 Three styles of the paradigm of rectangular cuts

and naturally draped and hung over the body of the wearer (Carrico and Kim 2014). Use of grains and types of fabric may critically function for a final look (Fig. 1).

Tunic. The most frequently emerged style (16 out of 31) was tunic, a poncho-like-style garment. The patterns of the tunic style showed a "T-shape," which includes a neck hole in the middle of the panel. While there are many modifications of the T-shaped style, pattern commonalities include minimal cutting and a neck hole. There is no traditional side or shoulder seam in the tunic and the main panel usually wraps around the body. The width of a cloth decides the length of sleeves and the fabric weight affects the overall clothing fit. In the examples of Holly McQuillan, one of the designers who started popularizing zero waste designs, the negative spaces of neck holes were used for neck facings. This type of pattern is usually used for simple tops or long dresses.

Djellaba or kaftan. These garments involve a piece of cloth that wraps around the upper body or lower body, but does not entirely cover the body, creating distinct drapes. A separate waist belt usually fixes the clothing fit. This type of patterns has been frequently used for wraparound jackets with or without hoods. The negative spaces were used for belts or pockets.

Sandwich. The sandwich styles, which are not seen in Hamre's analysis (1980), involve a rectangular front and back panels that are attached at certain fixed points such as shoulder or waist areas. The rest of the panels naturally hang down to the body, creating natural drapes and covering the body.

Paradigm of tailoring matrix. In the paradigm of the tailoring matrix, 33 cases were found. These garments included semi-fitted or contour-fitted silhouettes, which are closely related to the existing ready-to-wear styles. Their patterns were based on basic slopers and included dart manipulation, added fullness, or contoured shapes.

Jigsaw method. Most of the patterns (29 out of 33) were developed using the Jigsaw method, consisting of a "fixed area" and modifying the patterns to utilize the remaining fabric (Rissanen 2008). Many of the designs used remaining fabric for adornments, pockets, or facings. As following the basic sloper including inevitable curvy lines in neck holes, armholes, hood, or sleeve patterns in the paradigm of tailoring matrix, large pieces were positioned in the pattern layouts, and negative areas of the large pieces were filled with small pieces such as a collar, facing, or pockets. Design of Jigsaw method in three dimensions showed fitted silhouettes but selectively also showed loose fits (Fig. 2).

Fig. 2 An example of Jigsaw method (Julia Lumsden, 2010)

Transformational Reconstruction method. Alternatively, few of them (4 out of 33) seemed to adopt the Transformational Reconstruction (TR) method, creating a three-dimensional garment first and re-patterning it to two-dimensional patterns (Saeidi and Wimberley 2018). The patternmaking technique, developed by Shingo Sato, is an innovative tool for modifying conventional darts and seams into desired seam lines (Saeidi and Wimberley 2018). Therefore, the unique technique seems to be used for zero waste designs, allowing for combined pattern pieces shaped in rectangular cuts. Designs of TR method in three dimensions showed a fitted silhouette reflecting designers' desired shapes. However, inevitably, there could be certain negative spaces in the pattern layout, and spaces were used for facing and lining as the Jigsaw method showed (Fig. 3).

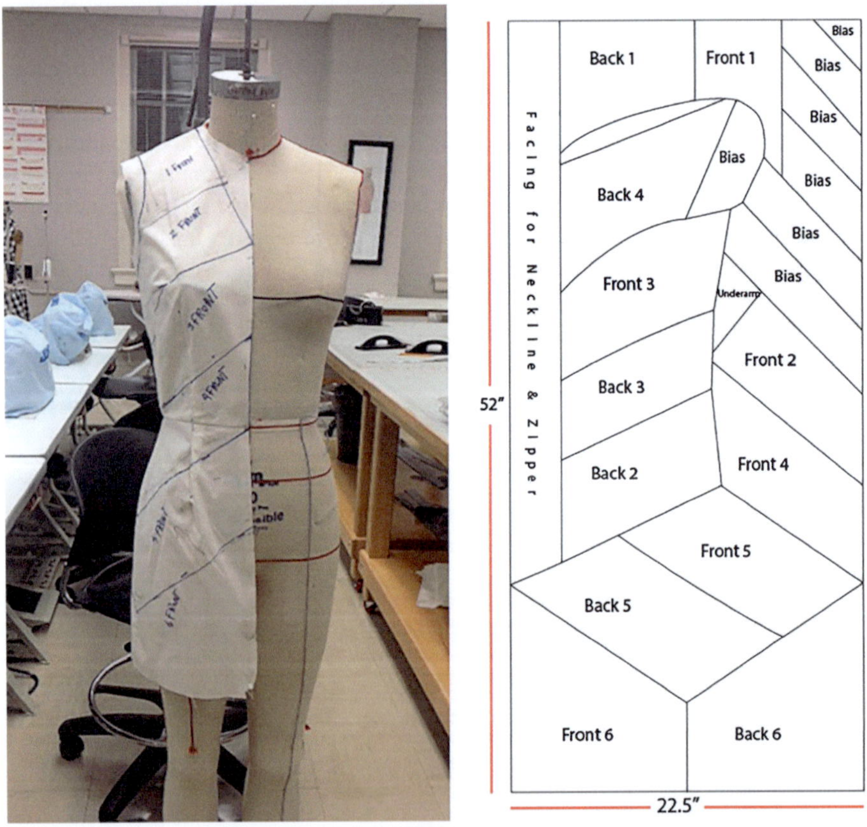

Fig. 3 An example of Transformational Reconstruction method (Saeidi and Wimberley 2018)

Zero Waste Patternmaking Strategies

In Stage 2, through the completed PMTSA assignments and related discussion, a number of patternmaking strategies applied by zero waste pattern designers to use the entire piece of textile fabric (i.e., no discard or waste) were identified. These strategies are discussed below.

Merge seams. In many zero waste patterns, side seams and shoulder seams were merged. This was a way to avoid the triangles of wasted fabric that is created when generating shoulder slopes or waist shaping. Unsewn darts were converted to flare.

Wrapping. Belts or sashes were often used to fit the garment to the body, as well as another way to utilize all the textile yardage. The need to contour the garment to the body is related, in turn, to the avoidance of shaping devices as a means of waste circumvention.

Create body openings with slits instead of curves. This strategy avoided the generation of the circular-shaped wasted fabric that is typically created along armscyes and neck openings. Neck openings may be a slit straight across the shoulders of a T-shape.

Creative use of cutout pieces. If leftover pieces were created through side seam shaping, armscyes, or neck openings, these pieces were creatively incorporated into the garment design. For example, neckline cutout pieces could be used as reinforcement fabric at stress points in the garment.

Tessellation. In this technique, repeated triangular- or rectangular-shaped pattern pieces are fitted together with no space or overlap as a way to eliminate waste in the cutting process. Tessellation refers to use of the same shape repeatedly to compose the entire garment silhouette. It must be a shape that interlocks to create a continuous 2D plane that corresponds to textile. However, the shapes must be able to be reconnected in a way that creates the desired 3D garment silhouette.

Zero Waste Patternmaking Garment Outcomes

The following garment outcomes were commonly found in zero waste garment designs.

Loose and boxy silhouette. Students found that most of the patterns they selected utilized the traditional patternmaking principle of *added fullness*, which resulted in loose and boxy garments. In the attempt to minimize use of shaping devices (seams, neck and armhole openings) so as to avoid textile waste generation, the garment silhouette commonly resulted in a loose and boxy shape. As observed through content analysis of publicly available zero waste patterns, fitted silhouettes are possible with the incorporation of numerous seams and dart manipulation.

Shoulder as the supporting point of the garment. Most zero waste design garments hung from the shoulders. Strapless garment designs were not seen. This could be due to the need to contour a garment closely to the body to create a strapless garment. To do so would require shaping devices and corresponding fabric waste.

Benefits of Zero Waste Patternmaking Approaches

Most importantly, zero waste patternmaking methods eliminate, or at least largely reduce, contribution of textile waste for the landfills. Another interesting outcome the authors observed was that zero waste patternmaking techniques most often resulted in garments that did not copy current silhouettes—resulting in unique design outcomes. For example, designs had seams in unexpected locations, such as diagonal, instead of vertical side seams. Such uniqueness may give apparel brands a competitive advantage. Additionally, some zero waste patternmaking methods, particularly those in the paradigm of rectangular cuts, have fewer seams than traditional garments. This may reduce apparel production costs in cutting and sewing aspects. In addition, zero waste patternmaking garments may feature universal design, allowing people of diverse sizes to wear the type of garments because of their loose and boxy silhouette.

Challenges of Zero Waste Patternmaking Approaches

Zero waste patternmaking methods are often size-constrained. It was found that most methods only produced a garment of a single size. For some students, creating their first samples of selected patterns was extremely challenging and required several iterations to achieve the design. This was because the instructions for the selected pattern were for a different size and did not translate well to the dress form size.

Textile width and length affect design. Textile widths are determined by fabrication equipment, such as a weaving loom, and cannot be changed. This is a problem if the design you are trying to produce is engineered for a different width fabric than what the designer has.

Difficulty finishing edges. Many zero waste designs tested by the students lacked consideration of how to finish the edges. This was particularly challenging when slits were cut into garments for neck or arm openings. Properly finished edges are important as they contribute to the longevity of the garment and may impact quality perceptions from a customer standpoint.

Off-grain. It was found that the grainline of the pattern piece to the fabric was often unusual, breaking traditional rules of aligning the length of pieces on the straight grain of the fabric. These off-grain and cross-grain situations typically occurred as the result of seam merging in the pattern design. However, in some cases pieces were purposely cut on the bias grain to help the garment wrap around the body.

Lack of mass production knowledge. Some of the published zero waste designs used solutions to eliminate waste that were not in line with mass production practices. For example, zero waste pattern designers tended to add additional adornments or pockets that might not be necessary to remove the negative spaces. While using these scraps as adornments eliminates waste, it also adds cost in terms of sewing the added elements to the garments. To increase use of zero waste practices, designers and producers must be aware of all design decisions on mass production feasibility and costs.

Insufficient directions. Available zero waste designs lacked clear and comprehensive pattern drafting directions. Insufficient development description, instructions, and visual references may result in a poor garment fit or waste generation. These insufficiencies in design communication make it extremely challenging for others to reproduce the method successfully.

Discussion and Implications

The findings of this detailed review of zero waste patternmaking methods provide practical knowledge for apparel professionals (Delong et al. 2013). Let us consider how professionals in the apparel industry and researchers may use this knowledge. Both apparel designers and patternmakers can benefit from these study findings. Further, it is clear that both designers and patternmakers must work synchronously to achieve a zero waste outcome. Alignment of development team members would require a change to current fashion industry processes (Fletcher 2012). Ultimately, the goal is for the fashion industry to increase the use of zero waste patternmaking strategies and reduce textile waste.

Zero Waste Design Approaches and Relationship to Existing Patternmaking Frameworks

In the paradigm of rectangular cuts, zero waste patternmaking garments feature tunic, djellaba or kaftan, or sandwich styles. The type of garments includes minimal cuts and least darts. These styles were developed by contemporary clothing designers such as Yeohlee Teng or Zandra Rhodes, who are pioneers of zero waste garments. Typically, rectangular-cut designs present loose silhouettes and are more likely to fit a variety of sized individuals. Variability of fabric widths may not affect the overall size of the garments. Designs in the paradigm of rectangular cuts do not relate the traditional patternmaking method (e.g., Joseph-Armstrong 2010). For example, designs do not show the typical sloper shape, and with enough amount of added fullness, darts of the basic slopers are ignored. Traditional shoulder or side seams are not used in pattern development. Instead, one or a couple of whole pieces may wrap across the body with being folded or manipulated without any seams. Experimental patternmaking is required in the rectangular-cut process.

In the paradigm of tailoring matrix, zero waste patternmaking garments feature existing ready-to-wear styles. Garments usually incorporate traditional seams and show basic sloper shapes. By adopting Jigsaw method or TR method, designs in the paradigm of tailoring matrix minimize negative spaces in the pattern layout. Strategies for the approach are modifying some curvy lines in the pattern or intentionally creating the negative spaces to put all of the pieces together. Therefore, it may require a lot of time and effort to achieve the designated pattern layout. In addition, there could be possible fit issues because the garments may be a fitted or semi-fitted silhouette. In addition, the garment patterns developed can only be associated with one particular size because overall dimension of textile (width and length) indicates the full puzzled layout. Apparel industry professionals may use the identified major zero waste patternmaking techniques and their relationship to existing patternmaking concepts, to easily comprehend and use zero waste patternmaking.

Strategies Used to Avoid Fabric Waste in the Patternmaking and Apparel Design Process

Similarly, apparel industry professionals may use the identified major zero waste patternmaking strategies as a roadmap to implement zero waste patternmaking in their own company structure. One simple and effective strategy is to eliminate garment seams by merging two pattern pieces together. As a result, the "extra" space between pieces becomes added fullness around the body. If such fullness was not desired, designers and patternmakers, as a team, could explore garment adjustability strategies (McKinney and Wei 2019) to control excess fabric in a creative way. Currently, zero waste patternmakers use wrapping and belting to fit the garment to the body, which is an easily implemented strategy and often in alignment with prevailing fashions. Another simple strategy is to convert circular neck and armscye openings into slits.

Tessellation is another interesting approach to eliminating fabric waste. This strategy would require some additional thought by designers and patternmakers but could result in innovative, as well as fabric-efficient, outcomes. First, the designer must consider how the desired 3D garment silhouette could be divided into 2D shapes, which could be rotated to fit together into a flat planar structure (i.e., textile yardage). As such, tessellation is a bit of a puzzle-solving exercise. As 3D virtual patternmaking continues to develop, it may become easier for designers to visualize and implement such patternmaking methods. This is an area that researchers could assist in exploration for future works.

Garment Outcomes Achieved with Zero Waste Patternmaking

Boxy garment silhouettes that hang from the shoulders are often the outcome of minimizing the use of shaping devices (seams, neck and armhole openings) to avoid textile waste generation. By knowing what outcomes are typically achieved, designers may design into these silhouettes. This knowledge gives designers a starting point to achieve a zero waste outcome more easily.

Conversely, the known outcomes give researchers and designers a challenge. How can zero waste techniques be used to create more fitted garment silhouettes or strapless silhouettes? Some researchers have already begun to tackle these issues. McKinney and Bennet (2015) created a strapless gown with minimal waste. However, pattern shaping used to fit the bodice resulted in some waste. Saeidi and Wimberley (2018) successfully created fitted zero waste designs using the TR method. Creativity is needed to develop means for creating zero waste contoured garments. To achieve fitted garment silhouettes, development of patternmaking methods is further needed besides the existing methods such as TR (Saeidi and Wimberley 2018) or Jigsaw methods (Rissanen 2008).

Benefits of Zero Waste Patternmaking in the Apparel Manufacturing Process

Equipped with knowledge of the benefits of zero waste patternmaking, designers and patternmakers may be more likely to integrate these techniques into their own practice. Due to the challenging nature of using 100% of textile yardage, designers must dig deeper into their own creativity to find a solution. This is beneficial as an impetus for creative unique garments in an overcrowded fashion marketplace. Another benefit of zero waste patternmaking may be a reduced number of seams provided by combining seams that may reduce cutting and sewing costs.

Challenges of Zero Waste Patternmaking in the Apparel Manufacturing Process

The identified challenges may serve as future research direction. Garment sizes increase more in circumference, than they do in length (Mullet 2009). Therefore, the proportion of the pattern shapes changes as sizes increase or decrease. For example, a sleeve pattern that fits neatly between two bodice pieces on the cloth may no longer fit if the size is increased (or decreased). Work is needed to develop zero waste patternmaking methods that work for a range of sizes. As indicated by Saeidi and Wimberley (2018), our findings confirm that work is needed to develop zero waste patternmaking methods that function for a range of fabric widths. Most published methods are engineered for a specific width of fabric; yet fabric widths can vary from 27 to 60 in (Hatch 1993). Fabric width can significantly affect the layout of pattern pieces and/or finished garment size.

Another challenge common in zero waste patternmaking is devising appropriate edge finishes for slit openings, due to their lack of seam allowances. This issue may be explored by researchers through laser cutting (Rorah 2016), pre-finished edges of designs using digital embroidery (McQuillan and Rissanen 2011), or other creative methods. Alternatively, designers may come up with creative strategies to use the circular pieces from shaping devices, as functional or decorative elements. Even more creatively, designers may consider cutting the neck and arm openings into shapes that correspond to pattern shapes needed elsewhere in the garment. For example, armscyes could be cut as square shapes, which could be used as patch pockets.

Another challenge of zero waste patternmaking is that, when the pattern shape is cut, the fabric often ends up off-grain once sewn into full garment form. This can create unwanted fit problems or twisting effects in the final garment. It can be difficult to sew darts in a fabric that is not on-grain. Future researchers may consider how to design without waste while keeping the fabric on-grain. In general, many zero waste methods are incompatible with commonly used mass production strategies. Majority of the previous studies of zero waste patternmaking approaches were focused on experimental patternmaking and limited production; however, only one design does not have to fill out the full dimension of fabric. Parts of designs from different fabrics (McQuillan 2011) or multiple styles within one layout (Carrico and Kim 2014) can be applied to achieve zero waste patternmaking in mass production. Further research is needed to identify and address such issues.

Due to the addressed issues, designers and patternmakers may start to consider seeking change of the traditional apparel production process. Zero waste pattern development process may be involved in the stage of textile production, which was usually considered the first stage of apparel production. For instance, Holly McQuillan et al. (2018) has introduced pre-finished edges of designs using digital embroidery on the fabrics for her zero waste patternmaking designs. All the edges were pre-finished using digital embroidery and desired print designs were digitally printed in the designs.

Finally, it was observed that many of the publicly available designs, particularly through online sources (e.g., blogs, Pinterest), lack full explanations about the patternmaking technique employed in their created design. This limits the usefulness of information. Future practitioners should prioritize completeness and replicability in the sharing of their work, to further the environmental benefits of zero waste patternmaking.

Conclusions

Zero waste patternmaking approaches can be divided into two major paradigms—rectangular cuts and tailoring matrix. Within each paradigm, there are subcategories of garment styles. This organizational scheme allows designers and patternmakers to develop an overall understanding of the different types of patternmaking approaches that exist within the zero waste paradigm. This research has uncovered certain commonalities among patterns in the paradigms. For example, a garment made with the rectangular-cut approach is likely to have a loose and boxy fit, hang from the shoulders, and require a minimal amount of cutting and sewing. Armed with this knowledge, apparel makers can make better choices about the approach to achieve their desired garment outcome.

This study has positioned zero waste patternmaking approaches within the realms of current and historical methods. Further, it has delineated commonly used strategies in avoiding textile waste. Finally, this study contributes knowledge of achievable garment outcomes, benefits, and challenges of zero waste patternmaking. This knowledge may help to de-mystify zero waste patternmaking and increase its use.

The knowledge contributed through this study also gives focus to future research efforts. Researchers may explore garment adjustability strategies (McKinney and Wei 2019) to take in the excess fullness generated by rectangular-cut approaches. Also, in line with sustainable clothing design, the concept of universal design may be applied with zero waste patternmaking because the method often features non-tailored silhouettes; therefore, people of diverse sizes may wear the garments.

Alternatively, designers may explore how zero waste techniques could be used to create more fitted garment silhouettes or strapless silhouettes. 3D virtual patternmaking software may help designers visualize and implement this research. Two additional research areas ripe for improvement are the development of zero waste patternmaking methods that work for a range of garment sizes and fabric widths. Additionally, research is needed for exploring appropriate edge finishes for slit neck and arm openings, how to design without waste, and ways to keep the fabric on grain, as well as making zero waste methods compatible with mass production. To continue to build the body of knowledge, it is essential for future practitioners to share their work using adequate details and images to ensure replicability. By using the findings from this study, as well as continuing to build the body of knowledge around zero waste patternmaking, apparel researchers, designers, and patternmakers may together reduce the negative impacts of the apparel industry on the environment.

References

Almond K (2010) Insufficient allure: the luxurious art and cost of creative pattern cutting. Int J Fashion Des Technol Educ 3(1):15–24. https://doi.org/10.1080/17543260903582474

Antanavičiūtė A, Dobilaitė V (2015) "Principles of slow fashion application in clothing collection creation." Environ Res Eng Manag 71.2 :54–59

Ataniyazova OA (2003) Health and ecological consequences of the Aral Sea crisis. In: 3rd World Water Forum, Regional Cooperation in Shared Water Resources in Central Asia, Kyoto

Bye E (2010) A direction for clothing and textile design research. Clothing Text Res J 28(3): 205–217. https://doi.org/10.1177/0887302x10371505

Carrico M, Kim V (2014) Expanding zero-waste design practices: a discussion paper. Int J Fashion Des Technol Educ 7(1):58–64

Creswell JW (2014) Research design: qualitative, quantitative, and mixed methods approaches, 4th edn. SAGE, Thousand Oaks

Delong M, Goncu-Berk G, Bye E, Wu J (2013) Apparel sustainability from a local perspective. Res J Text Apparel 17(1):59–69. https://doi.org/10.1108/RJTA-17-01-2013-B006

Fletcher KT (1999) Environmental improvement by design: an investigation of the UK textile industry. (Doctoral dissertation, Open University)

Fletcher K (2012) Fashion & sustainability, 1st edn. Laurence King, London

Fletcher K (2013) Sustainable fashion and textiles: design journeys, Routledge.

Gam HJ, Cao H, Farr C, Heine L (2009) C2CAD: a sustainable apparel design and production model. Int J Clothing Sci Technol 21(4):166–179. https://doi.org/10.1108/09556220910959954

Gwilt A (2013) A practical guide to sustainable fashion. Fairchild Books, London

Hamre I, Meedom H, Artman I (1980) Tyg och funktion:[att skapa kläder av raka tygstycken]. Wahlström & Widstrand, Stockholm

Hatch KL (1993) Textile science. West Pub, St. Paul

Hollander A (1994) Sex and suits, 1st edn. Knopf, New York

Joseph-Armstrong H (2010) Patternmaking for fashion design. Pearson Education/Prentice Hall, Upper Saddle River

Krippendorff K (2019) Content analysis: an introduction to its methodology, 4th edn. Los Angeles, SAGE

Lawless E, Medvedev K (2016) Assessment of sustainable design practices in the fashion industry: experiences of eight small sustainable design companies in the Northeastern and Southeastern United States. Int J Fashion Des Technol Educ 9(1):41–50. https://doi.org/10.1080/17543266.2015.1116616

Leonie B (2017). Is cotton conquering its chemical addiction? Just-style.com, Just-style.com, Oct 11, 2017

Lindqvist R (2015) Kinetic garment construction: Remarks on the foundations of pattern cutting. University of Boras, Boras

Lumsden, J (2010) A garment for the upper body with a collar and sleeves and buttons down the front: shirts designed through patternmaking: an essay presented in partial fulfilment of the requirements for the degree of Master of Design at Massey University, Wellington, New Zealand. Diss. Massey University

McKinney E, Bennet S (2015) Sparkling socialite. a snow-dyed silk gown using inventive patternmaking and construction techniques for minimal fabric waste. International Textile and Apparel Association, ITAA 2015 Design Catalog

McKinney E, Wei B (2019) Patternmaking methods for creating size-adjustable garments. In: Moore JG (ed) Patternmaking history and theory. Bloomsbury Visual Arts, London

McKinney E, Stanley AE, Plummer B, Thompson K, Rorah W (2016) Tacit magic: understanding the principles behind Nakamichi's creative pattern cutting. Int J Fashion Des Technol Educ 9(2):130–141. https://doi.org/10.1080/17543266.2016.1165740

McQuillan H, Rissanen T (2011) YIELD: Making fashion without making waste. Twinset, The Dowse

McQuillan H, Archer-Martin J, Menzies G, Bailey J, Kane K, Fox Derwin E (2018) Make/use: a system for open source, user-modifiable, zero waste fashion practice. Fashion Pract 10(1):7–33
Mullet KK (2009) Concepts of pattern grading: techniques for manual and computer grading. Fairchild Books, New York
Neuendorf KA (2002) The content analysis guidebook. Sage, Thousand Oaks
Pedersen EL (2007) Theory is everywhere: a discourse on theory. Clothing Text Res J 25(1):106–128
Quinn S (2015) Fibers: exploring healthy and clean fiber. In: Hethorn J, Ulasewicz C (eds) Sustainable fashion: what's next? A conversation about issues, practices, and possibilities. Fairchild Books, New York, pp 148–206
Reynolds PD (2007) A primer in theory construction. Pearson/A and B, Boston
Rissanen T (2008) Creating fashion without the creation of fabric waste. In: Hethorn J, Ulasewicz C (eds) Sustainable fashion: why now? A conversation exploring issues, practices, and possibilities. Fairchild Books, New York, pp 148–206
Rissanen T (2016) Zero waste fashion design. London: Fairchild Books, an imprint of Bloomsbury Publishing
Rissanen T, McQuillan H (2016) Zero waste fashion design. Vol. 57. Bloomsbury Publishing
Rorah W (2016) Laser cutting and etching textiles and apparel design: an experimental study on the implementation and documentation of laser cutting and etching in the apparel design creative process. Iowa State University, Ames
Saeidi E, Wimberley VSJI (2018) Precious cut: exploring creative pattern cutting and draping for zero-waste design. J Fashion Des Technol Educ 11(2):243–253
Townsend K, Mills F (2013) "Mastering zero: how the pursuit of less waste leads to more creative pattern cutting." International journal of fashion design, technology and education 6.2: 104–111
Vennström K (2012) Sustainable fashion consumption: An interactive system between consumers and institutions. Stockholm University

Factors That Affect Sustainability in the Textile Design Industry in Kadoma, Zimbabwe

Mercy Rugedhla, P. L. P. Disele, R. Moalosi, and L. C. Fidzani

Abstract As sustainability gains prominence in many spheres of life and disciplines of study, there is also a need to produce sustainable textiles, or ethical fabrics. Sustainability in textile design aims at producing fabrics or textiles that provide the greatest long-term benefits for all the users at any given time to sustain the textile design industry. This paper explores the need for sustainable approaches for sustainable textiles and materials in Kadoma, Zimbabwe. It also highlights the need to incorporate environmental, social and economic sustainability in textiles and the textile industry as they produce sustainable textiles. The paper also outlines the factors that affect sustainability in textile designing companies in Kadoma and explains how the textile industry can gain sustainability through the use of sustainable textile production and people-centred designing methods for each of the factors as they produce sustainable textiles. The paper ends by recommending sustainable textile design strategies for the textile design industry in Kadoma.

Keywords Sustainability · Sustainable textiles · Sustainable textile design · Sustainable textile material

Introduction

Sustainable design emphasises the need for the reduction of the environmental burden. This is in line with the United Nations' goals for sustainability. Declaration number 33 of the United Nations' 2030 for sustainable development agenda states that members should determine to conserve the natural resources in

M. Rugedhla (✉) · P. L. P. Disele · L. C. Fidzani
Department of Family and Consumer Sciences, University of Botswana, Gaborone, Botswana
e-mail: diselep@mopipi.ub.bw; fidzani@mopipi.ub.bw

R. Moalosi
Department of Industrial Design and Technology, University of Botswana, Gaborone, Botswana
e-mail: moalosi@mopipi.ub.bw

S. S. Muthu, M. A. Gardetti (eds.), *Sustainability in the Textile and Apparel Industries*, Sustainable Textiles: Production, Processing, Manufacturing & Chemistry, https://doi.org/10.1007/978-3-030-37929-2_3

order to promote social and economic development through managing the resources sustainably. For this reason, the textile industry has been challenged greatly in terms of sustainable practices on its production processes. Bonilla et al. (2018) commend the industry's contribution to social welfare through the supply of high-quality products. However, they note that the problem of the current production is not environmentally sustainable. Nerurkar (2016) alludes to the negative impact the clothing industry has resulting from the indigenous raw material production and disposal of such products among other things. Thus, producers are encouraged to find ways of using materials (resources) that could bring a positive change and reduce the challenge. Examples include the pollution of rivers from dyes and the depletion of natural resources from use. Product sustainability has been found to be the answer to this challenge hence the emphasis of this study to identify the factors that affect textile design and textile products in Kadoma in order to suggest sustainable strategies for the industry. Hence, the paper proposes the incorporation of sustainable textile design methods to enhance the textile design industry in Kadoma.

The Cambridge Dictionary (2019) defines sustainability as the quality of being able to continue over a period of time: or the quality of causing little or no damage to the environment and therefore be able to continue for a long time. Sustainability thus has to do with a number of factors such as the quality of performance over a period of time. It also has to do with the quality of resources and their durability. According to Walker (2013), sustainability has emerged as a central issue for contemporary societies all over the world, and many of the social and environmental concerns that are embodied in sustainability are related to design directly or indirectly. He adds that sustainability emerged in response to perceived public discontent over the products and services offered by those in business.

The United Nations World Commission on Environment and Development (1987) defines sustainable development as development that meets the needs of the present without compromising the needs of future generations. Echoes the same which shows that this notion is accepted and adopted by many researchers about sustainability. Concurring, Walker (2013) links the work of the designers to sustainability and that it helps to define how the environment is produced and used and how long it endures. Similarly, Chen, Li and Mo observe that sustainable design approaches consider the environmental impacts of products during their whole life cycle. The application of sustainable design is aimed at the development of more environmentally benign products and processes. According to them, the aims and scope of sustainable design should not be separated from those of traditional design but be an extension of traditional design. They highlighted that sustainable design without any cultural meaning would not be acceptable by the society. Thus, Tonelli et al. (2013) define industrial sustainability (IS) as the end state of a transformation process in which industry is part of and actively contributes to a socially, environmentally and economically sustainable planet.

Alluding to this notion, Bonilla, Silva, Silva, Goncalves and Sacomano (2018) indicate that companies now recognise the benefits and competitive advantage of sustainability which include the satisfaction of the stakeholders and the improvement of financial performance through new foreign markets. To achieve sustainability, different companies now employ the concept of the 3Ps, thus 'people, planet and profit'. Measures (2017) defines 3P as a business model that was developed to encourage social responsibility and sustainability among business worldwide. This model emphasises that an organisation should take steps to ensure that the company's operations benefit the employees. This should help the workers to find value in their work. At the same time, the organisation should engage in activities that promote and not harm the environment and as much as possible reduce negativity impacts in the environment. By so doing, the company will contribute to the well-being of the planet. In textile design the concept of 3Ps is employed in areas such as recycling, upcycling, slow fashion and sustainable product-service system. It is against this background that this chapter will discuss sustainable approaches in the textile industry as well as factors that affect sustainability in textile design industry in Kadoma. Furthermore, the study will also discuss the current practices in the textile design industry and textile designing and sustainability strategies that can be used to design sustainable textile.

Sustainable Approaches in the Textile Industry

The whole world has been experiencing climatic changes which have been linked to the various production methods from the various industries including the textile design industry. This calls for the industry to devise and implement sustainable approaches that make the industry sustainable and produce sustainable textile products. The textile design industry has implemented approaches to reduce pollution such as recycling and upcycling among other things to encourage sustainability.

Recycling of Textiles

Recycling is one of the major aspects of sustainability that are related to textiles. This term refers to reusing or reproducing. According to Worbin (2010), traditional textiles have a way of recycling material over time. By the same token, the industry is encouraged to use recycled materials such as fibre to reduce pressure on virgin resources and reduce pollution. In the same manner, consumers are also expected to participate through recycling, re-fashioning, restyling and swapping of investing in quality when buying. Some of the benefits of recycling are that it benefits the needy and the environment as well as creates a sustainable market. Fabrics from recycled materials have become more popular in the industry. One such example is the polyester which is made from recycled bottles. Recycling of fashion is usually done to save or preserve the resources. The other reason is to use materials to their maximum capacity before discarding them. Materials have been recycled in the textile industry to preserve the environment through dumping waste and avoiding intoxicating the environment. Textile products should be biodegradable for easy disposal.

Upcycling

Upcycling is about using waste material to do new products. Vadicherla, Saravanan, Ram and Suganya (2001) observed that textile waste can be used as the raw material for value-added products. Consumers can also assist by avoiding the depletion of the natural resources and through creating and maintaining an ecological balance. Upcycling is a very simple concept that could go a long way in saving our communities from environmental pollution. Many industries have a lot of waste with which they do not know what to do with it. Upcycling helps them get rid of their waste with some benefits.

Bonilla, Silva, Silva, Goncalves and Sacomano (2018) allude to a more concerned society, public sector and nongovernmental organisations demanding that the industrial sector should produce in an economically, environmentally and socially sustainable manner. Nerurkar (2016) cites design as the way of delivering the sustainable solutions to the environment. Saeed and Kersten (2019) state that organisations have adopted sustainable practices due to increased social and environmental awareness.

The Textile Industry in Zimbabwe

The African continent is characterised by a wide variety of climate, temperature and terrain. In many African countries including Zimbabwe land is of great significance. The country's position on the map of Africa made it easy to trade with the surrounding nations. Sithole et al. (2016) argue the textile and clothing industry are the key manufacturing subsectors in Zimbabwe. Mlambo (2017) confirms this and further explains that in 1980 Zimbabwe was the second most industrialised country in sub-Saharan Africa.

Sithole et al. (2016) highlight that the textile and clothing subsectors in Zimbabwe have played and continue to play a major role in the economy. The clothing and textile industry in Zimbabwe was wide and boasted on employing more than 20,000 workers. It consisted of a wide variety of companies which dealt with a wide variety of textile products. They included ginners, spinners and fabric and apparel producers. Kadoma Textiles was one of the first textile companies to be opened in Zimbabwe. Zimbabwe is agro-based and Kadoma is a city in the central part of Zimbabwe. Kadoma Textiles is named after the town. Farmers from and around Kadoma produced a lot of high-quality cotton; as a result the city housed a lot of clothing and textile companies in Zimbabwe. According to Mlambo and Phimister (2006), these companies experienced great success between 1950 and 1980 and attracted both the local and international market. According to Sithole et al. (2016), the textile and clothing industry in Zimbabwe started with the establishment of three ginneries in the then Gatooma town which is now called City of Kadoma which was funded by the British Cotton Growing Association (BCGA). The first cotton-spinning mill was built in 1941 in the then Gatooma town which is now called Kadoma.

Mlambo and Phimister (2006) advances that in 1954 the industry continued to grow and the number of companies increased with the major ones being Kadoma Textiles, David Whitehead, Security Mills, Textile Mills, Lancashire Cotton Corporation and Rhodesian Weaving Mills which manufactured most of the country's blankets.

A number of the African countries rely on their textile industry for economic growth. However, some such as Zimbabwe have not been able to sustain the textile industries to provide for their apparel as well as for their other textile needs continuously. Zimbabwe used to produce many textile products and fabrics through its textile industry, which according to Mlambo and Phimister (2006) flourished between 1948 and 1989. This has led to the closure of most of the textile industries including Kadoma Textiles which used to produce large quantities of fabric. Currently, most of the fabrics that are used are imported from the other continents and neighbouring African countries. Mlambo and Phimister (2006) also noted that the most difficult times were between 2000 and 2010 due to the depreciation of the Zimbabwean dollar.

Factors that Affect Sustainability in the Textile Design Industry in Kadoma

Zimbabwe used to have a very vibrant textile industry with Kadoma being the centre of textile activities. It housed most of the textile production and designing companies. These specialised in the cotton fibre which was produced locally. Unfortunately, these have now closed. The closure of the companies shows the lack of sustainability that is in most textile companies not only in Zimbabwe but across Africa. As observed by Tonelli et al. (2013), Bonilla et al. (2018), Mpofu (2012) and Mlambo (2017), the major factors that affect sustainability in textile design are resources, restrictive policies, cooperation between internal and external functions, current practices, consumer consumption, implementation of sustainable initiatives and the cost of labour.

Textile Resources

Tonelli et al. (2013) envision a future industrial system that delivers high value to international customers while using less than a quarter of the current resources. There is great need for the textile industry to reduce the amount of resources that are used in most textile designs and products. This issue becomes of great concern since some of the resources are natural and may not be recovered or replaced easily.

According to Bonilla et al. (2018), the rate of exploitation of natural resources should not exceed the rate of regeneration, and the rate of waste generation should not exceed the assimilative capacity of the biosphere of environmental stocks. This study assumes that some of the resources from plants, e.g. cotton, could be saved by changing the methods of growing them. An example would be to allow cotton to be harvested for a span of 2–5 years without destroying the plant to protect the environment as well as the people. According to Mpofu (2012), the social or economic position of the country also affects the operations of the industry. This has been true in the case of Zimbabwe where the companies' operations have been directly affected by the various laws and policies as cited earlier in this paper.

Laws and Policies

Laws and policies also impact sustainability in the textile industry. These could be local (from within the nation of operation or company) or global. They range from social, political or economic. Mlambo (2017) advances that laws and policies can promote or destroy the economy of a nation. Policies are binding once they are signed, and hence there is not much that the organisations and companies can do to safeguard themselves against restrictive policies. However, one way is to collaborate with other similar companies to avoid being limited to one place. An example of these policies is the Multifiber Arrangement (MFA). According to Kenton, the Multifiber Arrangement was an effort by the United States and the European Union (EU) to protect their domestic textile industries by restricting imports from developing countries. Africa is still developing and Zimbabwe is among the developing countries; hence the restrictive policies seem to be among one of the key factors affecting textile design in Zimbabwe.

Zimbabwe introduced many economic policies between 1989 and 2018. Most of these impacted negatively upon the textile industry. Mlambo (2017) cites the Economic Structural Adjustment Program (ESAP) which was implemented by the government of Zimbabwe as one of the government policies that had negative effects on the manufacturing sector of Zimbabwe with the textile manufacturing being affected most. The World Bank initiated the ESAP in 1991. Under this program companies that had been receiving subsidies from the government had to be given less than their usual amounts. There were no notices given and this forced many companies to reduce their production. Most of them had to devise strategies for survival instead of growing. The majority of them had to reduce production as well as the workforce. This to an extent also shows that many of them had no sustainable strategies in terms of avoiding termination and living long.

Mpofu (2013) observed that the textile industry was facing a great challenge for second-hand clothing, especially in light of the policy of duty-free finished clothing. Second-hand clothes sell cheap because the expenses are lowered by free duty. The pricing of products is also determined by the policies that govern the industry. Thus companies need to acquaint themselves with these in order to be informed by them in their operations. In 2017, the Reserve Bank of Zimbabwe observed that many cotton farmers were affected by the low prices for cotton and were abandoning cotton farming. This in turn impacted negatively on the textile industry in Kadoma because in the absence of locally obtained cheaper cotton, the companies have to import cotton. The Reserve Bank also notes that the other challenge that faces the textile industry in Zimbabwe is that the international prices of cotton are not constant making it difficult for the industry to plan and budget for business.

Cooperation Between Internal and External Functions

The need for cooperation between internal and external functions should never be underestimated. Tonelli et al. (2013) suggest that industrial sustainability requires cooperation between the external agents in the supply network and an organisation's internal functions. As seen in the example of companies in Zimbabwe, especially Kadoma being affected by external policies to the extent of closure, it is important to ensure understanding and cooperation between the external and internal forces. These functions include the workers, administrators, suppliers as well as consumers. Lack of cooperation at any level may lead to failure or a compromise of the standards. Internal policies refer to the company's laws or policies as well as those of the country where the company is operating from.

The Current Practices in the Textile Design Industry

Bonilla et al. (2018) state that the current practice in industry contributes to the depletion of nonrenewable resources, climate change and a loss of biodiversity, among other ecological impacts through its encouragement of increased production. Currently, cotton fields are supposed to be burnt after the cotton harvest to prepare the field for the next crop. This is not environmentally friendly. Tonelli et al. (2013) advance that the current approaches are focused on manufacturing technology, supply chain management and product-service systems. Zimbabwe has been greatly challenged by the poor economic situations, and of late the manufacturing companies have been failing to procure the needed modern technology and machinery. This makes them unable to meet their production targets in terms of quality and quantity. The supply chain for the companies in Zimbabwe has also been affected by policies such as Multifiber Arrangement which reduces the chances of exports resulting in limited profits. These observations are based on the understanding that the current practices focus more on profit-making rather than sustainability.

Consumers' Consumption

Consumers' consumption of products matters. The amount of products that consumers use determines the production rates. Companies often produce to meet the consumers' demands. There is a need then for consumers to be educated and encouraged to practice sustainable consumption. This means that they should be able to use the textile products taking into consideration aspects of sustainability such as saving materials and recycling. Bonilla et al. (2018) note an apparent inconsistency between attitudes towards sustainable consumption and actual behaviour in the actual purchase of sustainable products by consumers. Person product attachment (PPA) could help increase the period of time that a consumer will use the textile product for before discarding it. This will help to reduce the levels of waste since a number of textile products are not biodegradable and usually not easy to dispose of. Bonilla et al. (2018) also observe that consumers have not adopted the principles of sustainable consumption; hence their demand of products is far beyond their real needs. Nerurkar (2016) talks of the need for person 'product attachment' for consumers to be able to keep the textile products for a long time. Zimbabweans do not seem to emphasise much on product attachment, hence the need for the textile industry to produce products that have some relationship with being Zimbabwean. Furthermore there will be a need for the people to be well informed of the products and their values for the people to appreciate them.

Implementation of Sustainability Initiatives

Tonelli et al. (2013) propose that sustainable industrial activities should focus on product design and product end-of-life management that lead to industrial manufacturers exploring significant savings in energy, water and waste among other things. Nerurkar (2016) commented that the role of sustainable design is to consider materials, uses and after use rather than just aesthetics and, therefore, all-encompassing. Saeed and Kersten (2019) state that the implementation of sustainability initiatives not only improves the environmental and social performance of organisations but also provides them a competitive advantage by acquiring a new set of competencies. Sustainability can be incorporated into design during any and all the phases of the design process. Gwilt (2012) explored sustainability and fashion design and proposes the integration of sustainable design strategies during the design process in order to allow for a change in the way products are produced and discarded. The textile industry in Zimbabwe has to incorporate sustainability initiatives throughout the production process and especially incorporate foresight. There is a need to project the operations of the company over a period of time taking into consideration the current trends in the industry globally.

The Cost of Labour

The cost of labour highly impacts the sustainability of the textile industry. Labour costs are generally high in Africa as well as in Zimbabwe. This calls for well-calculated financial operations for the company to do well in business. However, although Zimbabweans are highly educated, the labour costs in industry have been relatively low for a long time, and this gave the textile design industry an advantage during its hay days. Mlambo confirms that one of the advantages of the textile industry in Zimbabwe was that the labour costs in the country are low. This probably helped the textile companies to gain large profits.

Another challenge that the textile companies or organisation may face is to engage unskilled labour which may be cheaper but compromise the standards of the textile products to be produced. Kenton gives an example of Bangladesh which was expected to suffer the most from the ending of the Multifiber Arrangement as a result of more competition from countries such as China but did not because they had an advantage of cheap labour.

Textile Designing and Sustainability Approaches

Nerurkar (2016) states that design contributes 60–80% of the life cycle of a product; therefore, there is need to impart sustainability components in the textile designs. Textile designing is a creation of designs or pattern through the use of fibres. The fibres could be natural or synthetic. Textile or fabric designs could be produced through a wide variety of techniques or skills such as bonding, knitting and weaving. Worbin (2010) observed that new kinds of building blocks have been introduced into textile design for creating dynamic textile patterns. This is done through the use of various computer-aided drawing software. Computer-aided design (CAD) refers to any design that is done using a computer. This software makes it possible to make very specific and precise illustrations of a very wide variety of designs.

Mitra (2014) indicates that in the textile industry, computer-aided design and computer-aided manufacture (CAM) systems cover a very large application area including bed covers, towels, plastic mats, carpets, dress materials, sarees, laces, table mats, labels, knitwears, suiting and shirting, printing fabrics, furnishing upholstery and many others. The software also allows for specific colour matching. The designer can, therefore, introduce all the colours they want and be sure to have them well matching. Many colours with different textures can be introduced at the same time. This is appropriate for the multi-coloured African designs. Mitra (2014) further indicates that the computer-aided design (CAD) or computer-aided manufacture (CAM) module for printing has more than 200 tools and utilities, which include but not limited to the number of colours and repeat or size of design creation. This is very appropriate for the production of African textile products because Africans tend to love and use many colours; hence the colours, patterns and sizes will be put into perspective without difficulty. This technology has many advantages including changing designs to any scale and saving time, hence the need to maintain competitive market through satisfying the clients (Jhanji 2018). Computer-aided design enables manufacturers to meet this need as it makes it easy to make adjustments at any point of the process.

Computer-aided manufacture results from computer-aided design. Designs generated from computer-aided design are naturally complicated and intricate. They cannot be done manually, hence the introduction of computer-aided manufacture (CAM) because they have to be manufactured through the aid of computers. The process of computer-aided manufacture facilitates the manufacture of multidesigns and thus affords the textile industry the opportunity to producing in bulk. These include patterns such as Bobby and Jacquard. Supporting this view, Mitra (2014) states that computer-aided design and computer-aided manufacture software have developed many modules such as the edit module, and Print module Jacquard module by combining an excellent collection of painting tools, and powerful retouching capabilities all-in-one easy to use Windows application. The company only has to choose the appropriate modules for the designs that are most appropriate for Zimbabwe. Thus, the application of the design process in Kadoma will enable the textile industry to incorporate the expectations of the Zimbabwean consumers into the textile products that they produce.

The Design Process

The elements and principles of design are used in the context of the design process. The design process in Fig. 1 is cyclic in nature to show that it can be repeated several times for better results. In current designs, the steps are followed according to the Western or American ways where the process will have been implemented. This is rarely done in Africa including Zimbabwe.

The design process is important in obtaining and maintaining sustainability in textile design. The process emphasises on the inclusion of the consumer which is essential in product acceptability and appreciation. It is sad to note that most textile products are generated and tested outside Africa and only come to Africa as final products. In the model in Fig. 1, the double arrows show that in the design process any of the stages can be repeated until the expected outcome has been achieved. An example could be where one needs more information on the clients' recommendations to ascertain the design. There would be a need to go back and ask for the required information before proceeding. That would mean redoing the process twice or thrice before proceeding. This will help the textile industry to produce textile products that are appealing and meet the expectations of the consumers.

According to the Chicago Architecture Foundation draft (2015), the process has six cyclic steps. The first one is about defining the problem. The second step involves collecting information. This information is meant to help broaden the designers' understanding. The company can collect such information through interviews and market research. Design strategies can then be set after analysing data. The final textile product will be expected to be people-oriented according to the Zimbabwe context. Following the steps of the design process will thus help the textile design

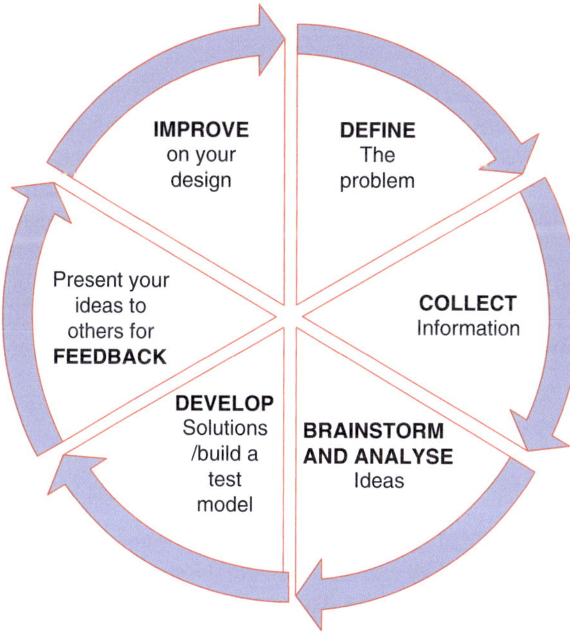

Fig. 1 The Design process Source: Chicago Archtecture Foundation Draft, April 2015

industry in Kadoma to identify the prospects of their industry and gain a better understanding of their market which should then inform their decisions for the industry. The result should be a sustainable textile design industry that can produce sustainable products and also be sustainable through the ability to maintain and sustain the textile design industry in Kadoma.

Recommended Sustainable Textile Design Strategies

Kanyenze (2006) expresses the need for the textile industry to implement strategies that promote industrial sustainability. Strategies are essential because they result from experience gained from practice. Strategies give direction to the work that should be done. The executions of well-planned strategies usually yield positive results. The strategies that are given in this study are based on the factors that affect the textile industry in Kadoma, Zimbabwe. These strategies should help to refocus the textile design industry in Kadoma in order to improve performance and become sustainable.

Be Resourceful

Kanyenze (2006) observes that there is need for government funding for the textile industry if it is to be resuscitated. However, it is important to note that funding should not be limited to the government only but also involve nongovernmental organisations. The companies also need to attract both local and foreign investors to get machinery and equipment as this can impact the textile industry positively or negatively. The current technology produces better products; modern machinery and technology is thus very essential. Zimbabwe's Ministry of Finance and Economic Development (2013) attributed the collapse of the clothing and textile industry of Zimbabwe to poor standard of technology as well as inadequate capital.

Collaborate

Adebanjo, Teh and Ahmed (2016) confirm the importance of integration and collaboration for managing social and environmental issues in order to achieve sustainability across the whole supply chain. There is also a need to become more integrated into the wider economy to avoid self-sufficiency or relying on limited sources. Life is not static, and society is affected by many factors such as natural disasters such as drought or floods which cannot be controlled by man. If an industry relies on one source for the production of its raw materials, it would shut down. The textile industry is highly impacted by agriculture since it gets some of the raw materials from there.

Adaptation of Sustainable Initiatives

The textile design industry needs to adopt sustainable initiatives that help transform their products and services according to the needs of the market so that they will not die at any moment. Sithole et al. (2016) observe that one of the challenges of the textile and clothing manufacturing sector is that most companies in Zimbabwe did not promote exports until around 1988. This was but one example of the challenges faced by the textile design industry in Kadoma. The adoption of sustainable design initiatives would help to enhance and promote the textile design industry in Kadoma.

Consumer Consumption

Nerurkar (2016) expressed the need for textile designers to simultaneously satisfy the consumer through appropriate design principles which ensures quality. Quality is one factor that attracts consumers to the various textile products. Thus, it is important for the textile design industry to be able to meet the consumers' needs and expectations. Currently, people change lifestyles according to the influences of technology around them. The textile industry needs to understand the changes in order to be able to provide products that attract and meet the expectations of the consumers. There is also need to observe that the consumers are varied and thus have varied interests and expectations. It is also vital to consider that consumers usually want quality at the cheapest possible costs. This has led to the further downfall of the textile industry in Kadoma as consumers are attracted to the second-hand market which provides cheap clothing. Failure to make the necessary considerations and implement sustainable design strategies may continue to lead to the downfall of the textile design industry in Kadoma.

Conclusion

Sustainable textile designs should make a great difference in terms of the growth of the textile industry and increase the acceptability of the textile industry as it also benefits the environment. This paper gives a background and context in which the textile design industry operates in Kadoma. The factors that affect sustainability in the textile design industry were highlighted and include resources, consumer consumption and the laws and policies of the company, state where the company is situated or international community among other things. Strategies for the production of sustainable textile design products were proposed, and these should be based on the implementation of the design process to ascertain person product attachment. This is anticipated to increase longevity of the textile design products that are produced by the textile design industry in Kadoma and thus achieving sustainable design and development. Such a sustainable approach can assist to revive the once vibrant Kadoma textile industry. This can only be achieved when all the stakeholders are involved and customers are educated on sustainability issues in order to empower them to make sustainable decisions when purchasing new products. The goal is to use the current resources sustainably and enable the future generation to also use the same. Finally, the paper recommends the adoption of sustainable methods and strategies in order to produce ethical textile and revive the textile industry in Kadoma.

References

Adebanjo D, Teh P, Ahmed P (2016) The impact of external pressure and sustainable management practices on manufacturing performance and environmental outcomes. Int J Oper Prod Manag 36(9):995–1013. https://doi.org/10.1108/IJOPM-11-2014-0543

Bonilla SH, Silva HRO, Da Silva MT, Goncalves RF, and Sacomano JB, 2018 Industry 4.0 and sustainability implications: A scenario-based analysis of the impacts and challenges. Sustainability, 10(10) 3740, https://doi.org/10.3390/su10103740

Gwilt A (2012) Integrating sustainable strategies in the fashion design process: A conceptual model of the fashion designer in haute couture. Doctor of Philosophy thesis, Design and Social Context, RMIT University, Australia
Jhanji Y (2018) Computer-aided design – Garment designing and patternmaking. In: Nayak R & Padhye R (Eds.). The textile Institute Book Series – Automation in garment manufacturing, Woodhead Publishing, Cambridge, pp. 253–290
Kanyenze G (2006) The clothing and textile industry in Zimbabwe. Herbert Jauch, Harare
Mitra A (2014) CAD/CAM solution for textile industry: an overview. Int J Curr Res Acad Rev 2(6):41–50
Mlambo A, Phimister I (2006) Origins and growth of colonial Zimbabwe's textile industry 1890–1965. His 52(2):145–175
Mlambo AS (2017) From an industrial powerhouse to a nation of vendors: over two decades of economic decline and deindustrialisation in Zimbabwe 1990-2015. J Dev Soc 33(1):99–125
Mpofu M (2012) Nhowe Zimbabwe Dress, Harare: Arts & Culture, Life and Style
Mpofu T (2013) Challenges faced by the clothing sector in Zimbabwe. J Bus Manag 1:83–84
Nerurkar O (2016) Role of psychosocial factors of fashion consumption and the challenges of design. Indian J Sci Technol, 9(15). https://doi.org/10.17485/ijst/2016/v9i15/91980
Saeed MA, Kersten W (2019) Drivers of sustainable supply chain management: identification and classification. J Sustain 11(4):1137. https://doi.org/10.3390/su11041137
Sithole L, Mutungwe E, Chirimuta C, Muzenda V (2016) Fashion designers and the shrinking customer base in Zimbabwe. J Humanit Soc Sci 21(12):44–55
Tonelli F, Evans S, Taticchi P (2013) Industrial sustainability: challenges, perspectives, actions. Int J Bus Innov Res 7(2):143–163
United Nations World Commission on Environment and Development (1987) Report of the World Commission on Environment and Development: Our Common Future. Oxford: Oxford University Press
Vadicherla T, Saravanan D, Muthu M, Suganya K (2001) Fashion renovation via upcycling. In: SS Muthu (Ed.) Textile and clothing sustainability, textile science and clothing technology pp 1–54
Walker S, Girard J 2013 (Eds.) Handbook of design for sustainability. Bloomsbury: London, pp.446–454
Worbin L (2010) Designing dynamic textile patterns. PhD thesis, The Swedish School of Textiles, University of Boras, Sweden.

Contributions to Sustainable Textile Design with Natural Raffia Palm Fibers

Feristah Unal, Arzu Yavas, and Ozan Avinc

Abstract Sustainability is to maintain the ability to be permanent while ensuring productivity and continuity of diversity. Sustainability is one of the biggest problems in the textile industry. Indeed, sustainable design is an indispensable element of today's production world from the ecology point of view. The solution is to take into account the ecological criteria in the process from the raw material selection of the textile product to the finished product. In other words, textile designs to be realized and the raw materials of these designs should be sustainable. Therefore, the use of sustainable natural fibers in the textile industry is increasing day by day. One of these is raffia palm fibers. Raffia fibers stand out because of their important performance characteristics and natural, renewable, sustainable, biodegradable, and environmental identity, and raffia fibers can be found in many different kinds of applications. Frankly, raffia palm plant is considered as a multifunctional plant due to its different usage areas. Generally, it is a plant used in food, cosmetics, medicine, and agricultural fields. It also has been traditionally used in the textile industry for many years. Since it is sustainable and renewable, it is becoming an alternative raw material in textile and different industries. Indeed, many different designed products can be generated from raffia palm fibers. For instance, many different designed products such as clothing (garment, dresses, shirts, ceremonial skirts, costumes, velvet tribute cloths, headdresses, cloaks, etc.), upholstery fabrics, blankets, carpets, mats, ropes, belts, hats, decoration products, baskets and basketry products, bags, shoes, women's accessories and jewelry, masks, knitted furniture, ornamental materials, art objects, rods, support beams and concrete reinforcement in the construction sector, ceiling panels and roofing sheets, geotextiles, and composites for different aims can be produced from raffia palm fibers. In this chapter, the structure, properties, production methods, and end use applications and designs of the raffia palm fibers are examined in detail.

Keywords Raffia palm fiber · Raffia · Sustainable · Design · Biodegradable · Renewable · Environment friendly · Natural fiber · Textile

---

F. Unal · A. Yavas · O. Avinc (✉)
Textile Engineering Department, Engineering Faculty, Pamukkale University, Denizli, Turkey
e-mail: oavinc@pau.edu.tr

S. S. Muthu, M. A. Gardetti (eds.), *Sustainability in the Textile and Apparel Industries*, Sustainable Textiles: Production, Processing, Manufacturing & Chemistry, https://doi.org/10.1007/978-3-030-37929-2_4

Introduction

The problems created by our habits of settlement and consumption are one of the common problems of the whole world. Global warming, environmental pollution, tabescent living spaces, and unhealthy societies are the most important problems of our time. The solution to all these problems is sustainability. Sustainability is not to harm people and the environment while ensuring social and economic development. Sustainability is to establish the balance between man and nature. Sustainability is to program the life and development of the future. In other words, sustainability aims to transfer natural resources to future generations. There is an increase in production and consumption with the increasing world population in the textile sector. Therefore, the use of recyclable natural raw materials is important for sustainability. With the help of developing technology and increasing environmental awareness, the use of natural, environmentally friendly, and recyclable plant-derived fibers is important not only for the textile industry but also for many different industries (Elenga et al. 2009; Chukwudi et al. 2015; Kocak et al. 2015; Fadele et al. 2017). Increased environmental pollution through the use of synthetic fibers increases awareness and increases interest in natural fibers, which are renewable, biodegradable resources (Elenga et al. 2009; Chukwudi et al. 2015; Dhas and Pradeep 2017; Kocak et al. 2015; Fadele et al. 2017). Particularly with the preference of natural fibers in composite structures, it is common to use flax, hemp, jute, kenaf, sisal, pineapple leaf, banana, abaca, raffia palm fibers, and many other plant fibers in fiber-reinforced composite structures (Elenga et al. 2009, 2013; Chukwudi et al. 2015; Anike et al. 2014; Johnson 2011; Obasi 2013; Dhas and Pradeep 2017; Rodrigue et al. 2017). Besides being environmentally friendly and renewable, being available in abundance and cheap and low density is another reason that increases the interest in the preference of plant fibers as reinforcing materials in composite structures (Chukwudi et al. 2015; Anike et al. 2014; Obasi 2013; Dhas and Pradeep 2017; Rodrigue et al. 2017; Kocak et al. 2015; Fadele et al. 2017).

Raffia fibers stand out because of their important performance characteristics and natural, renewable, sustainable, biodegradable, and environmental identity, and plant-derived raffia palm fibers can be found in many different kinds of applications. Therefore, in this chapter, it is aimed to draw attention to raffia palm fiber which is plant-derived sustainable and natural leaf fiber. Raffia palm is a species of plant belonging to the Palmae or Arecaceae family (Sikame Tagne et al. 2014). Therefore, it is similar to palm plant. The raffia palm which is used in textile is also used as a traditional beverage and edible oil in food sector. Also, it can be utilized for different usage types such as medication, cosmetic oil, building material, and broom and paper production. Therefore, raffia palm plant is in the category of multifunctional plants (Elenga et al. 2009, 2013; Johnson 2011; Akpabio et al. 2012; Abu et al. 2016). Raffia palm, which is mostly grown in Africa, has about 28 different species (Abu et al. 2016; Mert 2011). Raffia palm species used for textile purposes are known as *Raphia farinifera*, *Raphia hookeri*, *Raphia vinifera*, *Raphia textilis*, and *Raphia taedigera* (Mert 2011; Rodrigue et al. 2017; Fadele et al. 2017; Yazıcıoğlu

1999). These raffia palm species, which are cultivated in the same regions, are mostly grown in Central and West Africa, India, Singapore, Cameroon, Gabon, Congo, Cambodia, Nigeria, and Madagascar (Akpabio et al. 2012; Mert 2011; Wake 2006; Wiriadinata and Sari 2010; Odera et al. 2015; Musset 1933). Generally two types of fiber, piassava and raffia, are obtained from these raffia palm types used in the textile industry (Mert 2011; Raffia fibre 2019; Sikame Tagne et al. 2014). Raffia palm fibers are obtained from different parts of the raffia palm plant. Therefore, fibers obtained from the raffia palm species are named with different names (Mert 2011; Raffia fibre 2019; Sikame Tagne et al. 2014). Raffia fibers are obtained from the leaves of raffia palm species, while piassava fibers are obtained from the stems of the *Raphia hookeri* species, in particular, and thus the piassava fibers obtained are known to be coarser than raffia fibers (Akpabio et al. 2012; Mert 2011; Raffia fibre 2019; Sikame Tagne et al. 2014). Raffia palm fibers also attract attention with their elastic structure, break resistance, and easy dyeability (Mert 2011). Many different designed products can be generated from raffia palm fibers. In this chapter, the structure, properties, production methods, and end use applications and designs of the raffia palm fibers are examined in detail.

In this chapter, firstly, brief preliminary information about raffia palm plants and their cultivation is given. Then, the methods of obtaining fiber from raffia palm plant were classified. The physical and chemical structure and physical, chemical, and mechanical properties of the obtained fibers are reviewed in detail. Finally, different applications areas and designs of biodegradable, sustainable, and natural raffia palm fibers are examined in detail in order to raise awareness of this valuable fiber.

Raffia Palm Plant and Cultivation

Raffia is a type of palm with long leaves and large stems (Akpabio et al. 2012; Mert 2011; Raffia Palm 2017). Palm is the common name of the plants that grow in tropical regions that make up the palm tree family and have more than 200 genera whose leaves are mostly collected at the top of the stem. Its scientific name is known as Arecaceae (Sikame Tagne et al. 2014). The raffia palm plant is called monocarpic because it yields fruit only once (Figs. 1 and 2) (Mert 2011; Wiriadinata and Sari 2010; Odera et al. 2015; Raffia Palm 2017). Raffia palm, grown in the tropical regions of West and East Africa in Central and South America, is mostly produced in Madagascar. Also, it is produced in Congo, Central Africa Gabon, Cameroon, Nigeria, Guinea Bay, and Liberia (Akpabio et al. 2012; Mert 2011; Wake 2006; Odera et al. 2015; Musset 1933; Sikame Tagne et al. 2014; Sandy and Bacon 2001; Rodrigue et al. 2017; Fadele et al. 2017; Yazıcıoğlu 1999). Although the length of the raffia palm plant varies depending on the type of the raffia palm, it is generally in the range of approximately 10–25 m (Odera et al. 2015; Musset 1933). For example, plant lengths for some raffia palm species are as follows: *Raphia farinifera*, *Raphia hookeri*, and *Raphia vinifera* species can be around 25, 10, and 13 m, respectively (Mert 2011; Wiriadinata and Sari 2010). The leaves of the raffia palm

Fig. 1 View of raffia palm plant and its leaves (Wikimedia Commons 2018a, b)

Fig. 2 Raffia palm fibers

plant are quite large and consist of a large number of leaflets in the form of a feather about 150–200 pieces linearly arranged individually and in a regular manner (Mert 2011; Wiriadinata and Sari 2010; Raffia 2002). Leaf length of the raffia palm can be up to 15–25 m. Therefore, among the flowering plants, raffia palm is known as the broadest-leaved plant (Mert 2011; Odera et al. 2015; Musset 1933; Raffia 2002). Leaflets forming the leaves can be 1.5–1.8 m long and 4–5 cm wide (Fig. 1) (Mert 2011; Wiriadinata and Sari 2010; Musset 1933; Raffia 2002). The seeds of the raffia palm plant, which only bloom once, die after ripening (Odera et al. 2015; Musset 1933). Plant stalks die after the fruits are finished, but plant roots remain alive and produce new stems (Odera et al. 2015; Musset 1933).

The blooms of the raffia palm can be droop between the leaves or dense and clustered on the leaves (Mert 2011; Wake 2006). The type and cultivation conditions of the raffia plant affect the time from planting to flowering and fruiting (Mert 2011; Wiriadinata and Sari 2010). *Raphia vinifera* species need 8 years after planting for blooms. This period ranges from 3 to 7 years for *Raphia hookeri* (Mert 2011; Wiriadinata and Sari 2010; Tuan et al. 1985). *Raphia farinifera*, which is grown in Madagascar, blooms in 20–25 years and its fruits ripening takes 3–6 years (Mert 2011; Wiriadinata and Sari 2010; Tuan et al. 1985). Fruits of raffia palms are cylindrical-ellipsoid in shape, brown in color, covered with scales, single seed, and have a hard core structure (Mert 2011; Wake 2006; Wiriadinata and Sari 2010). The fruits of some raffia palm species can be edible (Johnson 2011; Wake 2006). Reproduction of raffia palm is usually provided by seed. Mostly grown in tropical rainforests, Savannah riverbanks, freshwater swamps, and wetlands, the raffia palm is very sensitive to salty soil conditions (Johnson 2011; Mert 2011; Wake 2006). According to research made by Otedoh, 18 raffia palm species living in the swamp were found (Johnson 2011; Wake 2006; Musset 1933; Otedoh 1982). Today, it is known that there are about 20 species of raffia palm, especially in Madagascar (Johnson 2011; Mert 2011; Wake 2006; Odera et al. 2015; Sikame Tagne et al. 2014; Raffia Palm 2017; Raffia 2002; Dransfield et al. 2008). According to paleoecological studies, it was suggested that a type of raffia palm was found in 2800 BC, and it is known that it was the first seen in Nigeria in the eleventh century AD (Mert 2011; Wake 2006). The international trade of the raffia palm is mainly carried out by Madagascar (Eicher 2001). In the 1950s, Madagascar exported 5000 tons of raffia fibers every year. In the 1980s and 1990s, the average annual export figure of raffia fibers was 2000 tons. Today, it is known that between 8000 and 10,000 tons are exported every year (Eicher 2001).

Raffia Palm Fiber Production Methods

Raffia palm fibers, commonly known as leaf fibers, are of two types, raffia and piassava (Johnson 2011). Raffia fibers are known as soft and durable fibers obtained from the leaves of *Raphia farinifera*, *Raphia hookeri*, and *Raphia vinifera* species. *Piassava* fibers are known as coarser fibers obtained from the leaf sheath and stems of the *Raphia hookeri* species (Mert 2011). The leaves of the raffia palm are mainly composed of thin-walled cells, vascular tissue strands, and longitudinally extending fibers (Mert 2011; Sandy and Bacon 2001). Its surfaces consist of epidermis and there is hypodermis underneath (Mert 2011; Sandy and Bacon 2001; Tomlinson 1961; Picton and Mack 1980). The hypodermis is composed of a plurality of fibers placed on the long axis of the parallel leaflets, and the upper hypodermis has more fibers. The upper surfaces of raffia leaves are generally used to obtain fibers (Mert 2011; Sandy and Bacon 2001; Tomlinson 1961; Picton and Mack 1980). Mechanical fiber extraction methods or retting methods are generally preferred in order to separate the raffia palm leaves from the woody structure or epidermis layer, to obtain usable and dyeable textile fibers (Abu et al. 2016; Mert 2011).

Production of Raffia Palm Fibers with Mechanical Methods

In this method, machines are generally used to extract the fibers. Moreover, different kinds of knives and manual stripping also can be utilized to extract the fibers as a kind of primitive method. The upper parts of the young leaves of the raffia palm plant are cut, and the raffia fibers are manually removed from the cut opening manually or with the aid of a knife (Elenga et al. 2013; Mert 2011). The extracted fibers have a strip-shaped appearance of 1–2 m long and 2–3 cm wide (Elenga et al. 2013; Mert 2011). The ribbon-shaped raffia fibers are tied at one end for drying at room temperature (25 °C). The fibers obtained are 15 μm thick and exhibit mostly creamy and yellowish color tones (Elenga et al. 2013; Mert 2011).

Leaf shells of *Raphia hookeri*-type raffia palm, where coarser fibers are obtained, are broken into pieces, and fibers are extracted quickly and efficiently with the help of these machines which can provide high pressure with crushing cylinders and blunt blades called decorticators or raspadors (Mert 2009). Although the energy consumption is high, the process time is short and the obtained fiber quality is high. Therefore, preferableness of raffia palm fiber production method with these kinds of machinery is increasing. However, in the most African countries, the fine fiber obtained from the leaflets of the raffia palm plant can still be produced by manual stripping (Mert 2009).

Production of Raffia Palm Fibers with Retting Method

Piassava fibers obtained from the leaf sheaths and stems of the *Raphia hookeri* plant, a type of raffia palm, can be obtained with the machines as well as with the conventional method of retting (Mert 2011). The leaf stalks of the *Raphia hookeri*-type raffia palm are divided into 3–4 pieces for the retting process, and it is ensured that the retting process is carried out in bunches in dew and pond or with chemicals. This process can take from a few weeks to 2–3 months. The fibers obtained in reddish and brownish tones are dried under the sun for a few days (Abu et al. 2016; Mert 2011). In dew retting method, plant leaves and sheaths are laid on the soil, fields. The laid leaves are kept in dew for fermentation. Thus, it is provided that the cellulosic structure on the fiber is removed from the fiber by soil fungi. The processing time in this method ranges from 4 to 6 weeks (Mert 2009). In the method of water retting, leaves prepared in bunches are kept for 14–28 days at the sides of the water, pools or rivers. Cellulosic substances on the leaves are removed by anaerobic and pectinolytic bacteria (Mert 2009). *Raphia hookeri*-type raffia palm leaf bundles are kept for 2–3 weeks in a still swamp pond on the shore of Homonica, and at the end of the process, reddish fibers are obtained (Chukwudi et al. 2015; Abu et al. 2016; Dhas and Pradeep 2017). The resulting fibers should not be exposed to direct sunlight. It is therefore dried in semi-shaded areas or at room temperature. The length of the fibers obtained varies between 400 and 1200 mm and the fiber diameter is 0.70 mm (Abu et al. 2016; Dhas and Pradeep 2017). In the chemical retting method, raffia palm leaves are softened with boiling acid or alkali in high pressure. The most commonly used alkali in this process is sodium hydroxide (NaOH) (Chukwudi et al. 2015; Mert 2009). In addition, the processing time is quite short in this method, which is an expensive method (Mert 2009). The quality, color, and physical and mechanical properties of raffia palm fibers may vary depending on production differences, growth conditions, harvesting time, and processing type for the fiber (Fig. 2) (Elenga et al. 2009; Mert 2011).

Gum Removal in Raffia Palm Fibers

The degumming method is a process which is applied after the raffia palm fibers are obtained by retting or mechanical methods. The degumming process is applied to the fibers in order to remove the gum substances remaining on the fiber bundles obtained. The fibers are treated with hot soap or alkali solution for this process. The nonremovable impurities on the fiber surface make the subsequent use of the fibers difficult, especially preventing the interfacial adhesion in the composite structures (Ganan and Mondragon 2004; Hetal et al. 2012; Alvarez et al. 2003; Nekkaa et al. 2008; Beg and Pickering 2004; Jähn et al. 2002). Although the presence of wax, oil, and impurities on the raffia palm fibers forms a protective layer on the fiber surface, it causes poor bonding with the polymer when not removed for composite applications. On the other hand, the alkali-treated fibers display a much cleaner surface although they are more rough when touched (Hetal et al. 2012; Alvarez et al. 2003; Fadele et al. 2017).

Thus, the morphological and chemical structure of the raffia palm fibers is changed, and impurities that block interface adhesion with the polymer are removed from the fiber. It also forms less hydroxyl groups (Beg and Pickering 2004; Jähn et al. 2002; Fadele et al. 2017).

Gum removal by treating the raffia fibers with NaOH is mentioned above (Xiao et al. 2011). But nowadays, ecological methods have gained importance with the increasing environmental concerns. Therefore, the use of enzymes in the degumming process is also increasing. Enzymes can be used also in the degumming process of raffia fibers. The use of enzymes in the degumming process is ecological. Thanks to this process, the natural properties of raffia fibers are preserved and more valuable fibers are obtained. Recently, new enzymes have also been produced for new applications. Biodegradable enzymes have replaced chemical substances in the degumming process. Pectinase enzymes are the most commonly used enzymes for degumming from raffia fibers. This process is carried out at 45–60 °C. The pectins in the structure of treated (for 1 h) raffia fibers are broken down and removed. Enzymes are powerful biocatalysts that accelerate reactions. Thus, degumming with enzymes is carried out under more mild conditions. Thus, thanks to the degumming process with enzymes for raffia palm fibers, significant energy savings and easier reaction controls can be acquired leading to more economical, sustainable, environmentally friendly, and better quality production (Xiao et al. 2011).

Physical and Chemical Structure of Raffia Palm Fibers

Raffia palm fibers generally consist of overlapping flat sheets, as opposed to other plant fibers with luminal cylindrical and cut shapes (Elenga et al. 2009; Baley 2002). The structure of the upper surface of the raffia fibers is formed by the parallel fibrils adhering to each other in the longitudinal direction (Elenga et al. 2009; Sandy and Bacon 2001). It was reported that the average width of these fibrils is 10 μm and the average thickness is approximately 5 μm. These fibers, which have a smooth structure, consist of a structure similar to pula or tile unlike other vegetable fibers (Elenga et al. 2009, 2013). This tile-like structure which is 6–20 μm long covers the entire width of the fibril. This morphological structure of raffia palm fibers makes it possible to use these fibers as relatively waterproof construction materials and as roofing materials (Elenga et al. 2009). The lower surface of the fiber, i.e., the surface in contact with the body of the leaflets, has a honeycomb-like structure (Elenga et al. 2009, 2013). It was reported in the literature that the diameter of the alveoli in the honeycomb network is between 6 and 13 μm. It is separated by a wall approximately 0.5 μm thick (Elenga et al. 2009; d'Almeida et al. 2006). These alveoli are uneven in shape and size and show a random distribution in the volume of material. These alveoli and tile-like scaly structures can act as mechanical bonding in composite matrices. It also increases the willingness of the fibers to bond with the matrix (Elenga et al. 2009).

Raffia palm fibers, which have a layered structure, displayed a tile-like layer separated by a vertical thin interface aligned with the longitudinal axis of the fiber when examined under SEM microscope (Elenga et al. 2009). The longitudinal axis of the raffia palm fiber is almost vertical (Elenga et al. 2009, 2013).

It was observed that the fibers were not round but longitudinal as a result of SEM analysis of raffia palm fibers (Odera et al. 2015). At the same time, the surface morphology of the fibers is very important for their use in composite materials (Odera et al. 2015). The raffia palm fiber is a lignocellulosic fiber and is composed of cellulose, hemicellulose, lignin, and pectin (Table 1). The cuticula contains mineral substances, gum, and resinous, oily, and waxy substances. Cellulose, hemicellulose, and pectins in the structure of raffia palm fibers add elasticity and bendability to the fibers, while lignin in the fiber makes the fibers hard and brittle. The lignin in the structure of the raffia palm is among the micelles of the cellulose structure in the fiber (Mert 2009). The proportions of the chemical components of raffia palm fibers are indicated in different amounts in different sources.

Table 1 Comparison of chemical components of raffia palm fibers with other cellulosic fibers (Elenga et al. 2009; Schuchardt et al. 1995; Fadele et al. 2017; Lewin 1998; Taj et al. 2007; Kalia and Avérous 2011)

Fiber type	Cellulose (%)	Hemicellulose (%)	Lignin (%)	Pectin (%)	Ash (%)	Moisture content (%)	Crystallinity index (%)	References
Raffia palm	53.36–54.4	12.32	23.78–28.6	–	0.8	14	51–64	Elenga et al. (2009), Schuchardt et al. (1995), Fadele et al. (2017)
Abaca	63.2–70.1	19.6–21.8	5.1–5.7	0.5–0.6	4.80	14	–	Lewin (1998), Taj et al. (2007), Kalia and Avérous (2011)
Banana	60–67.6	6–19	5–10	3–5	–	10–11	60	Lewin (1998), Taj et al. (2007), Kalia and Avérous (2011)
Cotton	82–96	2–6	0.5–1.5	5–7	1–1.8	6–8.5	–	Lewin (1998), Taj et al. (2007), Kalia and Avérous (2011)
Bamboo	26–43	30	21–31	–	–	–	–	Lewin (1998), Taj et al. (2007), Kalia and Avérous (2011)
Flax	71–82	18.6–20.6	2.2–4	2.3	3.4	7.7	80	Lewin (1998), Taj et al. (2007), Kalia and Avérous (2011)
Hemp	70.2–88	17.9–22.4	3.7–6.8	0.9	3.9	1.8	88	Lewin (1998), Taj et al. (2007), Kalia and Avérous (2011)
Jute	51–84	13.6–20.4	12–13	0.2	1.0	1.5	71	Lewin (1998), Taj et al. (2007), Kalia and Avérous (2011)
Ramie	68.6–80	13.1–16.7	0.6–0.7	1.9	–	–	58	Lewin (1998), Taj et al. (2007), Kalia and Avérous (2011)
Sisal	60–78	10–18.1	5.9–11	10	1.0	4.0	71	Lewin (1998), Taj et al. (2007), Kalia and Avérous (2011)
Kenaf	31–57	21.5	15–19	2	–	–	–	Lewin (1998), Taj et al. (2007), Kalia and Avérous (2011)
Pineapple	70–82	16–22.2	5–13	1–3	–	–	–	Lewin (1998), Taj et al. (2007), Kalia and Avérous (2011)
Coconut	36–46	0.15–0.3	41–45	3–4	–	–	–	Lewin (1998), Taj et al. (2007), Kalia and Avérous (2011)

The chemical components of raffia palm fibers are given as a 53.36% cellulose, 12.32% hemicellulose, and 23.78% as lignin in a performed study by Fadele, Opeoluwa et al. (Fadele et al. 2017). Raffia palm fibers contain 54.4% cellulose, 14% moisture, 0.8% ash, and 28.6% lignin which were reported in another study (Schuchardt et al. 1995). Accordingly, the excess lignin content in the structure of raffia fibers supports the leaf and makes the fiber harder (Schuchardt et al. 1995). Raffia palm fibers have elements, which confirm its organic nature, such as carbon (C), hydrogen (H), and oxygen (O) (Odera et al. 2015). Raffia palm has carbon (C) content greater than the amount of carbon present in a normal plant material (Schuchardt et al. 1995). Total hydroxyl groups are about 14% of a normal plant material (Schuchardt et al. 1995). The chemical components of raffia palm fibers were compared with different cellulosic fibers and are shown in Table 1 (Elenga et al. 2009; Schuchardt et al. 1995; Fadele et al. 2017). The chemical components of raffia palm fibers were compared with other cellulosic fibers, and it was observed that the crystalline index (51–64%) was higher than ramie (58%), close to cotton (60%), but lower than sisal (71%), jute (71%), flax (80%), and hemp (88%) fibers (Elenga et al. 2009).

Physical, Chemical, and Mechanical Properties of Raffia Palm Fibers

The raffia palm fibers (51–64%), which have a high crystalline index, have very good mechanical properties, strength, and hardness when compared to many other natural fibers (Elenga et al. 2013). When the various physical and chemical properties of raffia palm fibers are examined, the raffia fibers exhibit around 500–660 MPa tensile strength, 12.3–30 GPa starting modulus, and 2–4% elongation at break (Table 2) (Elenga et al. 2009, 2013; Sandy and Bacon 2001). In the literature, the crystalline size of raffia fibers is 9.6 nm and the cell wall density is 1.52 g/cm^3 (Elenga et al. 2013; Sandy and Bacon 2001). The angle of its microfibril varies

Table 2 Properties of raffia palm fibers (Elenga et al. 2009, 2013; Sandy and Bacon 2001)

Density (g/m^3)	Tensile strength (MPa)	Initial module (GPa)	Elongation at break (%)	Crystallinity index (%)	References
–	527–660	26	–	51	Elenga et al. (2013)
0.75	500 ± 97	30	2–4	64	Elenga et al. (2009)
	500	12.3	–	–	Sandy and Bacon (2001)
0.128–0.236	–	0.88–7.92	–	–	Rodrigue et al. (2017)

Table 3 Comparison of properties of raffia palm fibers with other fibers (Elenga et al. 2009, 2013; Aizi and Harche 2015; Rodrigue et al. 2017)

Fiber type	Density (g/m^3)	Tensile strength (MPa)	Initial module (GPa)	Elongation at break (%)	Hygroscopic (%)
Raffia palm	0.75	500–660	26–36	2–4	–
Raffia farinifera	–	500	12.3	4	–
Raffia textilis	0.75	148–660	28–36	2	–
Abaca	1.50	12–980	31.1–41	2.9–3.4	14
Banana	1.35	540–550	20	5–6	10–11
Cotton	1.60	300–590	5.5–12.6	3–10	8
Bamboo	0.8	220–1000	22.8	1.3	–
Flax	1.50	345–1035	27–39	2.7–3.2	8
Hemp	1.48	550–900	70	1.6	8–10
Sisal	1.41	511–635	9.4–22.0	2.0–2.5	11
Jute	1.3–1.46	393–800	13–26.5	1.5–1.8	12
Ramie	1.5	220–938	44–128	2–3.8	12–17
Pineapple	1.07–1.52	170–1627	6.21–82.5	1.6	11.8
Fiberglass	2.50–2.55	2000–3500	70.0–73	2.5–3	–
Aramid	1.38–1.47	3000–3150	63.0–67.0	3.3–3.7	–
Carbon	1.7–2.1	4000	230–240	1.4–1.8	–

between 29,81° and 48,65° (Rodrigue et al. 2017). *Raphia farinifera* and *Raphia textilis* fibers, which are raffia palm species, were compared, and it was found that both raffia palm fibers had the same breaking strength, but *Raphia textilis*-type raffia palm fibers were stiffer and exhibit less flexibility. Changes in the mechanical properties of raffia palm fibers depend on the raffia palm species as well as on the production methods (Elenga et al. 2009; Sandy and Bacon 2001).

When the mechanical properties of different fibers and raffia palm fibers are compared, it is shown in Table 3 that the initial module of raffia palm fibers (26–36 GPa) is three times higher than that of cotton fiber (5,5–12,6 GPa) and slightly lower than that of the flax fiber (27–39 GPa) (Elenga et al. 2009, 2013; Aizi and Harche 2015). This suggests that raffia textile fibers appear to be more flexible than flax fibers but stiffer than cotton fibers (Elenga et al. 2009). In general, elongation at break of raffia palm fibers is known to be around 2–4% (Elenga et al. 2009). Raffia fibers are similar to hemp, sisal, and ramie fibers in terms of elongation at break (Elenga et al. 2009). Fiber density of raffia palm is generally known as 0.75 g/m^3. This value was lower than all other plant fibers in the literature (Elenga et al. 2009, 2013; Aizi and Harche 2015; Rodrigue et al. 2017). It is known that the density of raffia fibers is approximately half the density of abaca (1,50 g/m^3), flax (1,50 g/m^3), and sisal (1,41 g/m^3) fibers (Elenga et al. 2009, 2013; Aizi and Harche 2015; Rodrigue et al. 2017). This difference shows that the structure of raffia fibers is different from that of other plant fibers and the mechanical properties of raffia palm fibers are generally superior to other plant fibers (Elenga et al. 2009, 2013; Aizi and Harche 2015; Rodrigue et al. 2017). Raffia textile fiber is a potential reinforcing element for composites with its unusual structure and high mechanical properties (Elenga et al. 2009).

Applications Areas of Raffia Palm Fibers

The contribution of sustainable raffia palm fibers to textile design is promising. Since, first of all, raffia fibers are completely environmentally friendly due to the fact that they are natural and can be re-evaluated in the production line by disassembling the produced products and moreover waste raffia palm fiber materials that cannot be used anywhere are biodegradable in nature (Yıldırım and İşmal 2011). Raffia fibers are therefore important fibers for sustainable development. The ecological aspect of the design, which has an integral connection with technology and adds value to the textile product, should also be evaluated. At this point, the most important task falls to fashion designers. Today, shopping preferences are influenced by fashion trends. If fashion shifts towards sustainable natural raw materials, interest in natural products and environmental awareness can spread beyond a small segment. Thus, more raffia-like fibers can be used in the design. While environmental sensitivity of products in other sectors is the subject of advertisiment, unfortunately this sensitivity is not yet fully available for the textile sector. However, with the widespread use of raffia fibers and various natural raw materials, our world will be less polluted, people can buy more ecofriendly products and natural products with high added value will provide also economic development (Yıldırım and İşmal 2011). Natural raffia fibers are flexible, strong and extremely durable and can be dyed easily. Therefore, raffia fiber is a very popular material in crafts and high fashion (Abury 2018). For instance, Madonna Kendona-Sowah uses raffia fibers to create beautiful and stylish designs (Africa Fashion and Style 2014). Even, Madonna Kendona-Sowah has made Raffia fibers as a main raw material used in her designs. In 2013, she established the Raffia brand. It has proven that beautiful things can be designed from raffia fibers that are rough and dry in the raw state (Africa Fashion and Style 2014). Furthermore, Alexander McQueen, who gained fame in design, made products from raffia fibers such as clothes, bags, shoes and hats (Cohan 2011). McQueen uses elegance and an uncomfortable asymmetry in products designed from raffia fibers. What is more, Gucci also designed an elegant blouse made of raffia fiber (Cohan 2011). Different applications areas and designs of raffia palm fibers are examined below in more detail.

Raffia palm is one of the plants used in many tropical regions for the daily needs of the population (Bussmann et al. 2015). Besides being economical, it is used in different sectors due to its soft, flexible, and strong structure (Elenga et al. 2009, 2013; Johnson 2011; Odera et al. 2015; Fadele et al. 2017). Raffia palm fibers are used in different fields such as food construction, cosmetics, and paper making besides textile sector. So it is said to be among the multipurpose plant fibers (Elenga et al. 2009, 2013; Johnson 2011; Odera et al. 2015). Especially in America, *Raffia*

Fig. 3 Woven fabrics made from raffia fibers

Fig. 4 Rug made from greige raffia fibers (on the left); Bag made from raffia fibers and colored with natural dyes (on the right)

taedigera species are used in food sector and even consumed as food. In addition, the raffia palm produced for oil and the fruit juices obtained by fermenting from the raffia palm are of great importance (Elenga et al. 2009; Johnson 2011; Akpabio et al. 2012; Wake 2006; Musset 1933; Balick 1989; Beck and Balick 1990; Haynes and McLaughlin 2000). Raffia fibers are used in ropes, belts, baskets, bags, hats, shoes, women's accessories, knitted furniture, decorative products, ornamental materials, and rods and support beams in the construction sector (Figs. 3 and 4) (Elenga et al. 2009, 2013; Johnson 2011; Akpabio et al. 2012; Wake 2006; Fadele et al. 2017).

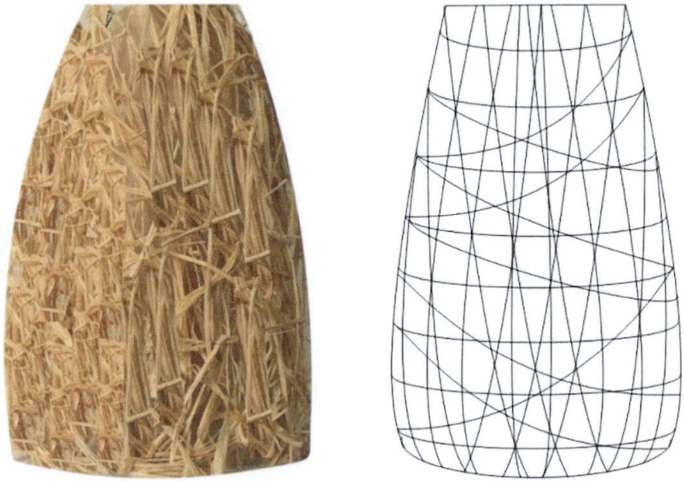

Fig. 5 Bell silhouette skirt made from greige raffia fibers (Gündoğan et al. 2018)

Although the use of synthetic fabrics affects the use of different textile fibers such as raffia, many designers in the fashion world today use straw-like natural raffia fibers in many designs, such as dresses, shirts, shoes, hats, and bags (Fig. 5) (Fadele et al. 2017). A dress made of raffia fibers was designed for Alexander McQueen in Paris, and raffia fiber was used for a stylish upper garment design for Gucci in Milan. Interior designer Michael Smith also used raffia fibers in decoration products, carpets, and upholstery fabrics (Cohan 2011). Raffia palm fibers have traditionally been used in African countries since the nineteenth and twentieth centuries. The use of raffia fibers is frequently encountered, especially in complex garments, carpets (Kasai Velvet), blankets, and art objects (Elenga et al. 2009). Today, the textile museum of George Washington University exhibits the textile products of the Kuba kingdom, which is one of the most outstanding and impressive work of African art. There are ceremonial skirts, velvet tribute cloths, headdresses, and basketry products, many of which are made from raffia palm fibers, in the museum. There are also more than 140 objects of the nineteenth and twentieth centuries, including many private collections (Weaving Abstraction: Kuba Textiles and the Woven Art of Central Africa 2011). In addition, the Horniman Museum, founded by Frederick Horniman, contains many African artifacts collected over the last 100 years. Many of the African objects in the collection are known to contain raffia fibers produced from the leaves of the raffia palm trees (Sandy and Bacon 2001).

Raffia palm fibers, which are very popular on the shores of the Gulf of Guinea and among some ethnic groups in Central Africa, are also used to make special ceremonial garments (Fadele et al. 2017). At the same time, the Pende people, an ethnic group in the South-West Democratic Republic of Congo, use masks, costumes, and dance clothing for traditional African folk festivals and Bapende ceremony with different animal figures made of raffia palm fibers (Fig. 6) (Haveaux

Fig. 6 Mask made from raffia fibers (Wikimedia Commons 2019)

1954; Van De Ginste 1946; Pende People 2019; Tribal African Art, Pende (Bapende, Phenbe, Pindi, Pinji) Democratic Republic of the Congo 2019; Fadele et al. 2017). In addition, raffia fibers are also used to bind tree vaccines, especially in Europe. Raffia fibers, which are widely used in textile and construction sectors, are also used today in the production of bags, hats, and shoes and are also used in composites such as other plant fibers, reinforcing concrete, panels, and geotextiles (Elenga et al. 2009; Odera et al. 2015; Musset 1933). Raffia palm fibers are used as reinforcing materials in composite structures due to its low cost and low weight. Thanks to raffia fiber, which is also used as a reinforcing material in cement mortar, composites used in roofing sheets are produced economically, sustainable, safe, and long-lasting environment-friendly building materials (Odera et al. 2015, 2011). Concrete, which is frequently used as construction material, is a material which is strong in compression but weak in tensile strength. These weaknesses of concrete are improved by using steel bars. However, due to their high cost, fibers, especially raffia palm fibers, are used instead of steel bars (Abu et al. 2016; Salau and Sharu 2004). Concrete becomes a homogeneous and isotropic material with the addition of raffia fibers to concrete. Fibers used as reinforcements stop crack formation and crack propagation and thus improve strength and ductility of materials (Abu et al. 2016; Wafa 1990).

In the literature, studies have been conducted to test the suitability of the use of raffia palm fibers in composite structures in the construction sector, and the thermal degradation behavior of raffia fibers has also been investigated. As a result of these studies, it has been suggested that raffia palm fibers are a potential composite reinforcing material and their thermal stability is improved by different methods (Odera et al. 2015). In another study, the use of raffia palm fibers as concrete reinforcement was investigated, and it was concluded that raffia palm fibers increased the tensile strength and toughness modulus of concrete (Abu et al. 2016). In Nigeria, the suitability of using raffia palm fibers as a ceiling panel in building design for the tropical region was investigated in terms of the current economic situation, alternative structural and thermally suitable building materials, and thermal properties. They concluded that the thermal properties of the raffia palm fibers were positively compared with those of other good insulators and that, if used appropriately, raffia palm fibers could be used as efficient ceiling panels for passive cooling building design (Abu et al. 2016).

Conclusion

The life of all people depends on natural resources and these natural resources are not infinite. At this stage, sustainability comes into play. Sustainability means creating a balance between nature and human. The creation of this balance is carried out, thanks to people again. Textile sector is one of the most polluting sectors in the world. Therefore, it is very important to use sustainable raw materials and production methods in textile. Raffia palm fibers are environmentally friendly, biodegradable, renewable, natural, and sustainable. Raffia palm fibers can be potentially used in many industrial fields due to their high strength properties. Raffia palm fibers, which date back to ancient centuries, are notable for being a preferred raw material in the food and construction sectors besides textile industry. Although raffia fibers are material that is used for almost all needs in these regions, which are better known and cultivated in African countries where it is grown, it is becoming a preferred raw material today especially in composite structures due to its superior performance properties.

Indeed, raffia fibers stand out because of their important performance characteristics and natural, renewable, sustainable, biodegradable, and environmental identity, and raffia fibers can be found in many different kinds of applications. Truthfully, raffia palm plant is considered as a multifunctional plant due to its different usage areas. Besides being economical, it is used in different sectors due to its soft, flexible, and strong structure. Raffia palm fibers can be used in different fields such as food, construction, cosmetics, medicine, agricultural fields, and paper making apart from textile sector. However, it has been traditionally used in the textile industry for many years. Since it is sustainable and renewable, it is becoming an alternative raw material in textile and different industries. Indeed, many different designed products can be generated from raffia palm fibers. For instance, many different designed products such as clothing (garment, dresses, shirts, ceremonial skirts, costumes, velvet tribute cloths, headdresses, cloaks, etc.), upholstery fabrics, blankets, carpets, mats, ropes, belts, hats, decoration products, baskets and basketry products, bags, shoes, women's accessories and jewelry, masks, knitted furniture, ornamental materials, art objects, rods, support beams and concrete reinforcement in the construction sector, ceiling panels and roofing sheets, geotextiles, and composites for different aims can be produced from raffia palm fibers. The use of plant fibers as reinforcing materials for renewable composite materials is increasing with sustainable production becoming increasingly important. Indeed, raffia textile fiber is a potential reinforcing element for composites with its unusual structure and high mechanical properties.

Today, it is necessary to use recyclable, sustainable raw materials and production techniques in order to reduce the damage caused by the textile sector to the world. Therefore, it is very important to increase the awareness of the raffia palm fibers which can be used in many different textile products and composites as well as in other sectors such as food, construction, cosmetics, medicine, agricultural fields, paper making, etc. and their different designed products in order to increase the utilization and preference of this valuable fiber.

Raffia fibers are environmentally friendly and high-quality fibers that can be used in all areas of textile. The choice of enzymatic methods for obtaining raffia fibers provides high energy savings. In addition to being a natural fiber, this fiber has a clean production technique. For this reason, it is in the form of a fiber sought by large companies. Famous brands and designers such as Alexander McQueen and Gucci used raffia fibers in their products. They proved with their aesthetic designs that raffia fibers are not coarse fibers. Studies conducted in this sense show that raffia fibers will be recognized by more designers and brands in the coming years and their usage in different designs will become widespread. Raffia and raffia-like fibers, natural fibers used in textile design applications, are predicted as a promising solution for increased environmental problems. Raffia-like fibers can be expected to replace nonbiodegradable synthetic fibers over the next few years.

References

Abu AK, Yalley PPK, Adogla F (2016) The use of raffia palm (Raffia Hookeri) piassava fibres as reinforcement of concrete. Int J Eng Sci 5(6):1–6

Abury (2018). https://abury.net/blogs/abury-blog/natural-raffia-fair-fashion. Accessed 12 Oct 2019

Africa Fashion and Style (2014) Raffia fabric in fashion. http://africashionstyle.com/raffia-fabric-in-fashion/. Accessed 12 Oct 2019

Aizi DE, Harche MK (2015) Extraction and characterization of *Retama monosperma* fibers. Afr J Biotechnol 14(35):2644–2651. https://doi.org/10.5897/AJB2015.14812

Akpabio UD, Akpakpan AE, Udo UE, Essien UC (2012) Physicochemical characterization of exudates from raffia palm (*Raphia hookeri*). Adv Appl Sci Res 3(2):838–843

Alvarez VA, Ruscekaite RA, Vazquez AJ (2003) Mechanical properties and water absorption behaviour of composites made from a biodegradable matrix and alkaline treated sisal fibres. J Compos Mater 37(17):1575–1588. https://doi.org/10.1177/0021998303035180

Anike DC, Onuegbu TU, Ogbu IM, Alaekwe IO (2014) The effect of alkali treatment on the tensile behaviour and hardness of raffia palm fibre reinforced composites. Am J Polymer Sci 4(4):117–112. https://doi.org/10.5923/j.ajps.20140404.03

Baley C (2002) Analysis of the flax fibres tensile behaviour and analysis of the tensile stiffness increase. Compos A: Appl Sci Manuf 33(7):939–948

Balick MJ (1989) Native neotropical palms: a resource of global interest. In: Wickens GE, Haq N, Day P (eds) New crops for food and industry. Chapman and Hall, London

Beck HT, Balick MJ (1990) Useful palms of the world. Columbia University Press, New York
Beg MD, Pickering KL (2004) Effect of fibre pretreatment on the mechanical properties of wood/polypropylene composites. In: The 2nd international conference on the structure, processing and properties of materials, Feb 2004
Bussmann RW, Zambrana NP, Kuhlman A, Rakotoarivony F, Razanatsima A, Rakotoarivelo N et al (2015) Palms and palm use in Ambalabe, a community in Eastern Madagascar. Ethnobot Res Appl 14:17–26. https://doi.org/10.17348/era.14.0.017-026
Chukwudi AD, Uzoma OT, Azuka UAA, Sunday EC (2015) Comparison of acetylation and alkali treatments on the physical and morphological properties of raffia palm fibre reinforced composite. Sci J Chem 3(4):72–77. https://doi.org/10.11648/j.sjc.20150304.12
Cohan S (2011) House and home – raffia end page. Susan Cohan Gardens. https://susancohangardens.com/house-and-home-raffia-end-page/. Accessed 29 July 2019
d'Almeida JRM, Aquino RCMP, Monteiro SN (2006) Tensile mechanical properties, morphological aspects and chemical characterization of piassava (*Attalea funifera*) fibers. Compos A: Appl Sci Manuf 37(9):1473–1479
Dhas JER, Pradeep P (2017) Mechanical property evaluation of palm/glass sandwiched Fiber reinforced polymer composite in comparison with few natural composites. IOP Conf Ser Mater Sci Eng 247(1):1–5. https://doi.org/10.1088/1757-899X/247/1/012010
Dransfield J, Uhl NW, Asmussen CB, Baker WJ, Harley MM, Lewis CE (2008) Genera Palmarum: the evolution and classification of palms. The University of Chıcago Press, Chicago
Eicher JB (2001) The art of African fashion by Prince Claus fund. Fashion Theory 5(1):103–110. https://doi.org/10.2752/136270401779045707
Elenga RG, Dirras GF, Maniongui JG, Djemia P, Biget MP (2009) On the microstructure and physical properties of untreated raffia textilis fiber. Compos A: Appl Sci Manuf 40(4):418–422
Elenga RG, Djemia P, Tingaud D, Chauveau T, Maniongui JG, Dirras G (2013) Effects of alkali treatment on the microstructure, composition, and properties of the raffia textilis fiber. Bioresources 8(2):2934–2949
Fadele O, Oguocha I N, Odeshi A, Soleimani M (2017) The effect of alkalization on properties of raffia palm fiber. In: Proceedings of the 26th CANCAM, Victoria, BC, 29 May–1 June 2017
Ganan P, Mondragon L (2004) Fibre-reinforced polyester composites: effects of fibre surface-treatment on mechanical behaviour. J Mater Sci 39(9):1573–4803
Gündoğan M, Kalaycı E, Avinc O,Yavas A (2018) Egzotik Doğal Çevre Dostu Bitki Liflerinden Etek Tasarımları. Alanya 4th International Symposium on Fine Arts, University of Alanya Alaaddin Keykubat, Alanya, 3-5 May
Haveaux GL (1954) La tradition historique des Bapende orientaux. Institut royal colonial belge, Marnixlaan
Haynes J, McLaughlin J (2000) Edible palms and their uses. University of Florida, Gainesville
Hetal S, Srinivasulu B, Subhas S (2012) The effect of surface treatment on the properties of woven banana fabric based unsaturated polyester resin composites. Int J Sci Eng Technol 1(3):86–90
Jähn A, Schröder MW, Füting M, Schenzel K, Diepenbrock W (2002) Characterization of alkali treated flax fibres by means of FT Raman spectroscopy and environmental scanning electron microscopy. Spectrochim Acta A Mol Biomol Spectrosc 58(10):2271–2279
Johnson VD (2011) Tropical palms. Food and Agriculture Organization of the United Nations, Rome
Kalia S, Avérous L (2011) Biopolymers: biomedical and environmental applications. Wiley, Chichester
Kocak D, Merdan N, Evren OB (2015) Research into the specifications of woven composites obtained from raffia fibers pretreated using the ecological method. Text Res J 85(3):302–315
Lewin M (1998) Handbook of fiber chemistry. CRC Press, New York
Mert M (2009) Lif Bitkileri. Nobel Akademik Yayıncılık, Ankara
Mert M (2011) Doğal Lif Kaynakli Minör Bitkiler. Nobel Akademik Yayıncılık, Ankara
Musset R (1933) Le raphia. Annales de géographie 42(236):190–193

Nekkaa S, Guessoum M, Chebira F, Haddaoui N (2008) Effect of fiber content and chemical treatment on the thermal properties of *spartium junceum* fiber-reinforced polypropylene composites. Int J Polym Mater 57(8):771–784
Obasi HC (2013) Properties of raphia palm interspersed fibre filled high density polyethylene. Adv Mater Sci Eng 2013:5. https://doi.org/10.1155/2013/932143
Odera RS, Onukwuli OD, Osoka EC (2011) Tensile and compressive strength characteristics of raffia palm fibre-cement composites. J Emerg Trends Eng Appl Sci 2(2):231–234
Odera RS, Onukwuli OD, Atuanya CU (2015) Characterization of the thermo-microstructural analysis of raffia palm fibers proposed for roofing sheet production. J Miner Mater Charact Eng 3(4):335. https://doi.org/10.4236/jmmce.2015.34036
Otedoh M (1982) A revision of the genus Raphia Beauv. (Palmae). J Niger Inst Oil Palm Res 6(22):145–189
Pende People (2019) Wikipedia Özgür Ansiklopedi. https://en.wikipedia.org/wiki/Pende_people. Accessed 29 July 2019
Picton J, Mack J (1980) African textiles. Trustees of the British Museum, London
Raffia (2002) Transport information service. http://www.tis-gdv.de/tis_e/ware/fasern/raffia/raffia.htm#informationen. Accessed 26 July 2019
Raffia fibre: the quintessence of high fashion. Fibre2Fashion.com. https://www.fibre2fashion.com/industry-article/7261/raffia-fibre-the-quintessence-of-high-fashion. Accessed 25 July 2019
Raffia Palm (2017) Wikipedia Özgür Ansiklopedi. https://en.wikipedia.org/wiki/Raffia_palm. Accessed 25 July 2019
Rodrigue STN, Ebénézer N, Dieunedort N, Didier F, Fogue M, Drean JY, Harzallah O (2017) Investigation of the physical and mechanical properties of raffia vinifera fibers along the stem. J Nat Fibers 14(5):621–633
Salau MA, Sharu AS (2004) Behaviour of laterized concrete columns reinforced with bamboo strips. West Indian J Eng 27(1):38–49
Sandy M, Bacon L (2001) Tensile testing of raffia. J Mater Sci Lett 20(6):529–530
Schuchardt U, Bianchi ML, Gonçalves AR, Curvelo AAS, Biscolla FC, Peres LO (1995) Piassava fibers.(*Attalea funifera*). 1. Chemical analysis, extraction and reactivity of its lignin. Cellul Chem Technol 29(6):705–712
Sikame Tagne NR, Njeugna E, Fogue M, Drean JY, Nzeukou A, Fokwa D (2014) Study of water absorption in raffia vinifera fibres from Bandjoun, Cameroon. Sci World J 2014:11. https://doi.org/10.1155/2014/912380
Taj S, Munawar MA, Khan S (2007) Natural fiber-reinforced polymer composites. Proc Pakistan Acad Sci 44(2):129
Tomlinson PB (1961) Anatomy of the monocotyledons, vol II. Oxford University Press, Palmae
Tribal African Art, Pende (Bapende, Phenbe, Pindi, Pinji) Democratic Republic of the Congo. http://www.zyama.com/pende/. Accessed 29 July 2019
Tuan D, Solomon W, Li Q, London IM (1985) The "beta-like-globin" gene domain in human erythroid cells. Proc Natl Acad Sci 82(19):6384–6388
Van De Ginste F (1946) Anthropometric study on the Bapende and Basuku of the Belgian Congo. Am J Phys Anthropol 4(2):125–152. https://doi.org/10.1002/ajpa.1330040210
Wafa FF (1990) Properties and applications of fiber reinforced concrete. JKAU: Eng Sci 2(1):49–63
Wake TA (2006) Prehistoric exploitation of the swamp palm (Raphia taedigera: Arecaceae) at Sitio Drago, Isla Colon, Bocas del Toro Province Panama. Caribb J Sci 42(1):11
Weaving Abstraction: Kuba Textiles and the Woven Art of Central Africa (2011) The George Washington University Museum-The Textile Museum. https://museum.gwu.edu/weaving-abstraction-kuba-textiles-and-woven-art-central-africa. Accessed 29 July 2019
Wikimedia Commons (2018a) https://commons.wikimedia.org/wiki/File:Starr-120522-5942-Raphia_farinifera-habit-Iao_Tropical_Gardens_of_Maui-Maui_(25116581256).jpg. Accessed 10 January 2020
Wikimedia Commons (2018b) https://commons.wikimedia.org/wiki/File:Starr-120522-6381-Raphia_farinifera-frond-Iao_Tropical_Gardens_of_Maui-Maui_(25143595515).jpg. Accessed 10 January 2020

Wikimedia Commons (2019) https://commons.wikimedia.org/wiki/File:Ndeemba_Mask_for_N-khanda_initiation_-_Yaka_people,_Bandundu_province,_Democratic_Republic_of_the_Congo,_early_20th_century,_wood,_raffia_fiber,_pigment_-_Brooklyn_Museum_-_Brooklyn,_NY_-_DSC08586.JPG. Accessed 10 January 2020

Wiriadinata H, Sari R (2010) A new species of Rafflesia (Rafflesiaceae) from North Sumatra. Reinwardtia 13(2):95–100

Xiao Z, Bergeron H, Grosse S, Lau PC (2011) Pectate lyases with increased thermostability and/or enzymatic activity. US patent 8,062,877, 22 Nov 2011

Yazıcıoğlu G (1999) Pamuk ve Diğer Bitkisel Lifler. Dokuz Eylül Üniversitesi Mühendislik Fakültesi Yayınları, İzmir

Yıldırım L, İşmal ÖE (2011) Tekstil Tasarımında Çevre Dostu Yaklaşımlar. Akdeniz Sanat Dergisi 4(8):9–13

Innovative Sustainable Apparel Design: Application of CAD and Redesign Process

Chanmi Hwang and Ling Zhang

Abstract This chapter aims to provide insights for designers, researchers, and educators seeking innovative ways to practice redesign activities within sustainable apparel design methods. The authors of the chapter present theoretical concepts for upcycling and applications of redesigning practices using innovative technologies, including laser etching and cutting, digital textile printing, and pattern digitizing through virtual computer-aided design (CAD) methods. The researchers further discuss a sustainable redesign framework that highlights the use of CAD and sequentially presents three design examples using (a) post-consumer textile waste (leather and silk) by an engineering laser cutting technique, (b) post-industrial waste (denim) by an engineering laser etching technique, and (c) post-consumer textile waste (denim) by pattern digitizing and engineering digital textile printing techniques. The presented redesign methods are sustainable as they reduce waste and increase efficiency for surface design and garment construction processes. The chapter concludes with some challenges of the redesigning process and suggested solutions suggestions for redesigning at micro and macro levels of the apparel industry.

Keywords CAD · Digital textile printing · Laser cutting · Laser etching · Pattern digitizing · Redesign · Upcycling

---

C. Hwang
Department of Apparel, Merchandising, Design and Textiles, Washington State University, Pullman, WA, USA
e-mail: chanmi.hwang@wsu.edu

L. Zhang (✉)
Department of Apparel, Events, and Hospitality Management, Iowa State University, Ames, IA, USA
e-mail: lzhang1@iastate.edu

S. S. Muthu, M. A. Gardetti (eds.), *Sustainability in the Textile and Apparel Industries*, Sustainable Textiles: Production, Processing, Manufacturing & Chemistry, https://doi.org/10.1007/978-3-030-37929-2_5

Introduction

A number of designers regularly practice redesign activities in the textile and clothing industry as part of their practical growth and development. The process of redesign adds value to discarded or used products and is associated with slow design in efforts to save resources, one of the important steps in the upcycling process (Aakko and Koskennurmi-Sivonen 2013). Currently, there are many innovative processes for new upcycling in the textile and clothing industry. Charitable organizations and small businesses are often the entities involved in remake activities where used clothes are converted into new clothing (e.g., I:CO, Isla Urbana and Triarchy Atelier, Goodwill, Brides for a Cause). Some apparel companies are also involved in the redesign activities for their surplus stocks, allowing suppliers or volunteers (e.g., student interns) to retransform unsold items to new clothing products (e.g., Reformation, Fabscrap, TRMTAB). Various upcycling and redesigning initiatives have been launched by leading clothing and textile brands (e.g., Patagonia, Nike) as well as luxury fashion brands (e.g., Elvis & Kresse, Charlotte Bialas), and upcycling workshops are supported throughout global attempts to achieve a more circular economy (e.g., Vaude supported by the German Federal Environmental Foundation). In redesign process, collected garments, such as jeans, can be reconstructed into children's wear, and woven scraps can be converted into elasticized sportswear by attaching strip of knitted clothes (Paras and Curteza 2018). Various types of surface designs and modifications, such as print or embroidery, are often incorporated into the designs to stimulate the interest of the user (Russell et al. 2010).

The role of designers and their creative practices in sustainable apparel design process has grown significantly in popularity in recent years. In order to redesign, repair, and recondition products to create something with added value, sufficient technical and creativity skills are required. Due to growing attention on the redesign process, an increasing number of designers have engaged in the redesign, and educators in the field of apparel design have developed curriculum to include upcycling design projects for their students (Lee and DeLong 2018; Park 2014). However, it is critical that designers and educators understand how redesigning differs from the traditional design processes and how to effectively use current technology to enhance the redesign process. Thus, this chapter aims to provide insights for designers, researchers, and educators seeking innovative ways to practice redesign activities within sustainable apparel design methods. The authors of the chapter present theoretical concepts for upcycling and applications of redesigning practices using innovative technologies, including laser etching and cutting, digital textile printing, and pattern digitizing through virtual computer-aided design (CAD) methods. The researchers further discuss a sustainable redesign framework that highlights the use of CAD and sequentially presents three design examples using (a) post-consumer textile waste (leather and silk) by an engineering laser cutting technique, (b) post-industrial waste (denim) by an engineering laser etching technique, and (c) post-consumer textile waste (denim) by pattern digitizing and engineering digital textile printing techniques. The presented redesign methods are sustainable as they reduce waste and increase efficiency for surface design and garment construction processes. The chapter concludes with some challenges of the redesigning process and suggestions of solutions for redesigning at micro and macro levels of the apparel industry.

Background

Textile Waste: Redesign in Upcycling Process

The current ready-to-wear (RTW) clothing, specifically the fast fashion system, makes cheap clothing available to the mass market and influences overconsumption behavior by producing resulting post-consumer textile waste. Currently, more than 15 million tons of used textile waste is generated each year in the United States alone, and about 10.46 million tons of this waste are sent to the landfill (Leblanc 2019). In terms of post-consumer textile waste, an average American throws away approximately 80 pounds of used clothing per person per year. Some common consumers' clothing disposal reasons are wear/tear, poor fit, and decreased interest in the fashion and boredom with the item (Laitala 2014). In addition, overproduction of fabrics or damaged products in the manufacturing sector that never reach the retail sector all contribute to the increasing textile waste (Williams 2011).

In order to consider ways to "reduce, reuse, or recycle in an effort to close the loop" of the textile waste, the upcycling process has been strongly encouraged in industry (Paras and Curteza 2018, p. 47). Upcycling is a mixture of *upgrading* and *recycling* including both adding value to and reusing products (Glaveanu et al. 2016). Glaveanu et al. (2016) defined upcycling in the book of *Creativity: A New Vocabulary* as "based on sustainable consumption" with the key notion to "revitalize old material by placing it into new constellations and by suggesting new ways of using it while, at the same time, keeping its essence intact as a main value-adding feature of the process" (p. 181). The concept of upcycling design is part of an effort to implement sustainability practices as a way of turning waste into valued products to protect and enrich the ecosystem (Delong, Alice Casto, Min & Goncu-Berk, 2017). In essence, the goal of upcycling is to prevent waste of potentially useful materials, thereby reducing the consumption of new raw materials when creating new products. The upcycling process creates higher value in a lower-value object by giving new value to discarded materials, and it increases the longevity of a product by redesigning the overall look.

Redesign could be considered an important step in the upcycling process. In the current manufacturing system, redesigning an existing product is considered to be relatively less important than manufacturing new products (Paras, Ekwall, Pal, Curteza, Chen & Wang, 2018). Redesign is a strategy for creating a second-generation product from existing unsold products or products in diminished demand. The aim of the redesigning process is to enhance product durability and longevity, as well as increase functional and aesthetic value of product, and is based on the technique of patternmaking and draping. The degree of redesign can differ from making minor changes to an overall transformation of the garment (Paras and Curteza 2018). Examples of minor changes include adding decorative trim or print and surface design modifications, such as the addition of embroidery. The overall transformation may even include completely changing the style or pattern of the original garment classification such as changing pants to a scarf. The redesign process is frequently a time-consuming process that requires critical analysis of the available products for determining their overall transformation into a new product.

To appreciate the relevance of redesign, it is helpful to first understand the foundation of design itself. In general, design can be defined as the process of creative problem-solving (Koberg and Bagnall 1981) and involves a set of highly organized procedures where various types of information are collected and synthesized into a consistent concept and finally transferred into a visual format (Cho et al. 2010). Specifically, apparel design involves the knowledge and arrangement of basic elements (i.e., line, color, shape, space, texture) and principles of design (i.e., proportion, balance, rhythm, emphasis, harmony), an understanding of the history of clothing, a knowledge of fibers and different fabrics, an understanding of body proportions and body/style interactions, and, finally, an understanding of garment making (Sumathi 2007; Keiser, Garner & Vandermar, 2017). For redesigning processes, these key design aspects should be considered for formulating a cohesive design aesthetic and function.

Despite the benefits of redesign, including an elongated product life and promotion of sustainable consumption, there are some complications that accompany the process. Paras and Curteza (2018) found that complex pattern and design, as well as multiple sizes, fabric types, and colors, were challenges in the redesigning process. However, the researchers suggest that solutions to these problems can be achieved through craftsmanship and innovation. The following section of this chapter presents concepts and applications of CAD technologies for surface design and garment construction to enhance the redesigning process.

Redesign Process and Framework

Thinking through a design planning is an iterative process, and, in general, designers move back and forth between decision-making that is explicit and intuitive and alternate between analysis and synthesis to develop effective designs (Watkins and Dunne 2015). Thus, understanding the relationship between materials, ideas, and systems is fundamental to comprehending the design process.

A three-step product development model by LaBat and Sokolowski (1999) can be broadly applied to designed products in the apparel field and includes (1) problem definition and research, (2) creative exploration, and (3) implementation. DeLong et al. (2017) suggest that the upcycling process can be performed during the creative exploration stage listed by LaBat and Sokolowski (1999); this is a stage that includes idea generation, conceptual design development, prototype development, evaluation, and refinement of designs and decisions for making the final product. Unlike typical design processes that start out with a combination of ideation and sketching, the starting point for the upcycling process starts out with

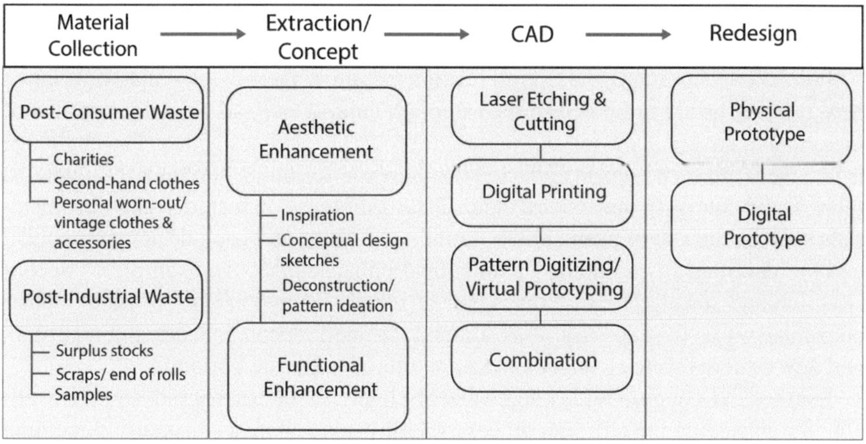

Fig. 1 A sustainable design framework: application of CAD and redesigning methods

the material. Similar to the concept of couture, a French term that means *sewing* and refers to high-quality apparel produced in small quantities, a fabric-driven design method is adopted by most of upcycling designers. Designers start their design process by selecting and manipulating fabrics to achieve their design goals (Delong, Alice Casto, Min & Goncu-Berk, 2017) and further make their design and construction decisions based on the yardage and pre-sewn construction limitations. According to these guidelines, this chapter focuses on the redesigning process within the exploration stage. As shown in Fig. 1, the process of redesigning can be divided into four generic steps: (a) material collection, (b) extraction/conceptualization, (c) CAD, and (d) redesign and creation of the final prototype. Redesigning processes have constraints based on the materials collected in terms of their properties and sizes; this framework highlights a creative redesigning process that integrates CAD to solve these constraints and develop unique garments with added values.

The first step of the redesigning process is collection of materials, namely, textile waste. This is critical and time-consuming compared to sourcing clean yardages of fabric at a store. The waste sources can be collected from post-consumer waste and post-industrial waste (Cassidy and Han 2017). Post-consumer waste includes worn-out (or never worn) or vintage garments and accessories, and they may be collected at charities and second-hand clothing or collected from friends and families. Post-industrial waste includes fabrics and yarns, such as end-of-roll materials or scraps from cutting and construction processes, and waste can be also collected at surplus stocks.

Once materials for redesigning are collected, they need to be sorted to maximize value that can be extracted out of the waste. According to Paras and Curteza (2018), the materials can be first sorted as waste or as valuable and then categorized by type and style to make this process more efficient. The authors further state that there are challenges to properly collecting textile waste in the current industry, especially at a mass production level, and recommend that more locations exist for the purpose of clothing and textile collection.

For the concept stage, the overall transformation of the collected materials into a new product needs to be determined through careful analysis of the properties of materials to ensure cohesion of the design properties and aesthetics. The challenge is that there are variations in collected textile waste including variant sizes, patterns, fabrics, and colors. In the concept stage, disassembly is conducted to examine materials such as the use of seam, stitch, lining, and trims, and each design component can be identified for aesthetic or functional enhancement. An example of deconstruction and sorting processes of a pair of jeans is shown in Fig. 2, where a post-consumer waste of jeans has been sourced, cleaned, deconstructed, and analyzed, and new patterns were arranged to explore minimum waste options.

For the CAD stage, a wide range of technologies in the current industry are available for enhancing the redesign process, thereby opening up a constantly expanding range of design possibilities and creating a complex, multifaceted set of decision points for designers. The applications of CAD tools allow for designers to explore their designs in relatively easy, quick, inexpensive, and sustainable ways. The following section discusses four main CAD alternatives as current technological advancements with potential for redesign: laser cutting and etching, digital textile printing, and pattern digitizing for virtual prototyping.

Laser Etching and Cutting

Laser technology in textile manufacturing offers new opportunities for cutting, surface adornment, and welding materials together. A laser is comprised of highly concentrated light energy that is funneled into a pinpoint (Nayak et al. 2008) and used via digital technology in the cutting and decorating of textiles (Baker 2016). Specifically, examples of surface treatment using a laser include hollowing out an area of fabric, engraving the fabric, creating patterns, engraving buttons, and embroidering fabric. Application of laser technology in apparel design emphasizes

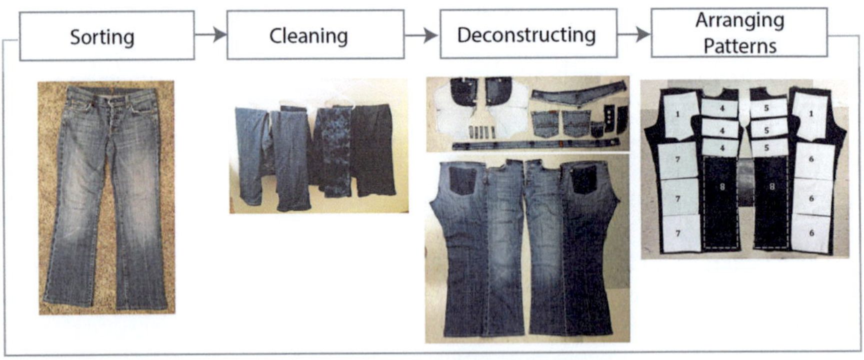

Fig. 2 Sorting and deconstruction processes of denim

more creative thought, removing the communicative barrier between the laser and designer (Goldsworthy 2009). The automated cutting method is now being used across a broad spectrum of industries (Yusoff et al. 2010), and because of the current affordability of laser cutters, apparel programs at universities have been implementing lower power cutters in the learning curricula (Viļumsone-Nemes 2012).

Laser cutting allows for a more creative approach when redesigning textiles since the laser cutters have vast capabilities in simulating desired aesthetics and surface patterns through cutting and etching processes. Similarly, laser cutting has created major advancements in movements toward sustainable design measures in terms of preventative solutions for textile application from chemical use in industry. Laser cutting applications can also be used in sustainable design formats such as upcycling projects and utilizing recycled elements left over or cutouts from one design use in another design. To apply this technique, raw edges of fabric are heated so fibers do not fray as easily, as hand-cut methods and textiles may involve the fusing of raw edges. Other key components to consider within sustainability are the use of laser cutting technology in the development of "zero-waste" garments as well as engineering applications to pattern cutting and CAD design programs (Trejo et al. 2016).

In addition to cutting, lasers can also alter the surface and look of textiles through etching or engraving. By altering the intensity of the laser pinpoint, designers may modify the look of the etching from slight damage on the surface to complete incineration of the textile fabric (Yuan et al. 2012). Fleece, suede, coated fabrics, and denim are best for use when etching and applying surface effects (Baker 2016). The testing of materials is important for achieving the desired aesthetic, as settings such as power, speed, and frequency levels may need to be adjusted according to fabrication and product design (Baker 2016). Fortunately, designers with limited technology skills can easily integrate laser cutting and etching tools into their redesign practices, since they require only basic skills in Adobe Illustrator or 3D modeling software. Figure 3 shows examples of patterns ranging from simple to complex that were created using vector-based graphics software (Adobe Illustrator) for laser cutting and laser etching. Using a Trotec Speedy 400 laser cutter, the applications of these techniques on cotton denim and lycra swatches (16″ × 16″) are shown in Fig. 4. A list of laser cutting and etching applications on both natural fabrics (e.g., 100% cotton satin and silk charmeuse) and man-made fabrics (e.g., polyester knit, lycra, and velvet) is presented in Appendix.

Pattern Digitizing and Digital Patternmaking

Pattern digitizing is a way to transfer a paper pattern to computerized form and is the patternmaker's method of scanning. The digitizing system and method can support and facilitate more cost-effective designs. Digitizing systems focus on the requirements of the apparel manufacturer, such as accurately generating paper patterns and marketing patterns to maximize the usage of the fabric. A pattern digitizing system is an effective tool for customizing and engineering prints or motifs

Fig. 3 Straight and curved lines for laser cutting (left) and etching (right): simple to complex patterns

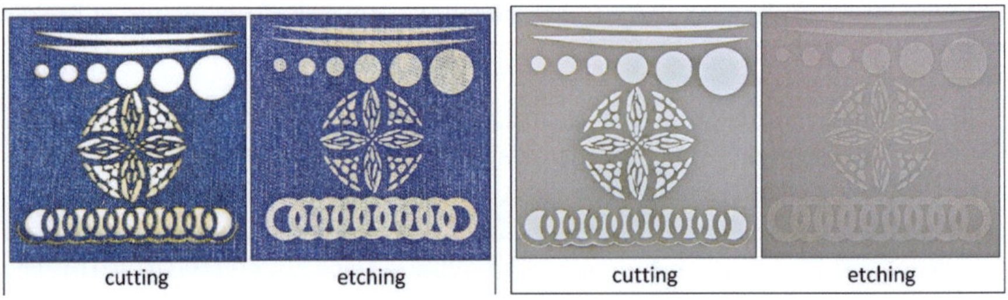

Fig. 4 Examples of laser cutting and etching on cotton denim (left) and lycra (right)

onto garment patterns for digital textile printing and laser cutting technologies. Especially for redesign, utilizing a digitizing system and CAD patternmaking can help designers minimize fabric waste and efficiently arrange the new garment patterns to fit existing deconstructed garment patterns.

Figures 5 and 6 show examples of how different pattern digitizing methods were applied when redesigning old T-shirts. The sourced T-shirts were washed and dried and carefully deconstructed along the sewing seams. The T-shirt pieces were then ironed flat and pined on the paper to ensure accuracy for digitizing. As shown in Fig. 5, flat patterns were created in Lectra Modaris software for a new design concept that will be made out of the old T-shirts. The measurements of the used T-shirts were also inserted and both Modaris flat patterns and digitized T-shirts patterns were converted to Adobe Illustrator. As shown in Fig. 6, the new garment patterns were arranged directly onto the digitized T-shirts patterns for the cutting lines. At the end, the stitching lines of the final garment patterns and texts were removed, and the outlines of the patterns were changed to the appropriate setting for laser cutting. Next, the new garment pieces were laser cut accurately based on the engineered patterns.

Figure 7 is an example of the results of utilizing existing garment seams to upcycle a new jumpsuit, with the technique simultaneously eliminating the usage of

Fig. 5 Garment flat patterns created in Lectra Modaris

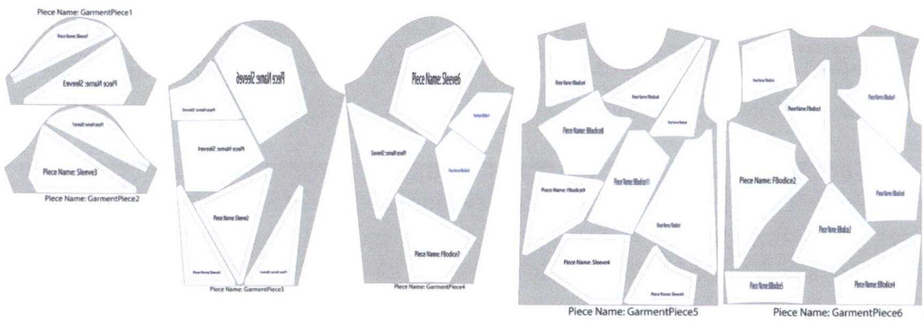

Fig. 6 New garment patterns (in white) engineered into digitized existing T-shirts patterns (in gray) to ensure minimum waste

additional threads. The plus-size men's jeans were deconstructed at the inner seams and the seams from the waist to the hip line. The Modaris flat patterns were converted into the Adobe Illustrator and placed on the digitized jeans patterns, and then the designer used the existing outer seams of the jeans at the center of the pant as structural decoration. Such digitizing systems can be executed using apparel patternmaking software like Lectra Modaris, Optitex, Browzwear, or Gerber AccuMark. All of the three redesign examples presented later in this chapter include a pattern digitizing process and describe how it can be utilized seamlessly throughout the sustainable design process through methods that reduce waste of fabric and increase efficiency.

Digital Textile Printing

Digital textile printing is another CAD application that can be integrated into the redesign process, since it offers greater flexibility for designers to make visual statements. Digital textile printing process works through use of the ink jet print system, whereas conventional printing works using a wet process. The conventional method consumes large amounts of dyes and water, which causes severe damage to the

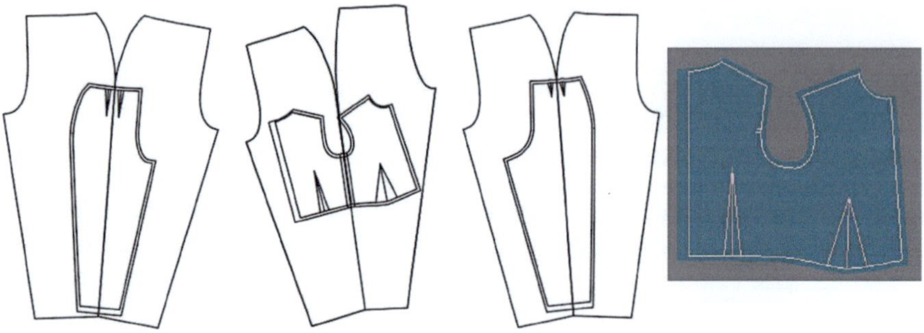

Fig. 7 New garment patterns (small size) engineered into digitized existing garment patterns (bigger size) for utilizing seams

environment (Civil Engineering Research Foundation 1997). For digital textile printing, various ink types (i.e., acid, disperse, reactive, pigment) are used to print on cellulous and protein-based fibers as well as synthetic fibers. Both these fibers and synthetic dyes, however, are detrimental to the environment because of textile dye waste. The amount of synthetic dyes produced in the world is estimated to be over 10,000 tons per year (Forgacs et al. 2004), and more than 700,000 tons of textile dyes are disposed to the ocean and rivers around the world (Bulut and Akar 2012; Fletcher 2006). The wastewater from synthetic dyes further contains a great degree of harmful chemicals that affect the soil, underground water, and atmosphere and lead to serious health issues for animals (Bulut and Akar 2012).

Digital textile printing, in the last decade, has made its way into the mass customization market and is slowly replacing original screen-printing processes; with this, the integration of print is becoming as vital to the designer's vision as the form of the garment. Fashion designers, such as Issey Miyake, Comme des Garcons, and Alexander McQueen, have continued to utilize and adapt digital textile design for powerful visual statements. This technique enables designers to create a repeat or engineered print that can be merged with upcycled garments and fit pattern pieces of a garment in a way that the design flows unbroken around the body. Some advantages of using digital textile printing over traditional printing (e.g., screen-printing, woodblock printing) include lessened impact on the environment, accelerated speed translating the design onto the fabric, and the ability to print intricate details and millions of colors.

The opportunities and advancements in digital textile printing likewise provide tremendous and continuous improvements on the customizable products such as home furnishings, apparel, accessories, and corporate promotional material (Bowles and Issac 2012). Currently, 60% of digitally printed textiles are produced in the developing countries such as Bangladesh, India, Indonesia, Pakistan, and Vietnam. According to the UK-based market research firm Smithers Pira report on "The Future of Digital Textile Printing to 2023," digital textile printing remains one of the

Fig. 8 Repeat (left) and engineered (right) prints using digital textile printing

fastest growing segments within the digital print ecosystem. There was $3.15 billion in 2018 which was more than double its value in 2013. The volume of fabrics printed digitally has risen from 1.40 to 2.42 billion m^2. For 2023, the future commercial of digital textile printing market is projected to increase to a total value of $5.46 billion and grow to consume over 4.54 billion m^2 (The Future of Digital 2018).

Digital textile printing is controlled by the CAD design process. The textile patterns are usually produced using creative software, such as Adobe Photoshop, Adobe Illustrator, Freehand, and Lectra Kaledo Print, and translated into a high-resolution image file. The image a designer has created could also encompass scanned, hand-painted artwork or photographs. Interactive functions, such as cleaning-up and defining the color contrast and brightness, adding filters and layers, combining images, and rotating and repeating the patterns, are the basic requirements. Repeat and engineered textile designs are two major surface design formats for digital textile printing. The amount of the colors and size of the repeats are not limited in the digital textile printing, though the width of the fabric must be considered. The first image in Fig. 8 is an example of a repeat print that was created based on two nature scene photographs using Adobe Photoshop. The repeat print can be used as garment panels or patch application to enhance redesign aesthetics.

Among the greatest advantages of using digital textile printing technology in apparel design is its application of engineered print, which integrates a garment form with "digitally manipulate surface imagery so that a printed design could be engineered to match across the seamlines" (Parsons and Campbell 2004, p. 90). The complexity of engineered designs must be aligned with the garment shape, as this allows large expanses of cloth to be used for innovative textile designs (Parsons and Campbell 2004). Figure 8 shows examples of engineered repeat prints that were developed using digital textile printing. The garment patterns were digitized and modified in the Lectra Modaris and converted into the Adobe Photoshop for engineering the prints. The repeat prints were applied on the digital garment patterns. In this representation, the prints were engineered to match across the shoulder seams. A single image can also be digitally developed and printed for garment patch application to enhance design aesthetics.

Applications of the Sustainable Design Framework

Following the redesign process and framework discussed in section "Redesign Process and Framework", three redesign examples using mainly upcycled leather, silk, and denim are presented in the next section. Each stage of the redesign framework (material collection, extraction/ concept, and CAD) is discussed with applications of laser cutting, laser etching, pattern digitizing, and digital textile printing in each redesign process.

Design Example 1

Material Collection, Extraction, and Concept

The objective of this design was to create women's couture two-piece suits from post-consumer waste and incorporate a laser cutting technique. The designers collected two second-hand, black leather jackets and a total of 16 worn-out men's silk neckties displaying a variety of prints. Two second-hand, women's leather jackets were disassembled along the constructional seam lines (i.e., along arm holes and side seams) to explore potential design strategies to use during redesign. The sleeves were disassembled along with the arm holes and the inner seam lines.

Once the materials were collected, designers defined inspiration of the design based on the aesthetics of the collected garments. Black leather jackets and ties were selected as main materials for the redesigned garment to represent the spirit of the Gothic Period. Thus, the initial inspiration of this apparel design was the aesthetics of Gothic architecture in the sixteenth century and expressed in the black and red colors and the materials of the garments; the collected ties contained geometric shape patterns with a majority of them being dark and red tones.

Black is the main color of the Gothic/Goth, which is a modern subculture that first became popular during the early 1980s within the Gothic rock scene (Eckart 2005). The purpose of including the red color on the men's neckties was to represent the Gothic theme, a representation of blood and roses related to life, death, despair, and passion that was prevalent during this era (Eckart 2005). As shown in Fig. 9, the placement of the laser cut patterns on the sleeves represented the *flying buttress* on the Notre Dame de Paris. The men's neckties on the wearer's right center front represented the pointed roofs, a common of Gothic architecture. The curve of the neckline on the wearer's left was inspired by the *Pointed Arch* in the ribs of Notre Dame Cathedral. Finally, the fitted waist and the peplum hem of the jacket was inspired by the silhouette of the women's dress during the Gothic period in sixteenth century in Europe.

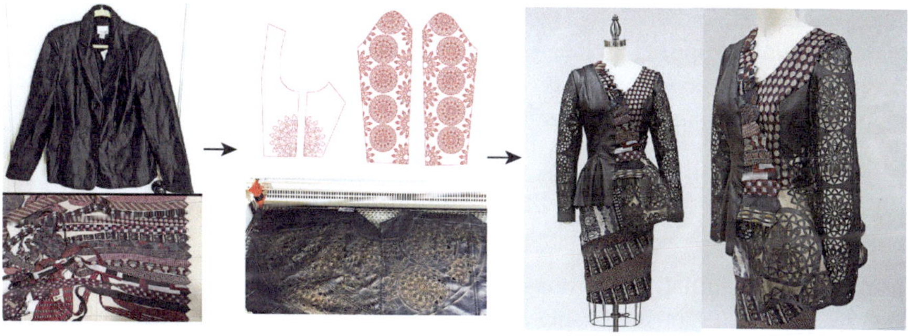

Fig. 9 A redesign process using post-consumer waste (leather jacket and silk ties)

CAD Application

The pattern of a rose window in Gothic architecture was the inspiration for the laser cutting pattern. The *Rose Window* patterns (pointed arch and flowers) were created in Adobe Illustrator. The sleeves and the wearer's left side hem were strategically placed and digitized through OptiTex PDS to ensure minimal waste and ideal fit on the body, and the two sleeves and hem panel were laser cut using Trotec Speedy 400. The 16 men's neckties were deconstructed using a piecing method, then the fabrics from three men's neckties were used on the jacket, and the narrow tails of 15 men's neckties were hand stitched and attached with the edge of the center front. The pencil skirt was completely created from the fashion fabrics of five deconstructed men's neckties.

Design Example 2

Material Collection, Extraction, and Concept

The objective of this design was to create a two-piece denim garment using post-industrial waste and incorporating the laser etching technique. The denim was collected from post-industrial waste and used end-of-roll scraps from a local textile retailer company. A total of three yards of denim was sourced and then was machine-washed in cold water and dried. Finally, the entire area of the fabric was hand-ironed so laser etching could be consistently applied on a clean surface.

The concept of the design was to use fashion as a medium to actively and creatively raise awareness and engage in the global environmental crisis, since dress is a powerful communication tool (Kaiser 2012). Specifically, this design was inspired by the cover image of an issue of the National Geographic, June 2018 edition. The cover image shows an iceberg-resembling a plastic bag partially submerged in the ocean and was photographed by artist Jorge Gamboa. The image was accompanied by the text, "PLANET OR PLASTIC? 18 billion pounds of plastic ends up in the ocean each year. And that's just the tip of the iceberg" (Gamboa 2018, p. 1). At first glance, the image appears to be an iceberg but instead depicts a plastic bag littering a pristine ocean, a powerful and proactive image part of a campaign issued to reduce global reliance on single-use plastics.

CAD Application

Inspired by this initiative, the image of an iceberg-resembling plastic bag was created by free-form sketching using a digital graphic drawing tablet. This image was converted from AI to DXF file format for laser engraving due to the laser engraving's sustainable design measure to prevent use of chemicals, such as those incorporated in industry's textile application. By altering the intensity of the laser pinpoint, the intended look of the engraved product, from slight damage on the surface to complete incineration of the textile fabric, can be achieved. This sustainable element of surface design, in combination with the efficiency and speed in design, makes the laser etching application a viable design option. Along with the plastic bag image, a geometric border portraying layers of the ocean floor with the written text phrase "fast fashion" was also developed (Fig. 10). These images were engineered into four panels and laser etched on 100% cotton denim. By adjusting the power levels of the laser beam, laser etching was performed to decorate the denim surface to create powerful visual effects. This visual statement of an iceberg and a plastic bag captures audience's attention and creatively raises awareness of the global environmental crisis.

The final design includes a two-piece dress with a sleeveless A-line top and an adjustable elastic skirt (Fig. 11). The top has two side panels with honeycomb smocking, a technique that involves creating a series of pleats and then stitching these pleats together at certain points in order to create diamond design repeats. This method allows the fabric to shrink and expand without elastic, and the addition of the honeycomb smocking to the side panels gives flexible ease and comfort to the wearer. The waist of the skirt has a ¾″ wide adjustable knit buttonhole elastic band that can fit a wide range of waist sizes (sizes 4–14). It is inserted inside of the waist band with two button holes attached so the size can be easily adjusted with a clean finish.

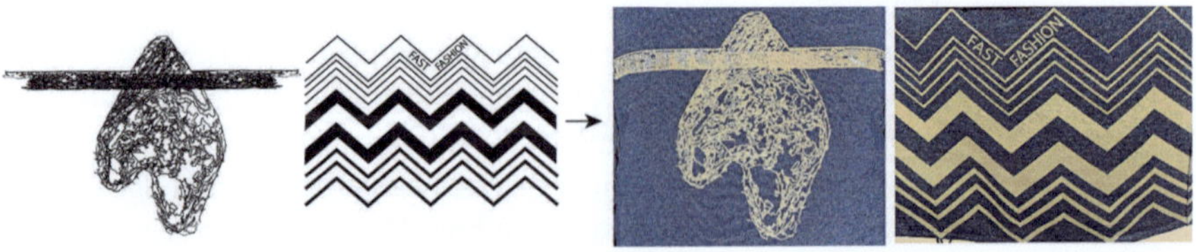

Fig. 10 Laser etching process on denim

Fig. 11 The final design: planet or plastic? Etching a global environmental crisis

Design Example 3

Material Collection, Extraction, and Concept

The objective of this design was to create a women's streetwear jacket from post-consumer waste and incorporate digitizing and digital textile printing methods. The design process also included combining traditional design techniques, such as hand painting and hand embroidering, with CAD methods. The designer collected three thrifted men's jeans that were made of 100% cotton to create a women's jacket. Three jeans were machine-washed and dried and were disassembled at the inner seams only. This design was inspired by Roksanda Resort 2015 Collection, a British fashion brand, and contemporary pop art aesthetics from Andy Warhol's Marilyn Monroe's artwork series. The new design features an oversized, but structural, silhouette with accent colors and prints to align with the design inspirations.

CAD Application

To undertake the pop art elements, the designer created a repeat print with the pop art figure made from an AI application, *Zepeto*, which is evenly spread upon the pointed-star background. Adobe Photoshop and Adobe Illustrator were used to create the repeat print, and the pointed-star shape was created using Illustrator with a stylized filter (Fig. 12). The figure portraits were placed on top of the single-pointed star motif; after merging those layers, the print was developed into a 52″ × 72″

Fig. 12 Digital textile prints development and applications

repeat in Illustrator. The portrait artwork shown in Fig. 12 was originally hand painted on canvas using colored tissue papers and was scanned into Photoshop to modify the colors and scale to a smaller size. The textile designs were digitally printed on a 100% cotton twill in order to match the texture.

The final garment includes an oversize jacket and a pleated mini skirt (Fig. 13). The pop art printed fabric was placed as the lining of the collar, bodice, and hem of the jacket for additional colors. The 8″ × 11″ printed portrait was placed at the back of the jacket as patch work. As for the closure, the skirt includes an elastic waistband for easy wearability and a flexible fit, and the jacket includes six metal buttons from the used jeans on the center front for a neat and professional finish. This playful and artsy ensemble communicates the wearer's personality and presents unique redesign aesthetics while still representing the potential for sustainable methods used and raising awareness of environmental issues.

Conclusions

This chapter provides useful solutions for understanding the redesign processes for designers to rethink and enhance redesign processes using CAD and upcycle textile waste into a new design solution. The redesign method is an important part of the sustainable design process in the apparel industry and adds value to textile waste for the generation of new clothing. As discussed in this chapter, the role of designers and their creative practices in the redesigning process has been emphasized, since it requires sufficient professional skills for patternmaking that minimize textile waste

Fig. 13 A redesign process using patching and digital textile printing methods

and innovation for combining different materials. The application of CAD discussed in this chapter enhances designers' efforts to creatively explore the redesign process with maximum technical and aesthetic solutions.

Both theoretical concepts of upcycling and applications of redesign practices using CAD were discussed to yield insights to designers, researchers, and educators who seek unique ways to practice redesign activities. The redesign framework includes material collection, extraction/concept development, CAD, and redesign prototype creation and contributes to a means for integrating innovative technologies and upcycling concepts into the contemporary apparel design process. The processes for sourcing the inspirations and applying them to contemporary apparel designs have demonstrated a unique way to utilize laser cutting and etching, pattern digitizing, and digital textile printing to accomplish the textile designs that combine the historical and cultural artistic inspirations. Laser cutting and etching techniques provide alternatives to add intricate surface designs on various types of textile waste. The pattern digitizing technique allows designers to engineer new garment pattern pieces into second-hand garment patterns with minimal waste, as well as utilize existing seams and stitches to maximize the use of existing materials. Integrating digital textile printing also adds value to textile waste, since powerful visual statements, like patches, can be flexibly incorporated into redesigned garments.

Upcycling is one of the most sustainable circular solutions in the waste hierarchy in textile production. The value of redesign has been increased in the industry since it requires little energy input and can eliminate the need for a new product (Ellen Macarthur Foundation 2017; Szaky 2014). Strategies were also proposed to strengthen the creative department and fashion design for small- and medium-sized enterprises to gain competitive advantages in the redesign sector (Cuc and Tripa 2018). Efforts toward remake activities have been typically associated with slow or couture design methods, but many upcycling solutions have been attempted in mass production in recent years (Cassidy and Han 2017; Cuc and Tripa 2018). The redesign system would furthermore generate a number of jobs, such as dealer, tailor, mending, and washing, though there are still limitations and challenges the apparel industry must overcome to ready the redesign process for mass production (Cassidy and Han 2017). A collection of large quantities of textile waste, along with resources for sorting, deconstructing, and cutting, would not be readily available for the larger manufacturers to adopt. Thus, redesigning processes that incorporate new technology need to be promoted to make the textiles system more sustainable and contribute to a circular economy. Designers' utilization and promotion of creative CAD applications in their redesign practices will increase efficiency in both technical and aesthetic design solutions in the industry.

Acknowledgments Thanks to Qiyao Xiong, Katelyn Otto, and Izabella Varela-Moran for providing the images of their designs and patterns.

Appendix

Applications of laser cutting and etching on natural and man-made fabrics

Laser cutting			Laser etching	
Cellulous fibers				
Cotton Denim (Blue)		P: 100 V: 1 PPI: 4000		P: 100 V: 65 PPI: 1000
Cotton Denim (White)	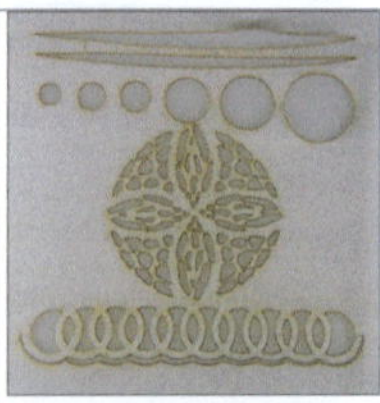	P: 100 V: 1 PPI: 4000		P: 100 V: 65 PPI: 1000
Cotton Satin		P: 75 V: 2 PPI: 5000		P: 20 V: 15 PPI: 500
Protein fibers				
Cotton Silk Charmeuse		P: 100 V: 2 PPI: 5000	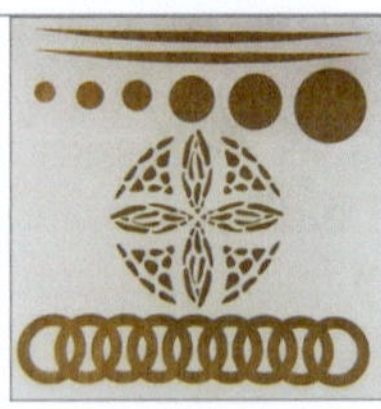	P: 13 V: 5 PPI: 500

Laser cutting			Laser etching	
Cotton Wool Felt	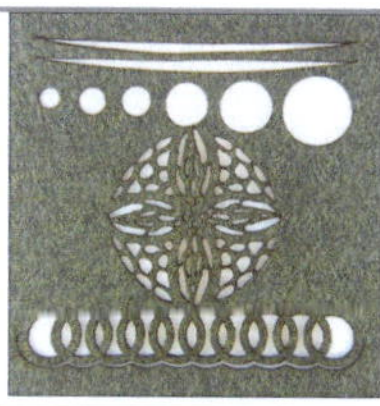	P: 100 V: 1 PPI: 4000		P: 100 V: 60 PPI: 100
Cotton Satin		P: 75 V: 2 PPI: 5000	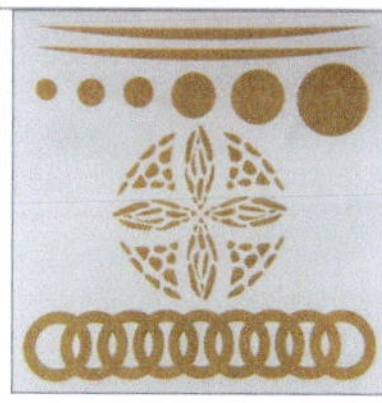	P: 20 V: 15 PPI: 500
100% leather				
100% Leather		P: 100 V: 2 PPI: 2000		P: 15 V: 5 PPI: 500
Man-made fibers				
Velvet		P: 100 V: 1 PPI: 1000		P: 35 V: 30 PPI: 1000
Polyester Knit		P: 100 V: 3 PPI: 5001		P: 30 V: 20 PPI: 500
Suede		P: 100 V: 3 PPI: 5001		P: 15 V: 10 PPI: 500

Laser cutting			Laser etching	
Lycra	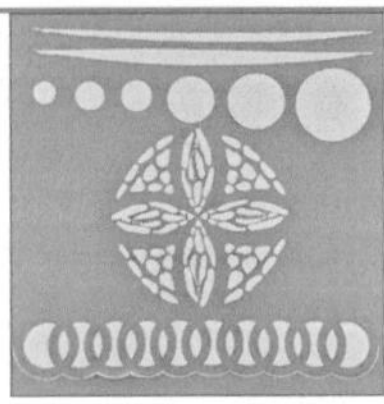	P: 100 V: 3 PPI: 5001		P: 20 V: 10 PPI: 500

Note: P indicates power, *V* indicates speed, *PPI* indicates power per inches

References

Aakko M, Koskennurmi-Sivonen R (2013) Designing sustainable fashion: possibilities and challenges. Res J Text Appar 17(1):13–22

Baker LB (2016) Laser cutting for fashion and textiles. Laurence King Publishing, London

Bowles M, Issac B (2012) Digital textile design, 2nd edn. Laurence King Publishing, London

Bulut MO, Akar E (2012) Ecological dyeing with some plant pulps on woolen yarn and cationized cotton fabric. J Clean Prod 32:1–9

Cassidy TD, Han SLC (2017) Upcycling fashion for mass production. In: Gardetti MA, Torres AL (eds) Sustainability in fashion and textiles. Routledge, London, pp 148–163

Cho G, Lee S, Cho J (2010) Review and reappraisal of smart clothing. In: Cho G (ed) Smart clothing: technology and applications. CRC Press, Boca Raton, pp 1–35

Civil Engineering Research Foundation (1997) Clean technologies in U.S. industries: focus on textiles. Info House. http://infohouse.p2ric.org/ref/06/05656.htm. Accessed 15 May 2019

Cuc S, Tripa S (2018) Redesign and upcycling-a solution for the competitiveness of small and medium-sized enterprises in the clothing industry. Ind Text 69(1):31–36

DeLong M, Alice Casto M, Min S, Goncu-Berk G (2017) Exploring an up-cycling design process for apparel design education. Fash Pract 9(1):48–68

Eckart G (2005) The German gothic subculture. Ger Stud Rev 28(3):547–562

Ellen MacArthur Foundation (2017) A new textile economy: redesigning fashion's future. https://www.ellenmacarthurfoundation.org/assets/downloads/A-New-Textiles-Economy_Summary-of-Findings_Updated_1-12-17.pdf. Accessed 27 April 2019

Fletcher K (2006) The circle of life, eco textiles: the past, present and hopes for the future. Selvedge 9:32–39

Forgacs E, Cserhati T, Oros G (2004) Removal of synthetic dyes from wastewaters: a review. Environ Int 30(7):953–971

Gamboa J (2018) Planet or plastic? National Geographic. https://www.nationalgeographic.com/magazine/2018/06/. Accessed 11 May 2019

Glaveanu VP, Tanggaard L, Wegener C (2016) Creativity, a new vocabulary. Palgrave Macmillan, London

Goldsworthy K (2009) Resurfaced: using laser technology to create innovative surface finishes for recyclable, synthetic textiles. In: Cutting edge: laser and creativity symposium. http://www.cuttingedgesymposium.com/pdf/kate-goldsworthy-paper.pdf. Accessed 14 June 2019

Kaiser SB (2012) Fashion and cultural studies. Bloomsbury, New York

Keiser S, Garner MB, Vandermar D (2017) Beyond design: the synergy of apparel product development. Fairchild Publication, New York

Koberg D, Bagnall J (1981) The design process is a problem-solving journey. In: The all new universal traveler: a soft-systems guide to creativity, problem-solving, and the process of reaching goals. William Kaufmann, Los Altos, pp 16–17

LaBat KL, Sokolowski SL (1999) A three-stage design process applied to an industry-university textile product design project. Cloth Text Res J 17(1):11–20

Laitala K (2014) Consumers' clothing disposal behaviour–a synthesis of research results. Int J Consum Stud 38(5):444–457
Leblanc R (2019) Textile and garment recycling facts and figures. Sustainable Business. https://www.thebalancesmb.com/textile-recycling-facts-and-figures-2878122. Accessed 20 May 2019
Lee YK, DeLong M (2018) Rebirth product development for sustainable apparel design practice in a design studio class. Fash Pract 10(1):34–52
Nayak R, Gon D, Khandual A (2008) Application of LASER in apparel industry. Man Made Text India 51(10):341–346
Paras MK, Curteza A (2018) Revisiting upcycling phenomena: a concept in clothing industry. Res J Text Appar 22(1):46–58
Paras M, Ekwall D, Pal R, Curteza A, Chen Y, Wang L (2018) An exploratory study of Swedish charities to develop a model for the reuse-based clothing value chain. Sustainability 10(4):1–19
Park J (2014) Upcycled parachutes project at Colorado State University. Fashion Pract 6(1):119–124
Parsons JL, Campbell JR (2004) Digital apparel design process: placing a new technology into a framework for the creative design process. Cloth Text Res J 22(1):88–98
Russell SJ, Morely N, Tipper MJ, Drivas I, Ward GD (2010) Principles of the recovery and reuse of corporate clothing. Inst Civil Eng Waste Resour Manag 163(4):165–172
Sumathi GJ (2007) Elements of fashion and apparel design. New Delhi, New Age International Ltd
Szaky T (2014) Outsmart waste. Berrett-Koehler Publisher, San Francisco
Trejo NK, Reyes CG, Sanchez V, Zhang D, Frey MW (2016) Developing composite nanofibre fabrics using electrospinning, ultrasonic sewing, and laser cutting technologies. Int J Fashion Des Technol Educ 9(3):192–200
Viļumsone-Nemes I (2012) Automated cutting of textile materials. Industrial cutting of textile materials. Elsevier, Amsterdam, pp 104–127
Watkins SM, Dunne LE (2015) Functional clothing design. Bloomsbury, New York
Williams S (2011) Sri Lanka upcycling factory makes waves in the fashion industry. The Telegraph. http://www.fashion.telegraph.co.uk/news-features/TMG8254833/Sri-Lankan-upcycling-factory-makes-waves-in-the-fashion-industry.html. Accessed 3 May 2019
Yuan G, Jiang S, Newton E, Fan J, Au W (2012) Application of laser treatment for fashion design. J Text Inst 103(1):48–54
Yusoff N, Osman NA, Othman KS, Zin HM (2010) A study on laser cutting of textiles. Int Cong Appl Lasers Electro-Optics 1:1559–1566

Bacteria Working to Create Sustainable Textile Materials and Textile Colorants Leading to Sustainable Textile Design

Fatma Filiz Yıldırım, Arzu Yavas, and Ozan Avinc

Abstract Sustainable textile design without harming the environment nowadays is one of the most important issues in the textile industry. Environmental concerns lead various researchers and artists to find and explore the applicability of more sustainable, renewable, and environment-friendly resources for fibers and colorants. In here, bacteria can help us. Microbial (bacterial) cellulose is an organic compound of the formula $(C_6H_{10}O_5)_n$ produced by certain types of bacteria. On the one hand, the material produced by the advancements in the detection and synthesis methods of bacterial cellulose, the resultant bacterial cellulose can be used in a wide range of commercial applications comprising textile, medical, cosmetic, and food applications and products. From the textile point of view, apart from bacterial cellulose production for textile substrates, bacteria can also be utilized for textile coloration. Indeed, bacteria can also produce colored pigments. It is now possible to produce some bacterial pigments for food, pharmaceutical, cosmetic, and textile applications. For instance, specific bacterial pigments can be used successfully as textile colorants. Therefore, bacteria are now working for us to create sustainable textile materials and textile colorants leading to more sustainable textile design. In this chapter, information about sustainable bacterial cellulose for textile substrates and sustainable bacterial pigments as sustainable textile colorants is given in detail.

Keywords Bacteria · Bacterial cellulose · Sustainable design · Dye · Pigment · Sustainable · Textile

F. F. Yıldırım
Research and Development Centre, Ozanteks Textile Company, Denizli, Turkey

A. Yavas · O. Avinc (✉)
Textile Engineering Department, Engineering Faculty, Pamukkale University, Denizli, Turkey
e-mail: oavinc@pau.edu.tr

S. S. Muthu, M. A. Gardetti (eds.), *Sustainability in the Textile and Apparel Industries*, Sustainable Textiles: Production, Processing, Manufacturing & Chemistry, https://doi.org/10.1007/978-3-030-37929-2_6

Introduction

Sustainable textile design without harming the environment nowadays is one of the most important issues in the textile industry. Today, our clothes can be produced using many natural or synthetic fibers. However, it is well known that the use of water to produce plants such as cotton, where the water demand is increasing considerably, can be quite high. On the other hand, it is a known fact that petroleum-based fibers may not be produced due to the depletion of fossil fuels in the future. These reasons lead various researchers and artists to find and explore the applicability of more sustainable, renewable, and environment-friendly resources for fibers and dyes.

Microbial (bacterial) cellulose is an organic compound of the formula $(C_6H_{10}O_5)_n$ produced by certain types of bacteria. Cellulose is the basic structural material of many plant substances but is also produced by bacteria from the genera *Acetobacter*, *Sarcina ventriculi*, and *Agrobacterium*. Bacterial or microbial cellulose exhibits different characteristics than vegetable-plant-based cellulose. These properties can be characterized as high purity, strength, moldability, and increased water retention. While bacterial cellulose can be produced in nature, a wide variety of methods are currently under investigation to increase cellulose growth from suitable cultures in laboratories for larger-scale commercial production. By effective control of the methods and conditions, it is stated that microbial celluloses with the desired specific properties can be developed. Therefore, nowadays, bacterial cellulose can be utilized in a wide variety of commercial applications such as textile, medical, cosmetic, and food applications and products with the material produced by the advancements in the detection and synthesis methods of bacterial cellulose.

The first discovery of bacterial cellulose was made in 1886 by A.J Brown (by *Acetobacter xylinum* bacteria). However, no intensive studies have been conducted on bacterial cellulose until the twentieth century. Later, other researchers have reported the formation of cellulose from other various organisms (such as *Acetobacter pasteurianum*, *Acetobacter rancens*, *Sarcina ventriculi*, and *Bacterium xylinoides* bacteria). In 1949, the microfibril structure of bacterial cellulose was characterized by Muhlethaler. Today, studies on bacterial cellulose continue in an increasing trend. Many researchers and artists began to look for alternative materials to cotton fiber. *BioCouture* is one of these alternatives. It is a radical project that expands the vision of the future, using nature, and is an idea from Suzan Lee's *Fashioning the Future: Tomorrows Wardrobe*. The project team examined the usability of microbial cellulose growing in the laboratory to produce clothing. The aim of the team was to grow microbial cellulose to form a complete garment on a liquid boat and to create the targeted textile garment in this way.

Additionally, interest and search in natural dye alternatives increases in textile applications due to the fact that synthetic dyes can be toxic to humans and nature, and also too much water is used in the conventional dyeing processes. However, due to some disadvantages about extraction difficulty of natural dyes from plants and the need for large cultivation areas for mass commercial productions, the utilization of microorganisms as a source of dyes and pigments comes on the agenda, and researches have been conducted on microbial dyes. In the industry, it is now possible to produce some bacterial pigments for food, pharmaceutical, cosmetic, and textile applications. Frankly, from the textile point of view, apart from bacterial cellulose production for textile substrates, bacteria can also be used for textile coloration. Since bacteria can produce colored pigments. Pigment-producing bacteria can produce many different colors from all colors of the rainbow to unusual colors such as black, white, brown, gold, silver, fluorescent green, yolk, or blue. Red and purple pigments (known as *prodigiosin* and *violacein*, respectively) were applied to different fabrics such as acrylic, silk, cotton, polyester, and polyester microfibers. All pigments were found to have high staining and dyeing ability on fabrics. All these findings indicate that specific pigments can be utilized successfully as textile colorants. New type of safe and effective natural dyes and pigments can be used potentially instead of harmful conventional synthetic dyes.

Therefore, bacteria are now working for us to create sustainable textile materials and textile colorants leading to sustainable textile design. In this review chapter, information regarding sustainable bacterial cellulose for textile substrates and bacterial pigments as sustainable textile colorants is given in detail. In the first part of this chapter, bacterial cellulose, its history, and bacterial pigments are introduced. In the second part of the chapter, extended information about bacterial cellulose and its interesting application areas are given. Finally, in the third part, coloration with bacterial pigments is examined.

Bacterial Cellulose and Its Application Areas

In the production stages of many natural or synthetic fiber-based textile materials, massive amount of water is used. However, it is an important fact that oil-based fibers will be eventually depleted as a result of depletion of fossil fuels. These reasons lead various researchers and artists to search more sustainable, renewable, and environment-friendly resources for tomorrows' textiles (Biocuturc 2014). Suzan Lee, one of these researchers, established BioCouture when she began to look for alternative materials to cotton fiber (Biocuture 2014). BioCouture is a radical research project that uses nature and sheds light on the vision of the future fashion and is an idea from Suzan Lee's book *Fashioning the Future: Tomorrow's Wardrobe* (Biocuture 2014; Growing Your Own Fabric is Possible 2014). The project team examined the usability of microbial cellulose growing in the laboratory to produce garments (Biocuture 2014). As aforementioned, the material produced by the advancements in the detection and synthesis methods of bacterial cellulose, the resultant bacterial cellulose could be utilized in a wide range of commercial applications comprising textile, medical applications, cosmetics, and food products.

Cellulose is one of the most common polymers in the world (Güzel and Akpınar 2018). Although cellulose of formula $(C_6H_{10}O_5)_n$ is the basic structural material of many plant materials, it can also be produced by *Komagataeibacter*, *Gluconacetobacter*,

Acetobacter, *Agrobacterium*, *Aerobacter*, *Achromobacter*, *Alcaligenes*, *Azotobacter*, *Pseudomonas*, *Rhizobium*, and *Sarcina* (Güzel and Akpınar 2018; Wikipedia 2014; Dursun et al. 2006; Tozluoğlu et al. 2015; Costa et al. 2017). This type of cellulose is the first product of bacterial metabolism and serves as a cell protector (Güzel and Akpınar 2018). Bacterial or microbial cellulose is distinguished from vegetable cellulose by its properties such as high purity, strength, molding suitability, and increased water retention ability (Güzel and Akpınar 2018; Wikipedia 2014). Bacterial cellulose is about 100 times thinner and porous than vegetable cellulose (Fig. 1). During the synthesis of this cellulose, glucose chains are secreted from the bacterial cell wall and nanofibril cellulose strips are formed. The diameters of these strips were reported to be in the range of 10–100 nm (Güzel and Akpınar 2018; Tozluoğlu et al. 2015; Costa et al. 2017). Bacterial cellulose production is also 40 times faster than vegetable-plant cellulose production (Dursun et al. 2006).

The comparison of the mechanical properties of bacterial cellulose (BC) with other fibers is given in Table 1 (Costa et al. 2017). Intermolecular and molecular hydrogen bonds hold the cellulose chains together, thus providing bacterial cellulose fibers with properties such as low solubility and high water retention, as well as high purity, mechanical resistance, elasticity, flexibility, and biocompatibility. BC membranes are elastic and flexible and can be sterilized. In the scientific world, bacterial cellulose is now being investigated and explored to find new potential usage possibilities in various fields (Costa et al. 2017).

Although the first discovery of bacterial cellulose with the utilization of *Acetobacter xylinum* bacteria by Adrean J. Brown was in 1886, the intensive works on bacterial cellulose have begun only after the twentieth century (Wikipedia 2014; Dursun et al. 2006; Costa et al. 2017; Tiboni et al. 2014). Brown vinegar fermentation studies observed a pellet-gelatinous heap on the surface of the liquid. Further analysis showed that this layer is actually cellulose (Dursun et al. 2006). Researchers reported that cellulose can actually be manufactured from various organisms such as *Acetobacter pasteurianum*, *Acetobacter rancens*, *Sarcina ventriculi*, and *Bacterium xylinoides*, and in 1949, the microfibril structure of bacterial cellulose was characterized by Muhlethaler (Fig. 2) (Wikipedia 2014; Costa et al. 2017). Today, bacterial cellulose studies are still continuing.

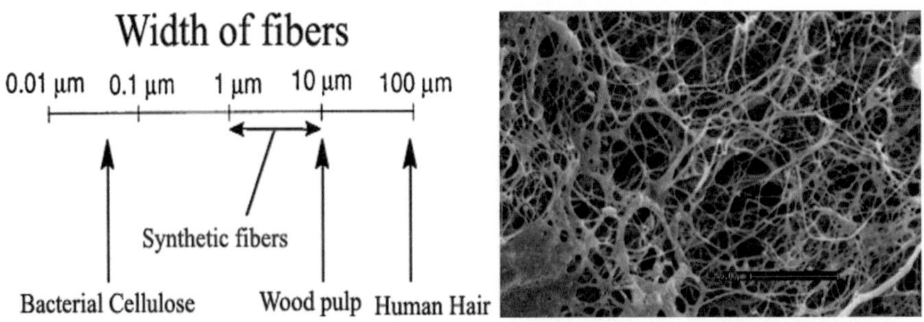

Fig. 1 Porous structure and width of bacterial cellulose fibers (Goh et al. 2012; https://commons.wikimedia.org/wiki/File:Sizes.png)

Table 1 Mechanical properties of bacterial cellulose

Material	Tensile strength (MPa)	Deformation (%)	Young modulus (GPa)
Polypropylene (PP)	30–40	100–600	1–1.5
Polyethylene terephthalate (PET)	50–70	50–300	3–4
Bacterial cellulose (BC)	200–300	1.5–2	15–20

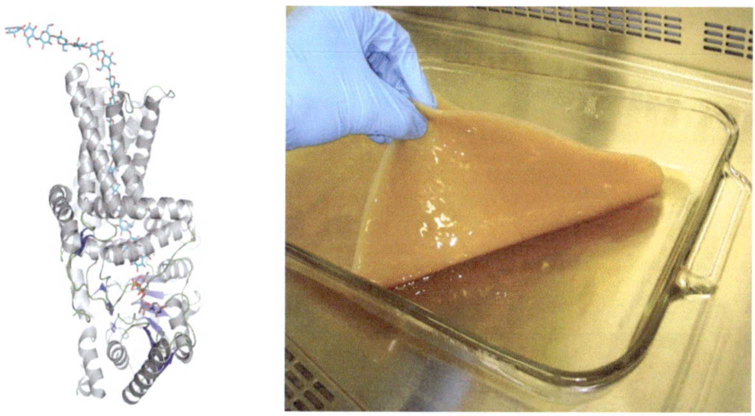

Fig. 2 Chemical structure of bacterial cellulose (https://commons.wikimedia.org/wiki/File:Structure_of_a_bacterial_cellulose_synthase.png)

It is very interesting that bacterial cellulose is used as a local food (Nata) in Southeast Asia. What is more, it is also used in the production of some functional drinks (kombu and Manchurian tea) (Güzel and Akpınar 2018). It is possible to produce bacterial cellulose in nature, but studies are still continuing in the laboratories in order to realize large-scale production. Various methods are being investigated to enable the production of bacterial cellulose in larger sizes than appropriate cultures (Wikipedia 2014). Today, *Acetobacter xylinum* cellulose is used as a model microorganism in biosynthesis studies (Dursun et al. 2006). Another type of bacteria used today is *Gluconacetobacter xylinus*. This bacterium was taken as a model of microorganism in biosynthesis and crystallization and in obtaining structural characteristics of bacterial cellulose. More than 100 pores present in the membrane of this microorganism allow the extrusion of cellulose to form a fibril element with a diameter of 3.5 nm. Approximately 46 contiguous fibrils join along hydrogen bonds to form a band having a width of 40–60 nm. The bands are wound to form the other entangled fibers dispersed in the culture medium. Its thickness depends on the culture time and can usually reach to 1 or 2 cm (Costa et al. 2017; Pecoraro et al. 2008; Wood 2007). Using a bacterium *G. xylinus and Gluconacetobacter hansenii*, 2% glucose (as a carbon source), pH 5–7, and a temperature range of 25–37 °C begin to form a viscous gel after 2 h (Costa et al. 2017).

Nowadays, bacterial cellulose could be used in a wide variety of commercial applications such as textile, medical, cosmetic, and food applications and products with the advancements in the synthesis of the material production and the detailed determination of the bacterial cellulose properties (Güzel and Akpınar 2018; Wikipedia 2014; Dursun et al. 2006). For instance, London fashion designer Suzan Lee (senior lecturer at Central Saint Martins Textile Fashion School in London) has been able to grow a kind of vegetable skin using green tea, sugar, bacteria, and yeast (Biocuture 2014; Growing Your Own Fabric is Possible 2014; Wood 2007; Lee 2014; 10 Eco-Fashion Garments Inspired by Nature and Biomimicry 2012; Alter Nature: The Future That Never Was 2011; Arcadia Boutique 2011). In kombucha tea, a kind of fermented tea, it is possible to produce BioCouture extract and dry cellulose with the help of bacteria, tea, and sugar which are used for fermentation (Fig. 3) (Biocuture 2014). Kombucha tea is a fermented beverage/drink prepared from tea, sugar, bacteria, and yeast (Wood 2007; Bauer 2014).

Yeast is added to the tub containing sweetened tea and bacteria to produce bacterial cellulosic fibers, which then turn into a skin-like state (thin bacterial cellulose leaves) (Biocuture 2014; Costa et al. 2017; Lee 2014). This material is a by-product

Fig. 3 Kombucha tea, microorganisms, and mother of kombucha tea (https://upload.wikimedia.org/wikipedia/commons/b/b8/Kombucha_fermenting_black_tea.jpg; https://upload.wikimedia.org/wikipedia/commons/4/4a/Kombucha_mushroom.jpg; https://upload.wikimedia.org/wikipedia/commons/f/f6/Kombuchacultsm.jpg)

of the fermentation process and can remain in the tub for about 2 weeks (under the controlled temperature conditions) to reach a sufficient thickness to produce garments. This layer is then taken and dried to form a final layer or dried to mold in order to produce jackets, clothing, or other types of end-use-product garments (Biocuture 2014; Costa et al. 2017). As the layers dry, the overlapping edges fuse and form seams. When all fibers dry, the garment can be colored with natural plant dyes or bleached if used as white (Biocuture 2014; Costa et al. 2017; Lee 2014; 10 Eco-Fashion Garments Inspired by Nature and Biomimicry 2012). Using this material, Lee has been able to produce many different types of garments such as jackets, dresses, shoes, kimono, shirts, etc. Shoes also were made as part of the BioCouture project (Fig. 4). In this study, nine green tea bags in plastic containers, 540 g sugar, 632 mL vinegar, and 100 g commercial combo yeast bacterial cellulose were created and used for Scarlett and Rhett designs (Nam and Lee 2016).

Having studied at Central Saint Martins Textile Fashion School, Erdem Kızıltoprak has combined fashion, textile, and science and technology in his collection. In his study, he aimed to design modern body jewelry for men. In Kiziltoprak's work, kombucha tea, kombucha yeast (mixture of yeast and bacteria), sugar content, and black/green tea mixture were used to grow its own fabric (Coleman 2011).

Another designer, Sammy Jobbins Wells, stretched the bacteria-made material to make wearable objects and placed them on previously prepared frames. Wells says her first inspiration was Suzanne Lee's BioCouture project. He used kombucha-cultured bacterial cellulose in his project. Wells stated that cellulose was very flexible and durable when wet and that it was very difficult to remove bacterial cellulose having a thickness of more than 2 mm. It was also stated that when the material dries, it retains much of its strength and exhibits a structure and attitude similar to old human skin. Wells kept a mixture of glucose, tea, and water at room temperature to produce the material. Tea is added as a nutrient for bacteria and gives color to the final material. Wells stated that the material has some defects. For instance, unfortunately, the material is not waterproof in its present form, and if it comes into

Fig.4 Shoes made by bacterial cellulose (Nam and Lee 2016)

contact with constant moisture, it will eventually return to its original wet state (Dezeen 2014; http://www.2014interiorideas.com/beauty-and-fashion/wearable-structure-of-bacteria-references-antique-animal-bone-corsets.html).

Rachel Trattles conducted a bio-jewelry study using bacterial cellulose. In Rachel's study, Rachel used green tea, yeast, bacteria, and vinegar to obtain bacterial cellulose (Trattles 2014). Some graduate students at the University of Iowa also obtained bacterial cellulose from the fermented green tea bacteria and yeasts, along with sustainable biopolymers from corn and soybean oil, which are agricultural plant products. This project team is one of the 7 teams to win awards among 40 teams at the National Sustainable Design Expo in Washington (National Sustainable Design Expo 2014).

According to Bunch, bacterial cellulose membranes are durable and can be dyed naturally. Therefore, clothes can be produced (Costa et al. 2017). In 2012, microbial cellulose produced from kombucha tea was characterized in a study. In order to characterize the produced microbial cellulose, SEM, FTIR, and X-ray diffractometry analyses were carried out, and also swelling properties were examined. The SEM results showed that the microbial cellulose layer exhibits an ultrafine mesh structure of compact cellulose. FTIR results also displayed that microbial cellulose does not contain contaminants such as lignin or hemicellulose (Goh et al. 2012).

In a study conducted in 2018, the current state of bacterial cellulose in the development of textile was examined, and innovative cultivation techniques were tried to adapt to different usage forms and forms of bacterial cellulose in various garments. Novel bacterial cellulose cultivation techniques shaped as cut in tailor not only improve the environmental values of this material but also improve the production method as a sustainable application. This organic material can be grown in any desired garment panel, providing less textile waste without the need to cut. Two different sowing techniques have been developed, called contact surface blocking sowing and panel-shaped sowing. The naturally degradable bacterial cellulose with its self-synthesized property can shed light on the development of sustainable new textile materials and production applications in the future (Chan et al. 2018).

In a recent study conducted in 2019, strength and thinness values of bacterial cellulose obtained from kombu tea were examined. Thus, the usability of bacterial cellulose in the fashion industry was also investigated. Owing to the presence of fermentation residues, the bacterial cellulose is sensitive to drying temperature. The best deformation characteristics are maintained when the bacterial cellulose material is dried at low temperature (about 25 °C). Bacterial cellulose material hardens at high drying temperatures due to rapid water vapor effect and tears when lower deformations are applied. Over time, it has been confirmed that the properties of bacterial cellulose have changed significantly and that the final products may have problems with the durability of this material (Domskiene et al. 2019). In 2019, a wallet was developed using leather-like textiles, namely, bacterial cellulose. As an outcome of this study, it was that the products that are tried to be developed by using bacterial cellulose may be more suitable for disposable plastics rather than textile applications (https://plantchicago.org/2019/05/29/scoby-textiles-part-2/). Apart from clothing, German designer Jannis Hülsen covered a stool with bacterial cellulose (Jannis Hülsen 2012). In addition to these applications, bacterial cellulose was also used in the medical field in a study conducted in Germany. In this study, bacterially synthesized nano-cellulose was tested for use in wound therapy by combining with various active ingredients (Views 2010).

Although the presence of bacterial cellulose produced by bacteria and yeasts was discovered in the late 1800s, it was expected to come to life in various innovative applications. Bacterial cellulose applications in the textile field have not yet reached an industrial scale. However, the potential future for bacterial cellulose is far beyond existing practices, especially if explorations can develop large-scale production processes using low-cost raw materials or agro-industrial waste. In addition to the economic potential of this biotechnology polymer, investment in research and industrial interest in the production of this high value-added polymer will not only reduce the disposal of by-products by reusing industrial wastes but also reduce global forest destruction (Costa et al. 2017).

In our world, where natural resources are about to be exhausted, the availability of renewable, environmentally friendly, and sustainable resources is becoming increasingly important. Therefore, with the interesting and improvable properties of bacterial cellulose, it makes it a promising potential biomaterial that plays an active role in the textile and medical applications of the future. From the textile point of view, apart from bacterial cellulose production for textile substrates, bacteria can also be used for textile coloration. Textile coloration with bacterial pigments is covered in the next section.

Coloration with Bacterial Pigments

Recently, since the problem of depletion of clean water resources is becoming more and more important, there is increasing interest and attention in the utilization of sustainable, renewable, and natural colorant resources instead of possible toxic and polluting synthetic dyes. However, significant water consumption in the extraction or cultivation of natural plant dyes and the need for large planting areas also necessitate the new searches for alternative colorant solutions. The use of microorganisms as a source of colorants comes to the fore and researches are carried out on microbial pigments. In the industry, it is now possible to produce some bacterial pigments for food, pharmaceutical, cosmetic, and textile applications (http://enneblog.com/; Venil et al. 2013; http://www.amm-mcrc.org/programmes/biotech/Pigments.html; Ökmen and Dilek 2014; Perumal et al. 2009). There is growing interest in the utilization of microorganisms as a color source because less labor and less soil is needed and bacterial colorants are more cost-effective than plant-based colorants (Perumal et al. 2009).

In the industry, it is now possible to produce some bacterial pigments for food, pharmaceutical, cosmetic, and textile applications. Indeed, bacteria can produce colored pigments (http://www.biotecharticles.com/Applications-Article/Colorful-Bacteria-612.html). Microorganisms produce a wide variety of stable pigments such as carotenoids, flavonoids, quinones, and verubramins and have higher yields than fermentation products of plants and animals (Gurcum and Ones 2018). Pigment-producing bacteria have many colors ranging from all colors of the rainbow to unusual colors such as black, white, brown, gold, silver, fluorescent green, yellow, or blue (http://www.biotecharticles.com/Applications-Article/Colorful-Bacteria-612.html). In the 2000s, yellow-orange, red, and purple pigments were obtained from bacteria such as *Chryseobacterium* sp., *Serratia marcescens*, and *Chromobacterium violaceum*, respectively (http://microbewiki.kenyon.edu/index.php/Industrial_Applications_of_Bacterial_Pigments). It has also been reported that textile materials are colored using these pigments. Red and purple pigments (known as *prodigiosin* and *violacein*, respectively) were applied as colorants to different fabrics such as acrylic, silk, cotton, polyester, and polyester microfiber. These pigments are told to have high staining ability on fabrics. This shows that bacterial pigments can be successfully used as textile colorants. New safe and effective natural dyes and pigments can substitute or take the place of potentially harmful synthetic dyes (http://microbewiki.kenyon.edu/index.php/Industrial_Applications_of_Bacterial_Pigments). For example, one of the most researched pigments of microbial origin is *prodigiosin*. Prodigiosins belong to the natural red pigment family. Prodigiosin was commercially produced for a short time before the synthetic chemical dyes were developed and utilized for dyeing silk and wool. Over the past years, interest in prodigiosins, as textile dyes, has been revived, and a small number of publications have been produced. It was suggested that biosynthetic prodigiosins can be utilized as a functional colorant for different textile materials (Kramar et al. 2014).

In addition to bacteria, fungal pigments are also alternative sources of synthetic pigments/dyes. Among the pigment-producing fungi identified, *Aspergillus rubber*, *Aspergillus glaucus*, and *Trichoderma viride* were reported to produce red, yellow, and yellow-green pigments, respectively. Fungi were obtained from barley in orange pigments from *Ganoderma lucidum*, *Coriolus versicolor*, and *Amanita muscaria* and applied to silk and cotton fabrics. Many fungi produce pigments from mycelium under natural growth conditions and when grown in a nutrient medium or body (Perumal et al. 2009).

As a result of the studies, the majority of bacteria are perceived (voting on a subject affecting the colony and reporting their presence to other bacteria), chemotactic signaling (movement of the organism in response to a chemical stimulus; detection of positive and harmful objects in the environment), and plasmid exchange (for transfer of antibiotic-resistant genes) (https://isea2011.sabanciuniv.edu/paper/communicating-bacteria; http://www.wellcome.ac.uk/News/2011/News/WTVM052287.htm). For instance, *Chromobacterium violaceum* is a common soil bacterium which produces striking purple colonies (https://isea2011.sabanciuniv.edu/paper/communicating-bacteria; http://exploringtheinvisible.com/page/4/). The color produced by these bacteria depends on the communication between the bacteria. If there is a small amount of bacteria, the color turns white, and if another bacterium receives notification, the color turns purple. If grown in the colony, individuals of these species constantly send and receive notifications to other bacteria, and as a result the colony turns purple (https://isea2011.sabanciuniv.edu/paper/communicating-bacteria; http://exploringtheinvisible.com/page/4/).

The advantages of using bacterial pigments as dyestuff (Gurcum and Ones 2018):

- Versatile and easy to produce compared to other sources.
- The ability to obtain different colors and shades by interfering with genes.
- Easy propagation and low cost for industrial production.

However, in addition to these mentioned advantages, it has been shown in the recent studies that industrial applications are needed for large-scale production of bacterial pigments produced by designers in small petri dishes (Gurcum and Ones 2018). The latest advances in molecular biology are utilized to make the production of bacterial pigment economically feasible. The production of these pigments can be increased by cloning genes responsible for the biosynthesis of a large number of pigments. Some scientists have succeeded in producing indigo on an industrial scale using the *Escherichia coli* bacteria in fermentation tanks. Furthermore, the process can be modified by biosynthetic means to form the molecular structure and hence the color of a bacterial pigment. *Streptomyces coel* producing blue pigment *actinorhodine* has been genetically modified to produce a *polyketite* which exhibits bright yellow color. Alternatively, studies have been conducted to produce orange or yellow-red colors. In order to use bacterial pigments in a wide range, they should avoid environmental factors, especially UV light. UV light initiates the reaction of free radicals leading to the deterioration of the pigments (Gurcum and Ones 2018). Studies on coloration textile materials by using pigments obtained from these bacteria and fungi have been going on since the beginning of the 2000s, and many artistic and scientific coloration processes were carried out.

For the last 30 years, Anna Dumitriu has been working with medical microbiologists (http://www.ecouterre.com/anna-dumitriu-embroiders-deadly-bacteria-antibiotics-into-textiles/anna-dumitriu-super-bug-quilt-11/). Dumitriu and Sue Craig investigated whether natural and sustainable bacterial pigments can be used in textile coloration. The starting points were *Serratia marcescens* red-pigmented bacteria and *Chromobacterium violaceum* purple-pigmented bacteria (Fig. 5) (http://exploringtheinvisible.com/2013/07/18/biodyes-2/).

Dumitriu also grew the MRSA bacteria (*Methicillin-resistant Staphylococcus aureus*) (Fig. 6) on fabric to produce colored fabrics (http://www.smithsonianmag.com/arts-culture/artist-dyes-clothes-quilts-tuberculosis-and-staph-bacteria-180949511/?no-ist).

After all these studies, Dumitriu produced a dress stained with *vancomycin* (an antibiotic species)-resistant *Staphylococcus* variant (VRSA), a bacterium more dangerous than MRSA (Fig. 7) (http://www.smithsonianmag.com/arts-culture/artist-dyes-clothes-quilts-tuberculosis-and-staph-bacteria-180949511/?no-ist). In another work, microbiologists Dr. Simon Park and Dr. John Paul, working with video artist Alex May, obtained and presented a dress as a result of their communicating bacteria project (https://isea2011.sabanciuniv.edu/paper/communicating-bacteria; http://www.wellcome.ac.uk/News/2011/News/WTVM052287.htm).

An autogenic textile design was carried out by inoculating bacteria growing on polyester/cotton fiber blended fabrics. In this study, the fabric was colored with completely natural bacterial pigments. The final design of the colored fabric is determined by the interaction of bacteria with each other. Red and purple bacteria

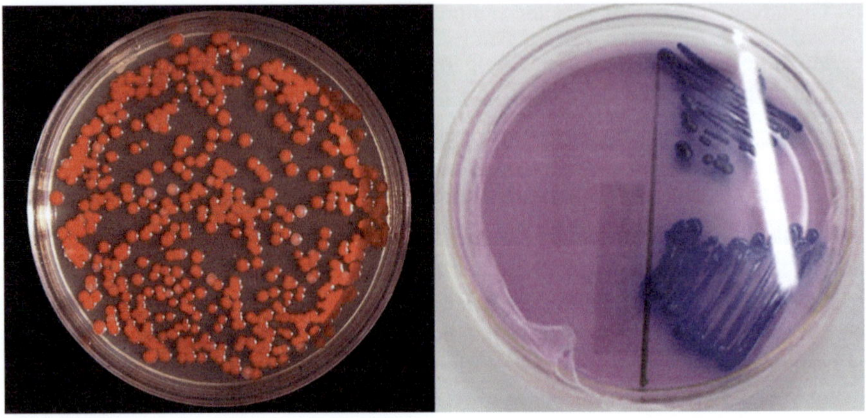

Fig. 5 *Serratia marcescens* and Chromobacterium violaceum bacteria in petri dish (https://commons.wikimedia.org/wiki/File:Serratia_marcescens.jpg)

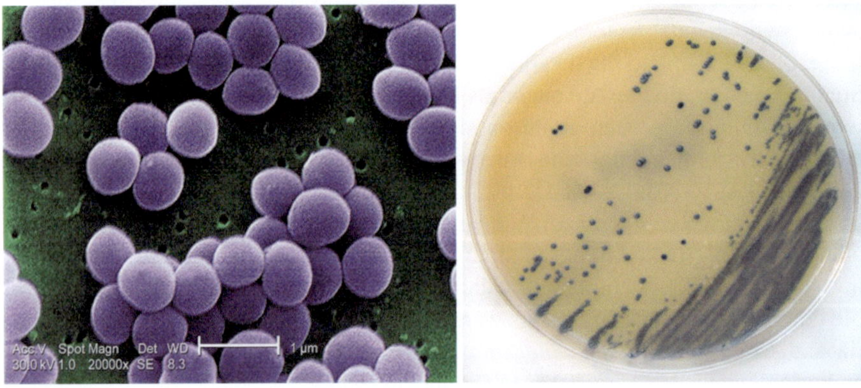

Fig. 6 *Staphylococcus aureus* and its color (https://commons.wikimedia.org/wiki/File:Staphylococcus_aureus_VISA_2.jpg; https://commons.wikimedia.org/wiki/File:MRSA_on_Brilliance_MRSA_Chromogenic_Agar.jpg)

are aggressive and want to cover a larger area on the fabric. Invaded regions also contain other bacterial species. Blue and orange-brown bacteria produce antibiotics and establish defense zones. Here, other bacteria die and invasion was prevented (http://exploringtheinvisible.com/page/4/). A similar method was used in the study of communicating bacteria.

Anna Dumitriu also exhibited a silk dress she designed using bacterial pigments, natural and synthetic antibiotics in an exhibition at the modern art museum (MOCA) in Taipei (http://annadumitriu.tumblr.com/post/68010439679/anna-dumitriu-has-created-a-new-work-the). Designer and researcher Natsai Audrey Chieza has also developed a new method to color fabrics with pigments produced from bacteria (Biocuture 2014).

Fig. 7 Dress dyed with bacteria (http://www.smithsonianmag.com/arts-culture/artist-dyes-clothes-quiltstuberculosis-and-staphbacteria-180949511/?noist)

Natsai Chieza also colorized a silk scarf with soil bacteria that produce beautiful colors such as dark blue, indigo, orange, and blood red (Fig. 8) (http://www.wired.co.uk/magazine/archive/2014/04/play/to-live-and-dye). According to Natsai Chieza, microbes have their own cellular color and can produce pigments of different colors. Chieza observed that these colors stain and color various types of silk fiber fabrics. As a result, Chieza wanted to modify the genetics of bacteria in order to produce the different colors demanded by fashion houses (http://www.wired.co.uk/magazine/archive/2014/04/play/to-live-and-dye). The basis of the research is that the *Streptomyces* bacteria commonly found in soil can be programmed to produce pigments when they grow under laboratory conditions. In this study, live bacterial isolates were inoculated into solids and directly into silk fibers. During a 7–14-day period in the laboratory environment, silk fabrics appear in terms of design according to environmental temperature, air circulation, and the effect of other factors affecting the speed and quality of pigmentation. Although the obstacles to the reproducibility of Chieza's results in a petri dish remain on a larger scale, bacteria-colored textiles may be a different alternative to sustainable fashion in the near future (Gurcum and Ones 2018).

Nidiya Vomiting, an Indonesian textile designer, began to work on the coloring of textile surfaces with bacteria and microfungus pigments, inspired by the *Aspergillus niger* microfungus, which caused black spots on moist clothing (Gurcum and Ones 2018). In the studies of Nidiya, she performed various experiments using the black *Aspergillus niger*, the orange *Monascus* sp., the white *Trichoderma* bacteria, and the red-pink bacteria *Serratia Marcescens* which can grow in the moist environment of the bathrooms. The fabrics were sterilized in an autoclave at high temperature and pressure to prevent infection while working with bacteria.

Fig. 8 Scarf dyed with bacteria (https://lovin.ie/entertainment/celebs/the-future-of-fashion-these-silk-scarves-are-dyed-with-bacteria)

Therefore, in the studies of Nidiya, coloration has utilized natural fabrics such as silk and cotton that are resistant to high temperatures (Gurcum and Ones 2018).

Amsterdam designers Laura Luchtman and Ilfa Siebenhaar examined the effect of sound frequencies on the growth of bacterial pigments and the controllability of this process with the "Living Color" project. With this project, it was tried to give antibacterial properties to the fabrics without using chemicals and consuming less water (Gurcum and Ones 2018). In a project led by Dutch designer Jelte van Abbema, alternative printing inks made from soy or natural plant pigments were used, and a series of bacteria were planted on fabrics to "enlarge" prints. As a result, it was a living bask pressure that developed over time, changed color, and died. This may not currently have a place in clothing, but is widely used as an art form (Wood 2007; https://lovin.ie/entertainment/celebs/the-future-of-fashion-these-silk-scarves-are-dyed-with-bacteria).

In 2013, designers Johanna Glomb and Rasa Weber developed a printing technique using microalgae, which they called an "Algaemy" to give an aesthetic approach to algae. In this technique, a wooden textile printing machine was developed, and rubber pattern molds were turned into printing rolls on the cylindrical surface. In order to obtain pigment, algae were planted in water and grown on sunlight and carbon dioxide, and then the resulting liquid solution was heated by filtration. Then a paste-like paint was created and printed with a printing roller (Gurcum and Ones 2018).

In addition to artistic studies, scientific studies have also been carried out on coloring textile materials using microbial pigments. A study conducted in 2014 focused on the identification of milk fungi from a compost pit containing cow manure and biodynamic herbal preparations. Pigments were obtained from these fungi and applied on cotton yarn. In this study, the colors obtained by pigments at different pH values were also examined. In Table 2, the colors of the extract at different pH values are given (Perumal et al. 2009).

After the pigments were extracted, pigment coloration was done on cotton fabrics using different chemical and natural mordants (Table 3) (Perumal et al. 2009).

Table 2 The effects of pH on pigment extracts (Perumal et al. 2009)

Samples	Color observed	pH value
1	Denatured color	2
2	Brownish orange	4
3	Pale orange	6
4	Pink	7.6
5	Brownish orange	8
6	Brownish orange	10
7	Pale orange	12
8	Denatured color	14

Table 3 The effects of mordants on dyeing cotton yarn (Perumal et al. 2009)

Samples	Color observed
Yarn (control)	White
Yarn + pigment (not boiled—control)	Light green
Yarn + pigment (boiled for 20 min—control)	Green
Pretreated yarn + pigment (boiled for 20 min—control)	Brown
Yarn + copper sulfate (treated with mordant)	Pink
Yarn + alum (treated with mordant)	Pink
Yarn + potassium dichromate (treated with mordant)	Green
Yarn + stannous chloride (treated with mordant)	Pink

Pigments obtainable from microbial sources have many important advantages. For example, they can be produced in the desired quantity, and production is not dependent on climate, geographic conditions, and the whims of nature. It is therefore advantageous to obtain pigments from microorganisms, and there are many species of bacteria and fungi that can provide high color yield (Perumal et al. 2009).

In another study conducted in 2014, *Streptomycetes* used as pigment sources were utilized, and pigments obtained from these microorganisms displayed different colors. Cultures developed in-house were stored in the refrigerator until the studies were performed (Ökmen and Dilek 2014). In addition, the effects of pH and temperature on *Streptomycetes* were also investigated. Different pH, medium, and temperature values provided different colors. The results of the study showed that *Streptomycetes* can produce pigments in various shades. Further studies have indicated that these pigments can be used in textile coloration (Ökmen and Dilek 2014).

Two new strains of *Streptomyces* (NP2 and NP4) were isolated to obtain deep blue and red pigments. Crude mycelium extracts of both strains were utilized as bio-colorants in conventional textile coloration. The thus obtained bio-stimulant yield was 62 and 84 mg/g per micelle for NP2 and NP4, respectively. With these pigments, polyamide and acrylic fibers were colored in the deepest shades, polyester and triacetate fibers are marked, but at a much lower color depth, while cotton and cellulosic fibers are colored in weak shades (Kramar et al. 2014).

Conclusion

Nowadays, one of the most important issues in the textile industry is sustainable textile design without harming the environment. Environmental concerns lead various researchers and artists to find and explore the applicability of more sustainable, renewable, and environment-friendly resources for fibers and colorants. In here, bacteria can aid us. Bacterial cellulose is an organic compound of the formula $(C_6H_{10}O_5)_n$ produced by certain types of bacteria. Cellulose is the basic structural material of many plant substances but can also be produced by bacteria such as *Acetobacter*, *Sarcina ventriculi*, and *Agrobacterium*. Although the presence of bacterial cellulose produced by bacteria and yeasts was discovered in the late 1800s, it was expected to come to life in various innovative applications. The material produced by the advancements in the detection and synthesis methods of bacterial cellulose, the resultant bacterial cellulose can be used in a wide range of commercial applications including textile, medical, cosmetic, and food applications and products. Bacterial cellulose applications in the textile field have not yet reached an industrial scale. However, the potential future for bacterial cellulose is far beyond existing practices, especially if explorations can develop large-scale production processes using low-cost raw materials or agro-industrial waste. In addition to the economic potential of this biotechnology polymer, investment in research and industrial interest in the production of this high value-added polymer will not only reduce the disposal of by-products by reusing industrial wastes but also reduce global forest destruction.

In our world, where natural resources are about to be exhausted, the availability of renewable, environmentally friendly, and sustainable resources is becoming increasingly important. Therefore, with the interesting and improvable properties of bacterial cellulose, it makes it a promising potential biomaterial that plays an active role in the textile and medical applications of the future. From the textile point of view, apart from bacterial cellulose production for textile substrates, bacteria can also be used for textile coloration.

The interest and search in natural dye alternatives increases in textile applications due to the fact that synthetic dyes can be toxic to humans and nature, and also too much water is used in the conventional dyeing processes. However, due to some disadvantages about the extraction difficulty of natural dyes from plants and the need for large cultivation areas for mass commercial productions, the utilization of microorganisms as a source of dyes and pigments comes on the agenda, and researches have been conducted on microbial pigments. As a result of these developments, bacterial pigments can be used in textile as alternative colorants. Indeed, bacteria can also produce colored pigments. It is now possible to produce some bacterial pigments for food, pharmaceutical, cosmetic, and textile applications. For instance, specific bacterial pigments can be used successfully as textile colorants. Pigments obtainable from microbial sources have many important advantages. For example, they can be produced in the desired quantity, and production is not dependent on climate, geographic conditions, and the whims of nature. It is therefore advantageous to obtain pigments from microorganisms, and there are many species of bacteria and fungi that can provide high color yield.

Indeed, bacteria are now working for us to create sustainable textile materials and textile colorants leading to more sustainable textile design. Although the usage of bacterial cellulose and pigments for textile materials may seem to come out of science fiction films, it has now been proven that this is happening with many different examples, and such samples are expected to increase day by day. In conclusion, it is an undeniable fact that this can open great horizons for people as a result of its development leading to more sustainable textile production, therefore more sustainable world.

References

10 Eco-Fashion Garments Inspired by Nature and Biomimicry. 2012 [cited 2014 30.09]. Available from http://www.ecouterre.com/10-eco-fashion-garments-inspired-by-nature-and-biomimicry/3/

Alter Nature: The Future That Never Was. 2011 [cited 2014 30.09]. Available from http://inspir3d.net/2011/02/06/alter-nature-the-future-that-never-was/

Arcadia Boutique; Inspiring Individuals of the Ecofashion World! 2011 [cited 2014 30.09]. Available from http://arcadiaboutiquephilly.blogspot.com.tr/2011/07/inspiring-individuals-of-ecofashion.html

Bauer BA. What is Kombucha tea? Does it have any health benefits. [cited 2014 24.10]. Available from http://www.mayoclinic.org/healthy-living/consumer-health/expert-answers/kombucha-tea/faq-20058126

Biocuture. [cited 2014 30.09]. Available from https://www.tumblr.com/search/biocouture

Biocouture by Suzan Lee. [cited 2014 30.09]. Available from http://www.ecouterre.com/u-k-designer-grows-an-entire-wardrobe-from-tea-fermenting-bacteria/bio-couture-2/

Chan CK, Shin J, Jiang SXK (2018) Development of tailor-shaped bacterial cellulose textile cultivation techniques for zero-waste design. Cloth Text Res J 36(1):33–44

Coleman M. Bacterial motility: a second skin that grows and moves. 2011 [cited 2014 30.09]. Available from http://fashioningtech.com/profiles/blogs/bacterial-motility

Costa AFS, Rocha MAV, Sarubbo LA (2017) Bacterial cellulose: an ecofriendly biotextile. Int J Text Fashion Technol 7:11–26

Dezeen: bacteria produce textiles for skin body adornments. 2014 [cited 2014 30.09]. Available from: http://www.dezeen.com/2014/08/12/skin-sammy-jobbins-wells-wearable-objects-bacteria-cellulose/

Domskiene, J., Sederaviciute, F. and Simonaityte, J. (2019), "Kombucha bacterial cellulose for sustainable fashion", International Journal of Clothing Science and Technology, Vol. 31 No. 5, pp. 644-652. https://doi.org/10.1108/IJCST-02-2019-0010

Dursun D, Yakar N, Çakır İ, Çakmakçı ML (2006) Mikrobiyel Selüloz Üretimi ve Gıda Sanayinde Kullanımı, Türkiye Gıda Kongresi, Bolu, 24–26 Mayıs 2006

Goh WN, Rosma A, Kaur B, Fazilah A, Karim AA, Bhat R (2012) Microstructure and physical properties of microbial cellulose produced during fermentation of black tea broth (Kombucha). II. Int Food Res J 19(1):153–158

Growing Your Own Fabric is Possible. 2014 [cited 2014 30.09]. Available from http://www.tds-blog.com/2014/02/13/growing-fabric-possible-see-suzanne-lee-discovered-ted-talk/

Gurcum BH, Ones A (2018) Bakteri ve mikroalglerin tekstil boyacılığında ve baskıcılığında kullanım olanakları. İdil Dergisi 7(46):701–709

Güzel M, Akpınar Ö (2018) Bakteriyel Selülozların Üretimi ve Özellikleri ile Gıda ve Gıda Dışı Uygulamalarda Kullanımı. Akademik Gıda 16(2):241–251

How to grow a Stool Cover Cellulose-Fibre Xylinum stool by Jannis Hülsen - Chair Blog. 2012. Available from http://www.chairblog.eu/2012/03/21/how-to-grow-a-stool-cover-cellulose-fibre-xylinum-stool-by-jannis-hulsen/

http://annadumitriu.tumblr.com/post/68010439679/anna-dumitriu-has-created-a-new-work-the

http://enneblog.com/

http://exploringtheinvisible.com/2013/07/18/biodyes-2/

http://exploringtheinvisible.com/page/4/

http://microbewiki.kenyon.edu/index.php/Industrial_Applications_of_Bacterial_Pigments

http://www.amm-mcrc.org/programmes/biotech/Pigments.html
http://www.biotecharticles.com/Applications-Article/Colorful-Bacteria-612.html
http://www.ecouterre.com/anna-dumitriu-embroiders-deadly-bacteria-antibiotics-into-textiles/anna-dumitriu-super-bug-quilt-11/
http://www.smithsonianmag.com/arts-culture/artist-dyes-clothes-quilts-tuberculosis-and-staph-bacteria-180949511/?no-ist
http://www.wellcome.ac.uk/News/2011/News/WTVM052287.htm
http://www.wired.co.uk/magazine/archive/2014/04/play/to-live-and-dye
https://commons.wikimedia.org/wiki/File:MRSA_on_Brilliance_MRSA_Chromogenic_Agar.jpg
https://commons.wikimedia.org/wiki/File:Serratia_marcescens.jpg
https://commons.wikimedia.org/wiki/File:Sizes.png
https://commons.wikimedia.org/wiki/File:Staphylococcus_aureus_VISA_2.jpg
https://commons.wikimedia.org/wiki/File:Structure_of_a_bacterial_cellulose_synthase.png; https://upload.wikimedia.org/wikipedia/commons/2/2f/Microbial_cellulose_pellicle.JPG
https://isea2011.sabanciuniv.edu/paper/communicating-bacteria
https://lovin.ie/entertainment/celebs/the-future-of-fashion-these-silk-scarves-are-dyed-with-bacteria
https://plantchicago.org/2019/05/29/scoby-textiles-part-2/
https://upload.wikimedia.org/wikipedia/commons/4/4a/Kombucha_mushroom.jpg ;
https://upload.wikimedia.org/wikipedia/commons/b/b8/Kombucha_fermenting_black_tea.jpg;
https://upload.wikimedia.org/wikipedia/commons/f/f6/Kombuchacultsm.jpg
Iowa State wins with leather-like vest at National Sustainable Design Expo _ College of Human Sciences _ Iowa State University. 2014 [cited 2014 30.09]. Available from http://www.hs.iastate.edu/news/2014/04/25/renewable-cellulose/
Kramar A, Ilic-Tomic T, Petkovic M, Radulović N, Kostic M, Jocic D, Nikodinovic-Runic J (2014) Crude bacterial extracts of two new Streptomyces sp. isolates as bio-colorants for textile dyeing. World J Microbiol Biotechnol 30(8):2231–2240
Nam, Changhyun and Lee, Young-A, "RETHINK II: Kombucha Shoes for Scarlett and Rhett" (2016). International Textile and Apparel Association (ITAA) Annual Conference Proceedings. 68. https://lib.dr.iastate.edu/itaa_proceedings/2016/design/68
Ökmen G, Dilek IŞIK (2014) Doğal Pigment Adayı Streptomisetlerin Gelişimlerine Çevresel Faktörlerin Etkisi. Biyoloji Bilimleri Araştırma Dergisi 7(1):37–41
Pecoraro, É., Manzani, D., Messaddeq, Y., & Ribeiro, S. J. L. (2008). Monomers, polymers and composites from renewable resources, Bacterial Cellulose from Glucanacetobacter xylinus: Preparation, Properties and Applications, pp 369–383. https://doi.org/10.1016/B978-0-08-045316-3.00017-X
Perumal K, Stalin V, Chandrasekarenthiran S, Sumathi E, Saravanakumar A (2009) Extraction and characterization of pigment from Sclerotinia sp. and its use in dyeing cotton. Text Res J 79(13):1178–1187
Tiboni, M., Grzybowski, A., & Fontana, J. D. (2014). Bacterial cellulose: a peculiar polymer for biotechnological applications. Cellulose and other naturally occurring polymers Research Signpost: Kerala, 1–10
Tozluoğlu A, Çöpür Y, Özyürek Ö, Çıtlak S (2015) Nanoselüloz üretim teknolojisi. Türkiye Ormancılık Dergisi 16(2):203–219
Trattles. [cited 2014 30.09]. Available from http://racheltrattles.tumblr.com/
Venil KC, Zakaria AZ, Ahmad WA (2013) Bacterial pigments and their applications. Process Biochem 48:1065–1079
Views C (2010) Bacteria help healing. [cited 2014 30.09]. Available from http://www.chemistryviews.org/details/news/886233/Bacteria_Help_Healing.html
Wearable structure of bacteria references antique animal bone corsets _ 2014 interior ideas. 2014 [cited 2014 30.09]. Available from: http://www.2014interiorideas.com/beauty-and-fashion/wearable-structure-of-bacteria-references-antique-animal-bone-corsets.html
Wikipedia. Bacterial cellulose. [cited 2014 30.09]. Available from http://en.wikipedia.org/wiki/Bacterial_cellulose
Wood JE (2007) Are microbes the future of fashion? The Microbiologist 18(2–4). https://e-space.mmu.ac.uk/618497/3/Microbiologist_June_Jane%20Wood%20%286%29.pdf

Sustainable Clothing Designs for Fashion: Design Strategies and Its Implementation Possibilities

R. Rathinamoorthy

Abstract Sustainability is the keyword among the researchers, irrespective of their research area. Textile and fashion sector is being well known among the researchers for its higher polluting nature. Many research studies were focused on the sustainable manufacturing process in different aspects. However, many researchers pointed that the steps that are taken for improving the sustainable aspects of clothing still have a huge gap in commercial implementation. With respect to clothing and fashion industries, several steps like organic and natural raw material usage, sustainable manufacturing steps, eco- or green processing, etc. were adopted to reduce the environmental impact of the clothing. But, the prevailing fast fashion concept and more frequent trend change in the market act as a barrier in implementing those concepts.

Sustainable designing is yet another concept which is meagerly addressed by the researchers due to its depending nature on customer preferences. This chapter details the importance of implementing the sustainable strategies in design stage along with the role of designers. Further various sustainable design strategies, namely, design for disassembly (DfD), design for zero waste, design for longevity, design for co-design, design for end of life (EoL), etc., were discussed with respect to the fashion applications. The chapter also details existing and most popular sustainable business model and their progress as a summary.

Keywords Sustainability · Sustainable fashion design · Design strategies · Challenges · Opportunities

---

R. Rathinamoorthy (✉)
Department of Fashion Technology, PSG College of Technology, Coimbatore, India

S. S. Muthu, M. A. Gardetti (eds.), *Sustainability in the Textile and Apparel Industries*, Sustainable Textiles: Production, Processing, Manufacturing & Chemistry, https://doi.org/10.1007/978-3-030-37929-2_7

Introduction

The Brundtland Report defines sustainability as the development that solves the requirements of the present without compromising the needs of the future generation's requirements (World Commission on Environment and Development 1987). But in reality, it is a different world with different requirements. Chouinard mentioned that "There has never been, nor is there now, sustainable business or a sustainable fashion on this planet" (Chouinard 2008). When it comes to the fashion industry, Chouinard's words are completely true. The complexity of the material and manufacturing process makes the textile manufacturing a challenging process in sustainability. Even though the raw materials are organic and sustainable, the materials are contaminated either in one or in another process during the production of the textile fabrics. A research report by the Ellen MacArthur Foundation mentioned that in the last 15 years, the clothing sales increased double the time from 50 billion units in 2000 to more than 100 billion units in 2015 (Ellen MacArthur Foundation 2017). The report also indicates that irrespective of the economic status of the country, the clothing utilization percentage is significantly reduced than the previous year. Many researchers indicated that the increased consumption of clothing and textile rapidly increased due to several reasons like affordability due to the economic level of the society, fast-changing trends, and availability of the cheaper products in the market. A recent research report mentioned that even after implementing so many sustainability practices in the clothing and textile industries in the UK, the total waste production in the specified sector increased to 26.2 million tons in 2016 compared to 24 million tons in 2012 (Waste and Resources Action Programme [WRAP] 2017).

The phrase fashion was defined by researchers as a practice or style adopted by a group of people within a limited period of time (Easey 1995). However, the dynamic nature of fashion always makes fashion unstable. The industry focuses on fast-changing trends and consumer requirements, and based on that, it always strives for novelty, producing new garments with different styles. But on the contrary, the word sustainability talks about longevity and maintenance up to a possible level. Basically, this definition is obtained from the natural ecosystem which assists themselves over periods of time (Thorpe 2007). Ultimately the goals of sustainability are open-ended and multi-faceted, considering the triple bottom line of environmental, social, and economic benefits (Elkington 1997) for current generations and the needs of future generations (World Commission on Environment and Development 1987). The recently adopted concepts of sustainability in the fashion industries are inspired by the report by an environmental architecture designer and a green chemist Braungart and McDonough (McDonough and Braungart 2002). They suggested the cradle-to-cradle concept over the cradle-to-grave concept. Cradle-to-grave concept urges linear way of consumption of energy, resource, and raw material, meaning that this concept needs resources to create a product, in use phase and also in the disposal. In this way, any product produced through the cradle-to-grave concept consumes resources or energy during its full cycle of life. Instead, Braungart and McDonough suggested a cradle-to-cradle concept, in which the researcher tried to mimic the regenerative cycle of nature. In nature, if some waste is created like the plant, tree, or animal dies, with time, that waste is converted into a nutrient for another process. This way the cradle-to-cradle system suggests that all products are nutrients of either biological (natural) or industrial (technical) cycles. Both can be reused in their own ways. In cradle-to-grave system, the main aim is to minimize the waste, but in the cradle-to-cradle system, it is to eliminate the waste (McDonough and Braungart 2002).

The important issue in the apparel and fashion industry is overconsumption of apparel products in recent years. Many research studies identified that the fast fashion and lack of consumer knowledge on the environment are the major reason for the increased purchase of apparels. In earlier times, the apparel manufacturers produced garments with standard styles in mass that were sold throughout the year. At that situation, the consumer's interest and knowledge on the fashion and style were very less, and the season lasts for a year. The manufacturing companies were predicting their styles in several seasons ahead (Brooks 1979). Later in the 1990s, the market became customer-driven, and the introduction of new supply chain methods like just in time increased the movement of the product by reducing the response time (Tyler et al. 2006). In the millennium, the consumer possesses more knowledge on their apparel and fashion product due to the increased accessibility through the Internet. As fashion brands want to hold their customers, they started to manufacture the product the customer wish, and that is how the fast fashion trend is introduced into the market.

Fast fashion products are the clothes that are inexpensive and produced in a shorter time with cheaper quality raw material. In general, the fashion clothing are copies of high fashion items, and it is very advantageous as it reaches the consumer very quickly compared to the old method. The major disadvantage of the fast fashion product is the very nature; the products are highly disposable as it is manufactured by focusing on the short life span or with cheaper raw material. The fast fashion concept is dominated by consumption, fast-changing trends, and low quality; leading consumers buy more clothes because they are affordable but discard these after only one season or sometimes one use (Fletcher 2008). The system forces the consumer to refresh their wardrobes very often. A study conducted by McKinsey & Company revealed that the environmental impact of the clothing manufacturing companies will increase 80% in 2025 if the fast fashion trend continues and the world market consumes to the same Western per capita level consumption (Remy et al. 2017). In results, the apparel retail giants like Zara and H&M are offering more collections per year to attract their customer, where they previously offered only two or three seasons per year. As a consequence of the fast fashion strategy, the manufacturing industries started to produce more, and so they contribute significantly to the increased level of environmental impact (Darshitamodi 2013). In this chapter, various non-sustainable and sustainable aspects of the fashion industry and the role of designers in developing sustainable products are detailed. The major emphasis is given for the role of design and its importance on the sustainability of apparel products. Further, the chapter details the various strategies the apparel industries can adapt during the designing process to develop a sustainable apparel product.

Fashion and Sustainability

Sustainability in the fashion industry is one of the widely discussed topics among researchers. The fashion industry had seen different terminologies in decades such as eco-, green, slow, and ethical fashion before the emergence of the term sustainable fashion. Before the term sustainable fashion, these words are used interchangeably (Carey and Cervellon 2014). At the beginning of the introduction of the concept, the term sustainable fashion had some negative connotation, however, that has been understood later based on their importance and recently accepted most widely. As per the definition of the International Institute for Sustainable Development (IISD), sustainable fashion can be defined as (IISD, International Institute for Sustainable Development 2015):

"Goods and services that respond to basic needs and bring a better quality of life, while minimizing the use of natural resources, toxic materials and emissions of waste and pollutants over the life-cycle, so as not to jeopardize the needs of future generations."

For fashion being sustainable, various researchers reported different methods to be adopted in the various stages of the product life cycle from the conceptual design to the disposal phase. Among them, Fletcher, the frontier scientist in the field of sustainable fashion products, mentioned four different aspects through her research that are essential to identify a product or fashion item being a sustainable one (Fletcher 2008). They are:

1. Child labor-free production – Any fashion product produced must be child labor-free across the entirety of the supply chain.
2. Harmless resources – The raw materials used within the production process should be less harming to the natural environment, recycled, upcycled, and/or more durable.
3. Ethical production locally – The production process must be in an ethical and socially responsible manner, paying workers fair wages and, ideally, creating jobs in the country of origin by producing locally.
4. Fair trade – Development of long-term relationship through sustainable practices across the supply chain. This will help in arriving fair prices and wages among the consumers and manufacturers.

The following sections will detail the widely adapted sustainable practices in the area of fashion and other factors influencing sustainability in the fashion industry.

From the raw material cultivation or development to the production of a complete garment involves numerous operations with a significant amount of carbon footprint to the environment. With respect to the raw material, in the fast fashion industry, cotton is the major fiber used along with synthetic; specifically polyester is the most commonly used material. As the general opinion, people believe that synthetic fiber has a significant impact on the environment as it is produced chemically, but at the same time, cotton does not have. But in reality, the cultivation of the cotton also significantly pollutes the environment. The research report says that cultivation of 1 kg of cotton requires approximately 8000 liters of water in comparison to the production of 1 kg polyester, which uses little or no water (McDonough and Braungart 2002). So, are synthetic fibers sustainable? The answer is no, as the water consumption is almost nil, but the energy consumption from the production is almost twice when compared to the cotton fiber. In the case of the production phase, even though the energy consumption is unavoidable, the materials used in each stage can be used as an eco-friendly or sustainable material. Localism is another topic which is widely discussed in sustainability initiatives. Logistics or transport is another main reason for the higher carbon footprint generation in the apparel industry. As the brands focus on fast fashion, they would like to buy products for a cheaper price; hence they source and produce material in different places of the globe where labor, material, and manufacturing costs are comparatively cheaper. Hence, the manufacturers use a lot of energy on transporting those goods from one part of the world to other parts; this ultimately increases the carbon footprint. Hence producing products in the local area reduces the transportation and associated carbon footprints. This is just one part of sustainability. The local manufacturing further has different advantages like economic benefits, societal benefits like employment for the local labors, and also the cultural and aesthetic diversity to the design (Fletcher 2008).

Consumer interest and purchase behavior is another important factor which encourages the fast fashion system and so the unsustainability. In general, the first motivation for the consumer's purchase is emotional needs (Strähle 2017). It is also noted through the research that the purchase behavior of the consumer mostly depends upon the brand and designer name; this is a strong indication for emotional motivation. Further, research results found that people use fashion as an identity and to adapt the social values and norms of a society (McNeill and Moore 2015). This is because identity is extremely important for fashion consumers, and sometimes the factor can outweigh other important factors, such as being ethical, sustainable, and functional (Hofstede et al. 1991). Byun and Sternquist identified in their research that consumers are mostly affected by short fashion cycles and limited supply (Byun and Sternquist 2008). They had mentioned that these factors strongly influence the buying pattern. It can be noted that the abovementioned two factors are the attributes of fast fashion; hence, the consumers are always motivated to buy new clothing and revamp their wardrobe often. At this kind of situation, the consumer did not bother about the negative effects of the product they are purchasing, and this also leads to overconsumption of the apparels.

Previously it is estimated that the knowledge about the environmental impact of the product is one of the important reasons for consumer's unsustainable choice in the purchase. The level of environmental knowledge had a significant influence on their purchase behavior (Thøgersen 2000). Similar results were noted by Henion, who reported that the consumer who provided with proper information and knowledge on the environmental impact of the product was most likely to purchase the safer products than the uninformed consumer (Henion 1972). But at the same time, studies also reported that the higher level of consumer's environmental knowledge and concern does not consistently translate into their activity or purchase (Borden and Schettino 1979; Rathinamoorthy 2018a). The consumer's emotional needs and feeling and their fashion consciousness with respect to the fashion items are the major reason for their higher consumption of apparels.

To address this fast fashion issue, other researcher details the importance of slowing down the fast fashion process called as slow fashion. The concept of slow fashion is not only slowing down the production phase of the fashion items; thereby it creates an account on the social, environmental, and economic impact of the process. The researcher mentioned that the slow fashion can be best described as having a conscience and slowing down the speed of production, the life cycle of the product, and the consumption of the resources (Friedman 2010; Johnston 2012). But at the same time, it does not imply a non-fashionable outfit. Slow fashion clothing can also be fashionable and trendy. However, the important point to be considered is the slow fashion garments are designed in a "classic" style, so that the style does not go out of trend and so the life of the product can be extended. Indirectly, the slow fashion concept urges the manufacturers and brands to bring a minimum number of collections/seasons per year, typically two, and a maximum of four collections like in the 1990s. The fast fashion products are designed for planned obsolescence and manufactured for a reduced number of wearing. The fast fashion industry focuses on the volume of the product over the quality. This is the main reason for the higher negative environmental impact of fast fashion. Joy et al. mentioned that consumer behavior is one of the main reasons for the generation of the fast fashion system. The researcher mentioned that the consumers who diligently select food products based on ethical, health, and environmental concerns may have no such attitude while selecting fashion products (Joy et al. 2012). The major reason for the lack of consumer awareness is attributed to the role of media in spreading awareness. Morgan and Birtwistle mentioned that the media downplay the social and environmental issues related to apparel production (Morgan and Birtwistle 2009; ENDS Report 2006). They also mentioned that the consumers also lack in their interest in knowing the impact of their new clothing on the environment.

Another important concept which is generally used along with sustainability is recycling and reusing the apparel for a different purpose after the end of life. Recycling is a process of converting a post used apparel into a useful product. If the produced product possesses a higher value than the original product, it is known as an upcycling process. If the recycled products' value reduced after recycling, it is known as downcycling (Rathinamoorthy 2018b). The recycling process by the abovementioned methods is more sustainable when compared to the complete recycling process, where the fibers were extracted from the fabric and reused. In the latter case, the energy consumption and the carbon footprint trails are very equal to the new manufacturing except for the resources used. The other potential option under the recycling process is clothing swap and use of secondhand clothing. The fundamental concept of clothing swap is a get together among the group of people, where they are arranged to exchange their clothing with others. The basic mantra of the clothing swap is "new to him/her." Based on the increased awareness and its environmental benefits, many charity organization and online stores are now organizing the clothing swap in different places (Rathinamoorthy et al. 2019). In these cases, the participant should register by paying a nominal fee to attend the meet, and they can swap as many items as they want. There are several fashion forums in the USA and UK doing the clothing swap meetups. The event takes place in different categories like men's swap, women's swap, all sizes swap, plus size swap, etc. (Kim 2013). As these meets are giving new life to the old products, the life of the clothing is further extended under a different owner. As far as the quality and aesthetic appearance are good, the clothing can be swapped in these meets. The clothing swap is one of the successful models in sustainable practice in the fashion industry.

In this recycling category, the use of secondhand clothing, known as vintage clothing, is recently trending in Western countries (Bramley 2019). As consumers become increasingly aware of the environmental impact of fast fashion, they are looking for a more sustainable way to shop. One of the recent options is buying famous secondhand vintage clothing, which is mainly driven by the influence of celebrity and media. A recent report stated that around 64% of the women were ready to buy pre-owned clothing in this year (2019) compared to 45% in 2016 (Bramley 2019). The main reason behind the increased awareness of vintage clothing is not only sustainable awareness but also the fit of the vintage clothing into the wider mood of the modern age. Other than vintage clothing, many well-established specialty online secondhand clothing stores like Thredup, Poshmark, and Grailed were also available for customers. A study reported that in the USA alone, there is a 25% increase in the count of the women who preferred the secondhand clothing in 2017 compared to the previous years. They also predicted a 15% of the market growth in the forthcoming year. A recent UK-based startup Depop, who sells the secondhand clothing through mobile application, has ten million users as of now. Depop has crossed an annual turnover of 400 million US dollars in the year 2017. Interestingly, their data revealed that 80% of their customers are aged between 13 and 24 years and buy an average of 20,000 items per day (Butler 2018).

However, the major drawback with the existing sustainable practices (like the abovementioned) is the products were designed and produced according to the fast fashion and frequently changing trend, meaning that the products are produced to have a very short life span, low in quality, and cheaper in cost by considering customer preferences. But at the same time, the industry cannot compromise the fashion and clothing requirements of the society, as they play an essential role in various parts of our life. The major drawback in following sustainability practices in different sections of the manufacturing is that even though the product is manufactured with eco-friendly and sustainable nature, the design of the product is not helping for that. Researchers named the fast fashion system as "planned obsolescence," which encourages us to give up or throw away perfectly good things in favor of others simply because consumers were led to believe that they are no longer useful or fashionable (Shedroff 2009). Hence, addressing the sustainability issues in the design stage is more reasonable rather than talking about the different phases of production and usage. Many researchers believe that sustainable design practice should be the typical approach to designing new sustainable products and services (Chapman and Gant 2007).

Design, Designer, and Sustainability

As things get complicated with the talk "sustainability" in the fashion industry, researchers identified that the better way to implement complete sustainable practice is only possible by implementing sustainable aspects in the design stage itself. The sustainable fashion garments like handmade products will give a sensorial experience to the consumers, and so it creates an emotional attachment with the customers. The consumer will feel that purchase as an investment and concerning to the products, the functional longevity and attractiveness beyond the fashion season makes them feel satisfied (Fletcher 2008; Clark 2008). The design is the only possible way out of an unsustainable situation, dominating and addictive patterns, and social behavior (Ehrenfeld 2008). The design is most of the time created only based on the needs and functionality requirements from the customer. The criticality of the design is also mentioned by Esty and Winston that the amount of environmental impact a product has can be estimated at this design stage itself (Esty and Winston 2009). Moreover, most of the designers believe that sustainability is an obstacle for the designing process and it restricts the innovation and newness of the design. This is mainly due to the unfamiliarity of the fashion designers with sustainability concepts. A report from the Centre for Sustainable Fashion mentioned that many of the designers are not aware of sustainability activity when it is offered to them and so they are more reluctant about it. But it is essential for the designers to understand the concepts of sustainability and act on their process (Centre for Sustainable Fashion 2008), because the designers are one of the major critical links in the fashion and apparel supply chain. They are the people who have space, time, and more specifically the correct opportunity to develop a design which produces very less impact on the environment and society. The design of any product is more important than any other process as it is responsible for 60–80% of life cycle impacts on the environment, and so, it is essential to reduce the environmental impact of any design at the concept design stage itself (Donato 2014). A report developed by the Danish Environmental Protection Agency after a series of research had suggested a seven-step method to reduce the environmental impact of any product in the product design stage itself as mentioned in Table 1. In this method, they had targeted at product developers and designers. Application of these steps will help to enhance environmental thinking in designers (McAloone and Bey 2009).

Many researchers reported different strategies for sustainable practices in the fashion design sector. Tischner and Charter reported that as of now all the modifications are performed in the operation level of the product to improve its sustainability. But the new solution must focus on the design to reduce the environmental impact and increase the sustainability of the product. They had suggested four approaches for the sustainable design development process, namely, repair, refine, redesign, and rethink. They specifically reported that the last two approaches redesign and rethink in the design stage itself will radically change the future lifestyle and also develop a sustainable product by satisfying all consumers' needs (Tischner and Charter 2001). Fuad-Luke had mentioned that the typical characteristics of the

Table 1 Seven-step method to reduce the environmental impact of any product in the product design stage

Steps	Context	Descriptions
1.	Use context	How is the product used? By whom? For how long? The other issue is to measure the environmental impact related to the product's functionality for the user
2.	Overview	View on how the product is manufactured? Distributed? And disposed of? And relevant environmental impacts
3.	Eco-profiles	Classify the environmental impacts to better measure and to identify the origins of the impact
4.	Stakeholder network	Illustration of stakeholder network which holds influence on the product. And identify the connections related to the environmental impacts of the product at different places
5.	Quantification	Evaluate and mention the products' environmental impacts Create possible alternative designs, methods, process, and life cycle. Consider likelihood scenarios
6.	Conceptualization	Try to remove or reduce the environmental impact of the product through a solution for the different life cycle approaches. By using the eco-design principle, design or conceptualize the product design
7	Eco-strategy	Make an action plan for reduced environmental product development

sustainable design should be a triple bottom line (TBL). Through TBL strategies, author balances sustainability with the profit, the economic issues; people, the social issues; and the environmental issues. The author came up with three different concepts in the designing process, namely, eco-design, sustainable design, and design for sustainability (Fuad-Luke 2009). According to Fuad-Luke, the eco-design represents the economically viable and ecologically stable design, which he called ecologically efficient design. In the case of sustainable design, he included social equity along with the economic and ecological aspect. In his "design for sustainability" concept, he further added institutional policy and strategy along with the previously mentioned TBL approach (Fuad-Luke 2009). Fuad-Luke developed a model called "sustainability prism," and he reported that as a foundation for design-led activism. Figure 1 represents the sustainability prism as reported by Fuad-Luke (Fuad-Luke 2009).

Similar research was performed by a different researcher, who represented the sustainable design aspects with the help of a milking stool. The researcher mentioned that the three legs of the stool stand for people, profit, and planet and the seat represents the sustainable platform. The researcher stated that even if anyone of the leg is defective, there is no sustainability in the design (Farrer 2011). In general, a design is considered as a human characteristic, and researchers believe that it has the power to change the people's conception. That is the reason why the design has such importance on sustainability. Chapman mentioned in his study that the design is a human phenomenon that ultimately shapes and modifies our culture. He suggested that improving the current design practices into an emotionally durable one will help to improve the sustainability of the new products (Chapman 2015). The researcher also states that design is generally based on various types of knowledge

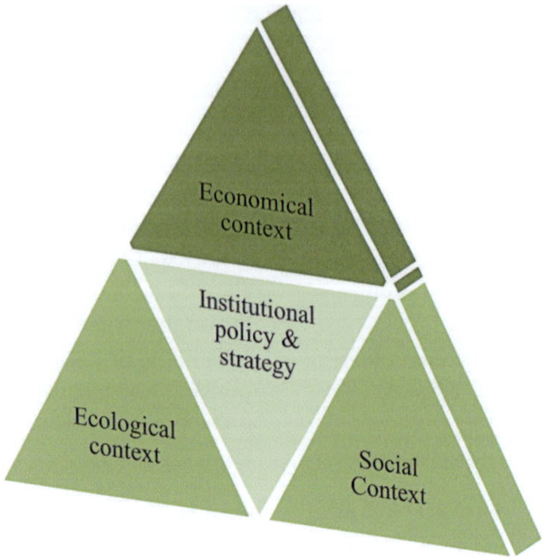

Fig. 1 Sustainability prism

like economic, social, medical, and cultural aspects of human life (Chapman 2015). The traditional designs were lacking in the sustainable aspects due to the nature of the quality assessment. As the visual and functional aspects of the designs only considered defining the quality of the design, the sustainability in it is found less. Researchers also mentioned that the practice of collaborating with other people from different domain has a potential effect on the sustainable aspects of design (McLennan 2004).

Few researchers mentioned that sustainable design practices should be closer to nature and the ecosystem. They have reported that efficiency, cooperation, and symbiosis are the key criteria which we need to adapt from nature (Fletcher 2008). However, Walker mentioned that sustainable development in the industrial world is a myth, and according to him, it is not possible to reach forever. He mentioned that as of today the sustainability was never achieved in any business or enterprise and it will never be (Walker 2007). Out of the literature, there are so many theoretical suggestions obtained, and they are summarized here. One of the good suggestions provided by Tischner and Charter is a suggestion to redesigning and re-imagining the parameters of a product to develop an eco-design (Tischner and Charter 2001). Further, research reports are suggesting industrial symbioses represent a closed-loop working system in the design stage itself to reduce the consumption of the resources and to reduce the carbon footprint (Ellen MacArthur Foundation 2017). Esty and Winston reported a concept called design for environment (DfE) thinking at the product development phase. This will reduce the pollution in the starting point itself, and they suggested to follow it till the end of the supply chain (Esty and Winston 2009). Similar factors were also reported by Thackara, who supports that the environmental impact can be reduced only from the design stage by using closed-loop or design for sustainability (DfS) strategies (Thackara 2006).

Sustainable Strategies for Fashion Design

In general, the sustainable practices employ ethical and social dimension to the manufacturing and designing process along with the user and disposal phase. The practices emphasize the use of renewable resources; even in nonrenewable resources used, it should be minimized to the possible extent. All the material should be recycled, and waste also must be avoided in all stages of the manufacturing. The important environmental principles for the sustainable manufacturing and designing as mentioned by the European Commission are (SEC 2009) use low-impact materials whenever it is possible, focus on resource efficiency, invest in high quality and durability, reuse, recycle, and renew. Sustainable design should have multiple life cycles by considering the use phase and end of life thinking so that the design will ensure the use of the product even after the end of life. A recent research by Sandy Black provided a set of guidelines for the fashion designers, and she mentioned to design the entire garments' life cycle including the user phase and disposal phases by using the following guidelines, namely, reuse waste materials; recycle, upcycle, repair, and remodel garments; recreate; reduce; use ecological materials; use monomaterials; use new technologies; create longer-lasting products; design multifunctional clothes; and design for delight (Black 2011).

The researcher suggested that fashion designers must understand and follow sustainable design strategies in their regular design practices. They have to introduce sustainable design strategies in all activities by thinking the parallel method to work out different designing concepts, so that the designer will be able to design, plan, and create a new garment design by integrating sustainable strategies (Lawson 2006). Gwilt followed the other researcher's parallel line concept and created step-by-step mapping for the design production process. Gwilt suggested the designers work with local people to collaborate and to understand their needs and adapt their cultures or designers engage in any of the low-waste or zero-waste method for the design development or to develop a durable high-end garments for longer life and designer must educate the consumer to consume the apparels slowly and accordingly the designer also should develop a classic style which can be used for multiple seasons (Gwilt 2009). The strategies mentioned by Gwilt were discussed in detail in the following sections.

Design for Waste Minimization/Zero Waste

Design for waste minimization, the concept urges the designer to waste less for their design or design possibilities with zero waste is the aim of the process. In the apparel manufacturing process, most of the waste is generated in the cutting process. The amount of the waste generated in the apparel manufacturing can be classified based on:

1. The garment style in relation to fabric width.
2. The number of garment sizes marked together in one marker.
3. The expertise of the marker-maker (whether manual or computer-aided).

However, literature reported that an average of 13.57% of the virgin fabrics is wasted in the cutting process itself before moving to manufacture. The results were obtained from various industries of Bangladesh (Rahman and Haque 2016). In another study, the waste generation was estimated in 25 different factories in Bangladesh and found that cutting section contributes a maximum of waste than any other manufacturing department in the industry (Tanvir and Mahmood 2014). A similar study was conducted among the 20 knitted garment manufacturing industries in India, Tirupur, and it is noted that approximately 23.57% of the fabrics are wasted in the cutting department alone (Rathinamoorthy et al. 2019). All these findings represented the poor design aspects of the product. The designers are more focused on mass production, and they were not considered the sustainable aspects in those designs.

Sustainable design considers the impact of the design on environment, economy, and society. Sustainable fashion is not a trend; it is the way of designing the styles as per positive future demands (Dickson et al. 2012). The first and foremost important stage in waste management is preventing the waste from being generated through economic design and fabric execution. It is also noted that zero-waste fashion designs already existed in the ancient culture. It can be found in the many countries' traditional styles like Japanese kimono and the Indian sari. Design elements often included gussets or gores, minimal arm shaping, rectangular sleeves or pants, and people engineered garment pieces to match the available fabric's length and width (Gay 2011). In line with the above statement, researcher Rissanen stated that reduction in the number of seams used in a garment potentially reduces the labor and raw material cutting and saves the resources. The lesser the number of seams, the lesser will be the wastage of the material (Khar and Ayachit 2013; Rissanen 2013). In the fashion design process, the opportunities to reduce the wastes come from the style feature, through the pattern making process. Hence, the control in the pattern making process is a vital factor for the zero-waste design generation. There are different methods reported by different researchers; however, the minimum wastage or zero-waste pattern mostly begins with the pattern making process along with the style and fabric width. It is necessary to integrate the pattern making and designing process to achieve a minimum wastage in the production process. Hence, a simultaneous pattern making and designing process helps us to achieve zero wastage in the designing process. Previous researchers reported different methods to develop zero-waste design as shown in Fig. 2 (Rissanen 2008; Niinimäki 2013; Carrico and Kim 2014; Black 2002; McQuillan and TimoRissanen 2013; Shakya 2016).

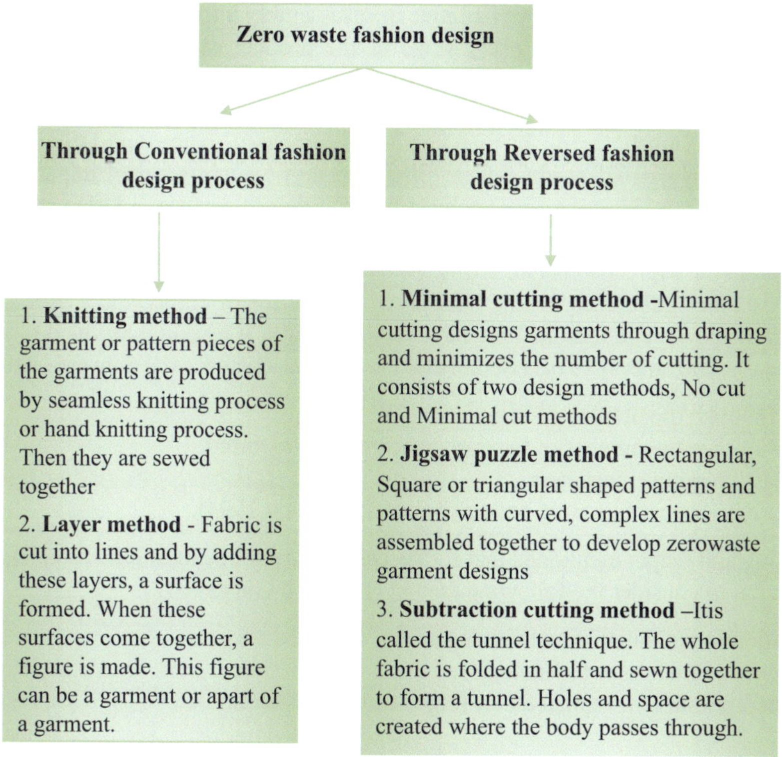

Fig. 2 Various methodologies to develop a zero-waste design garment

Design for Disassembly (DfD)

Design for disassembly (DfD) is a method, which promotes the designing process by considering the repair, refurbishing, and recycling kind of future needs at the designing phase itself. This will help the product to be utilized as much as possible and increases the lifetime of the raw material without sending it to landfills (Diener 2010). It is a type of green manufacturing strategy used to establish environment-friendly design practices for developing products. By using this method, the products can be developed in such a way that will allow the separation of components and materials from used product to encourage recovery and reuse (Galantucci et al. 2004). The concept of design for disassembly was first established during 1970; however, the concept gained momentum in recent years due to the increased awareness about the environmental impacts and resource depletions (Bogue 2007). The fundamental idea of the DfD concept is to increase the effectiveness of the product by asking serious questions about the product at different stages like before, during and afterlife. The questions like "Will a product need to be repaired?", "Which parts will need replacement?", "Who will repair it?", "How can the experience be simple and intuitive?", "Can the product be reclaimed, refurbished, and resold?", and "If it must be discarded, how can we facilitate its disassembly into easily recyclable components?" will be asked at every stage of the product design and development (Diener 2010).

Products produced and designed through DfD method are more valuable and are able to give much saving in manufacturing compared to the non-DfD product including the significant amount of difference in the environmental impact (Kerr and Ryan 2001). However, researchers reported that to perform a DfD process in design, few parameters must be considered during the designing phase itself as mentioned in Fig. 3 (Soh et al. 2014).

Few other guidelines for the DfD process, in terms of ergonomic aspect, were listed by Dowie and Simon. They have reported that it is essential for the designers to incorporate an easy to disassemble factors into the designs, and they are, namely:

1. Minimize the number of fasteners.
2. Minimize the number of fastener removal tools required.
3. Easy to remove fasteners.
4. Easy to access fastening points (Dowie and Simon 1994).

Ling and Gupta mentioned that for successful development of DfD product, it is essential to consider the following factors (Ilgin and Gupta 2012):

1. Adoption of suitable methodologies.
2. Implementation of technologies.
3. Incorporation of human factor (ergonomic) consideration.

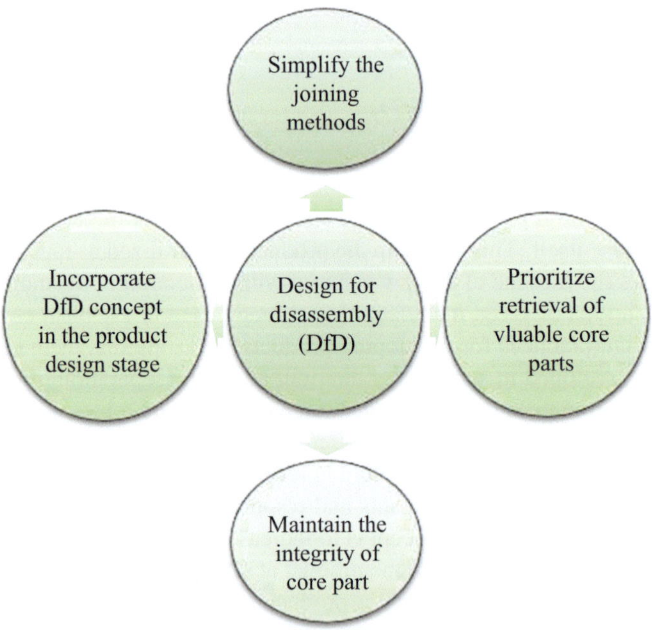

Fig. 3 Factors influencing the design for disassembly concept

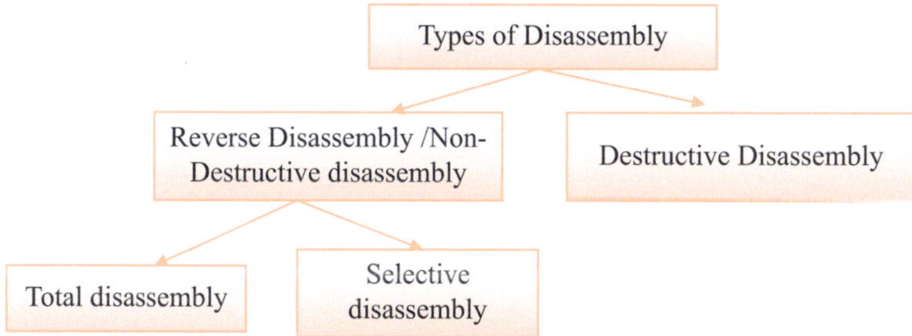

Fig. 4 Classifications of various disassembly methods for apparel and fashion industry

They reported that the third factor, which deals with the human and ergonomic aspects of the product design, is the important factor, as the process of disassembly is usually carried out by the humans. They also reported that the avoidance of any of the abovementioned factor may mislead and complicate the disassembly process and reduce work efficiency (Ilgin and Gupta 2012). A similar study was also performed by Bogue, and he reported that a successful design for disassembly requires thorough consideration of three critical stages: selection and use of materials; design of components and product architecture; and the selection and use of joints, connectors, and fasteners. Figure 4 illustrates the different disassembling processes types reported in the literature (Mule 2012).

The disassembly process is described as removing or taking each part of the product apart without damaging the parts by means of different operation. This process can be performed in two ways, namely total disassembly, where all components were removed, and another type selective disassembly. In the latter process, the complex parts will be converted into sub-components, and then the individual parts are removed from the product. Out of this, the total disassembly is mostly not possible due to economic reasons. The second type is disassembly; in this case, it is basically pulling apart the components or cutting them from the product. This can be performed at any stages of the product life cycle. This kind of disassembly is applied mostly in user phase after the product is being rejected by the customer for the repair and maintenance purpose or to recycle the materials.

McDonough and Braungart also mentioned the concept of "design for disassembly" as one of the sustainable strategies in their report *Cradle to Cradle* (McDonough and Braungart 2002). They have reported that the strategy can be successfully used for material recovery, reuse, recycling, or composting process. They have reported that the products that are designed for easy disassembly can be easily broken down into separate biological and technical nutrients. With respect to the fashion product, DfD is one of the key aspects of the sustainable fashion product. HaeJin Gam et al. analyzed the disassembling possibilities of an apparel (men's jackets) product through the conventional process. Their findings reported that conventional manufacturing and design made very difficult to disassemble the product and recover the material. Out of their research, they had suggested to designers to consider a few details before designing a jacket. They requested to use similar material for the sewing process and reduce the assembling steps. Further, they found the fusible interlining material is difficult to detach, so they advised going with the sewing process over fusing. They also advised larger stitches to sew biological nutrients (outer shell) and technical nutrients (lining) together. These points will reduce the disassembling time to reduce the material cross-contamination, and so the materials can be safely recovered and reused for a different purpose (Gam et al. 2011). This process will ultimately reduce material consumption and environmental impact.

Design for Slower Consumption/Longevity

Design for longevity or slower consumption is another important design strategy that should be adopted in the designing process for the introduction of sustainability in any product. Specifically with respect to the fashion, if their apparels have extended usage or life, they do not require any replacement. Lesser the quantity of products disposed of represents the reduced consumption of the resources (Ramirez 2014). The design stage is one of the best opportunities to increase the life of the garment. A report by WRAP (Waste and Resources Action Programme, a nonprofit organization supported by UK governments) mentioned that by addressing four fundamental areas as reported in Fig. 5, the design changes can bring an extended longer lifetime for the fashion clothings (Waste and Resources Action Programme [WRAP] 2016).

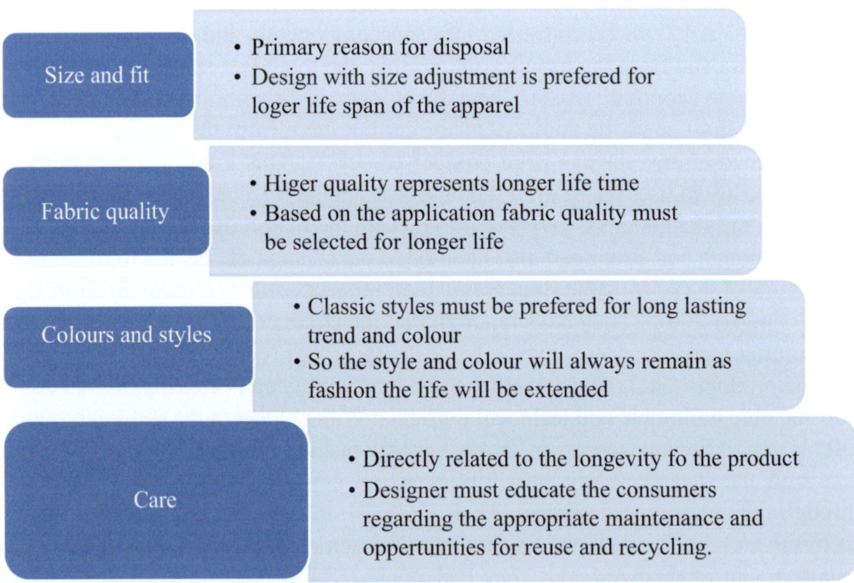

Fig. 5 Fundamental reasons for product obsolesce

Similar to longevity, the concept of slow design also represents the same meaning, which was introduced by researcher Fuad-Luke in 2002. The slow design theory focused on individual, sociocultural, and environmental well-being and emphasizes the value of time and experience (Fuad-Luke 2002). The main objective of the slow design is to encourage the local resources, create a richer experience, and involve the end user. Fuad-Luke represented some basic principles for slower design and mentioned design tool for the designers to design ideas, process, and outcome. The principles of the slow design process as represented by Fuad-Luke are as follows (Strauss and Fuad-Luke 2008):

- Reveal – Slow design provides space and experience in day-to-day life. Materials and process can be overlooked for creation.
- Expand – Slow design provides real-time expansion beyond their perceived functionality, physical attributes, and life spans.
- Reflect – Slow design provides environments that induce contemplation.
- Engage – Slow designs are collaborative, relying on sharing, cooperation, and transparency. This will help in future growth.
- Participate – Encourages users to be a part of the design process and increase their social responsibility.
- Evolve – Slow design will provide richer experience over time over the needs of the present.

The researcher reported that longevity is one of the most important alternative models in society with disposable culture. To accept the policies of the longevity, it requires a cultural shift, which will ultimately re-educate the consumer knowledge (Waste and Resources Action Programme [WRAP] 2016). Once the attribute is accepted and seen as an essential part of the product, then the product requirement will increase, and the price will reduce. The technical report by WRAP mentioned that there are approximately 15 decisions related to the life of the product that were taken in the design stage; hence it is very essential to introduce good design practice to increase the life of any product. The report also mentioned that the designers are instrumental in deciding these special characteristics of the garment based on their application requirements (Waste and Resources Action Programme [WRAP] 2016). The report also suggests that the semi-tailored garments had more lifetime than the ready-made one. Specifically, garments like oversized knits and kimono are the garments which possess higher lifetime due to its comfortability with different sizes. The report also suggests classic styles like a black dress, tailored shirts, pencil skirt, chino-style trousers, and V-neck jumpers, and basic colors like black, white, navy, gray, and red can be used for a longer time as they have acceptability over multiple seasons (Waste and Resources Action Programme [WRAP] 2016). The organization also released a longevity practice guidelines for different category of apparels for the public and manufacturers (Waste and Resources Action Programme [WRAP] 2016).

Several researchers reported different strategies for clothing longevity including the sustainable action plan launched by WARP in 2016 (Waste and Resources Action Programme [WRAP] 2016). Research reports of Alison Gwilt and Kate

Fletcher were also addressing similar sustainable strategies to reduce the environmental impacts of the apparel products (Gwilt 2014; Fletcher 2012). A report from Anja Claire Crabb mentioned various life extension strategies for apparel product that can be adapted for various products based on their application requirements. The strategies were provided as follows (Crabb 2014):

- Trans-seasonality – Represents a classic style and color which can be used for multiple seasons and which eliminated a phenomenon that can be described as "aesthetic obsolescence."
- Multifunctionality – Removable components and requirements of the user were carefully considered so that the user will have better control over repair, cleaning, and replacement.
- Modularity – Development of design in such a way that the garment can be reconfigured almost in all shapes and forms. This also supports the conversion or morphing of garments into other products in the hands of the user.
- Alterability – Provides possibilities to remodel the clothing, by introducing various manufacturing procedures and design guidelines like large seam allowances – which ultimately helps in extending the garment to newer size or to refashion it based on the requirements.
- Physical/emotional durability – The design should encourage strong physical manufacturing method. A report mentioned 90% of the product is discarded due to physical failure (Fletcher and Grose 2011). However, it is very important that the customer should keep the product for a longer time. Instead, there is no meaning in creating physically durable clothing. Hence, the researcher suggested creating an emotional attachment between the product and customer (Chapman 2009).

Recent research by McLaren et al. reported the consumer's mindset about the longevity in their study. They had performed a survey among different group of people who usually buy fast fashion items (youngsters), a group which prefers classic items (professionals) and parents having children. In another research, they had reported that the fast fashion consumers felt that the products are not long-lasting as they expected. They had mentioned that the fast fashion consumers expressed their feeling as trapped by the cheap, fast system of short-life garments, which "obliges" them to frequently buy new. In contrast, the slow fashion consumers reported that their clothes were up to their expectation (McLaren et al. 2015). A similar finding was also reported by Langley et al., who reported that the demand for long-lasting products was higher among the youngsters as they mostly buy fast fashion product (Langley et al. 2013).

Design for Social Well-being

Well-being is defined by the researcher Veenhoven (Veenhoven 2011) as:

"the degree to which an individual judges the overall quality of his/her own life-as-a-whole favourably."

In his definition, the term happiness represents a positive appreciation. Similarly, in earlier the well-being was defined *By* Eid and Diener (Eid and Diener 2004) as it:

"refers to one's multi-dimensional evaluation of their lives, including cognitive judgments of life satisfaction as well as affective evaluations of moods and emotions".

In general, the well-being is a state of human which designates that they are happy or flourishing; by doing so they are representing that their life is going well. The researcher reported that it is one of the highest values in which the other values can be subsumed. It is the one which makes one's life good and that possesses the highest value for many people (Brey 2015). In 1980, William James linked fashion with well-being. He reported that our psychological happiness is with our clothing system, and he mentioned that our clothing was an extension of ourselves (Watson 2004). Desmet and Pohlmeyer reported that the well-being state of human by using his positive design ingredients, namely, design for pleasure, design for personal significance, and design for virtue, which altogether stimulate a subjective well-being, and they denoted it as the sweet spot. That sweet spot is the point where all other three attributes intersect and that is where the people get flourish (Desmet and Pohlmeyer 2013).

Concerning to the design, there are questions such as whether the design for well-being is actually possible. Brey reported that the technical parameters can be reliability correlated with the consequences beyond its intended function and so it can be predicted. These consequences may be desirable, and sometimes it may not be. Hence, the researcher reported that it is possible to design a product by avoiding undesirable consequences and by encouraging the desired consequence, and Brey mentioned that in principle it is potentially possible to design a product for well-being (Brey 2015). To evaluate the influence of fashion on well-being, Rebecca Smith and Julia Yates conducted a study and identified five features, namely, positive emotion, engagement, relationships, meaning, and achievement, as attributes that create flourishing as authentic and sustained happiness and well-being (Smith and Yates 2018). They reported that fashion plays a significant role in flourishing among the participants, and their research results were in line with the previous findings (Seligman Martin and Csikszentmihalyi 2014). Sasha Walton mentioned in her article to fashion conversation that the findings of the researchers like Rebecca and Julia were very crucial for the designers and innovators. In the current fast fashion trend, both the consumers and designers were not able to build a bond on their designs and clothing due to the prevailing throwaway system. If we properly design and chose, then the clothing that we wear can help us to moderate our relationship with the outside world (Walton 2016). By adding values in the design stage, the role of designer with the clothing can be more personal than ever. Further, the designs will encourage the consumer's relationship with the clothing, and those can be emotionally connected with the consumers. A recent study was conducted among the fashion-conscious and style-conscious customers by a group of researchers. The results unveiled the negative relationship between materialism and subjective well-being. This is due to the fact that the materialistic people believe that the product comes with pleasure and associated signals. However, they also made them feel as they want things again and put them in a lesser well-being situation. Additionally, when we talk about the hedonic pleasure, both the consumers expressed satisfaction, delight, and form of entertainment and recreation, but the fashion-oriented consumers felt less well-being than the style-oriented. As the style-oriented consumers are interested in longevity, authenticity, and uniqueness to express, they do not need to acquire new product very often (Gwozdz et al. 2017).

Design for User Participation (Co-Design)

Design for user participation is another design concept, which popularly known as "co-design." In this concept, the design process must motivate to design the product along with the consumers participation. Researchers reported that co-design is a process of offering the opportunity to multi-stakeholders and actors to collectively define the concept and solve the problem. This process ultimately improves the design purpose effectively (Carroll 2006). However, the practice of co-design has presently existed with architectures and urban design planning than other design fields (Bell 2004). The co-design process is sustainable as the process involves the customer voices in it, and it is known as open innovation in opposite to the conventional practice of designing which is commonly known as closed innovation. Based on these definitions, by using the consumers' involvement in the development of the new product, companies started to capture their market by exposing their customer's experiences (Prahalad and Ramaswamy 2004). The degree of accessibility of the co-designer (may be a customer) to the product information is mentioned as one of the major issue with the aforementioned design concept, as the design concept of the new product is maintained under intellectual property rights. However, the risk of information theft can be avoided through the customer's loyalty and ratification with a specific brand; further, it generates an emotional bond between the companies and customers (Bowen et al. 1989). The co-design process connects the designer to the end user or manufacturer directly, and so the complex problems of the manufacturing and consumption were discussed in the design stage itself and designs prepared accordingly. This process allows the designers to get knowledge about every stage of production and usage. This further directs the designer to work on sustainability at every stage of the product. Due to the recent advancements, co-creation is an integral part of the fashion business.

Threadless is one of the perfect examples in the fashion field, the company invites consumer's design for the T-shirt and conducts voting and selects the top five designs. The selected designs will be used for the garment production and Threadless markets those fashion products. This is one of the best examples of the co-design process application in the fashion industry, and this is possible only due to the online platform and the Internet. Similarly, the world leading brand also Nike launched NikeID, a program for customers to design their own shoes and apparels (https://www.nike.com/nike-by-you). However, the first effort for mass customization in the apparel industry was launched in early 1994 by Levis, and it became a huge flop due to the immature technical support and unmet economic order quantity (Daamen 2015). The apparel retailer Burberry launched their label Burberry Bespoke in 2012, a program that allows people to design and purchase their own, personalized version of the company's iconic trench coat (Chowney 2011). The problem for the designer is to bring their end user into the designing phase, but the major restriction is the fear of mistakes. Once the participant is educated, their involvement can be motivated. Easily reachable entry level and reducing the fear of mistakes is assumed to be a key factor. Garments that are planned for co-designing allow easy customization, so they can be changed over time, and at its best, they reveal their life story by altering their beauty (Lee 2008). Another best example for the user participation is halfway items also known as semi-stitched apparels. A researcher reported that the halfway garments are the products which are intentionally unfinished to let the end user make the rest so that they can be customized for the individual. This is how these halfway garments can be used to create an emotional attachment with consumers due to their participation in the designing process (Lee 2008). Researcher represented the halfway designs as (Chapman 2015):

"the designer creates a piece in such a way that offers the opportunity to make changes within the original design."

Fletcher reported that the halfway products can satisfy the consumer needs, namely, individuality, self-expressing, creativity, and personal style customization. The personal participation of the customer in the design process provides space for the consumer's personal involvement in product development both physically and mentally. This creates a higher emotional attachment with the clothing and makes the product more valuable (Langley et al. 2013). This kind of personal style creation and individualistic creations adds personal values to the garment and increases the garment shelf-life beyond the seasons and trends.

Design for Product/Service System (PSS)

Being one of the highly polluting industries in the world, the fashion and textile industry is already looking for some alternative strategies and design methodologies to reduce its environmental footprints throughout its product life cycle. The higher water consumption, carbon footprint, and waste development are the major challenges in front of the fashion industry (Business for Social Responsibility 2009). The phenomenon of product/service systems (PSS) is emerging in recent time due to the increased market-driven consumption pattern. Due to the overconsumption, the service activities provided along with the product may increase the lifetime of the product. In the sustainable aspect, the life of the product increases, and the resource consumption reduces further. Hence, the application of PSS is one of the potential and viable options for the clothing and fashion industry.

Goedkoop et al. defined PSS as "PSS is a marketable set of products and services capable of jointly fulfilling a user's need" (Goedkoop et al. 1999). In commercial term, PSS is a mix of products and services designed to deliver utility rather than personal ownership of a product. The concept emphasizes interaction with consumers to meet needs and a life cycle approach to reduce environmental impact (UNEP 2002). This PSS is considered as one of the new selling concepts where the product as a material and service as an immaterial are sold together to fulfill the need of the customer. Hence, the PSS should initiate from the design stage of the product. Thus it is important to develop product and services within an integrated design process in order to achieve a sustainable design (Business for Social Responsibility 2009; Maussang et al. 2006). In general, PSS are always viewed as environmental-oriented system as they address the level of material consumption. Further, the system provided service to the consumer and keeps the product without lowering the value. Hence, the requirement of the new product is either delayed or not required. Agri et al. reported that the advantages of the PSS are (Agri et al. 1999):

- Decreased use of virgin materials in the production.
- Increased lifetime of each part of a product.
- Minimized number of times the materials pass through the production cycle.

PSS may offer any kind of services which add values to the product like repair/maintenance, return/exchange plans, or participatory/co-design options. Thus the PSS helps the product owners to utilize the product for different application or in a new form or for a lifelong. Sometimes it also helps to remove the ownership of the product by using the service options renting, swapping, or consultancy (Maxwell and van der Vorst 2003). However, the difficulty is noted in the acceptance of the system, and a new mindset among designers, manufacturers, and consumers is needed in order to find more sustainable ways to fulfill consumer needs and to attain sustainable improvements in the relationship between production and consumption (Chapman 2015). To develop a design associated with the PSS, designers should think beyond the scope of the company's existing product. Through this PSS, the company is going to capture increased values for their product and also implements the sustainable production in the manufacturing; hence, the designers should recognize context behind general customer needs; social issues such as health care and labor right can be involved. Hence, the designers should learn and develop their knowledge on the sustainable aspects in advance to the design process (Joore and Brezet 2015). Chong-Wen Chen suggested a four-step guideline principle to help the designers to create a PSS design in the planning stage as in Table 2 (Chen 2018).

With respect to the clothing industry, the PSS can be classified into two categories, namely, (a) product-oriented service and (b) use-oriented service (Armstrong et al. 2015). The authors reported that even though a complete PSS implemented in apparel and clothing industry is less, some product- and use-oriented PSS are already prevailing in the market to improve the products' lifetime and longevity. The details of those systems were shown in Fig. 6.

Table 2 Guideline for the designers to develop a PSS-oriented design

Step no.	Step	Tasks
1	Find the problems	• Select an issue that fails to meet the sustainability requirements
		• Describe all problems associated with this issue
		• List the problems under the TBL context
2	Recognize the context	• Explore the relationships of all listed problems
		• Link the problems to generate a network of causal relationships
3	Explore the possibility	• Classify highly related problems into a group, and propose inquiries for these problems in each group
		• Seek out whether there are solutions used to deal with these problems
4	Built the network	• Summarize the proposed questions into several keywords, and link their relationships
		• Define potential participants/partners and ideal goals for new solutions according to the organized keywords

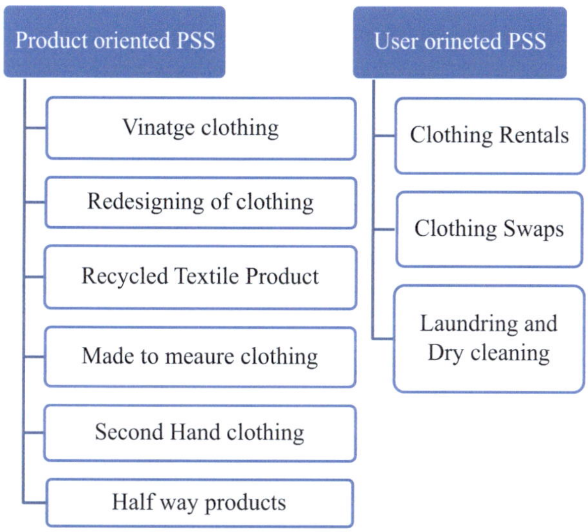

Fig. 6 Types of PSS in the fashion and apparel industry

Design for End-of-Life (EoL) Strategies

Designing products for the end-of-life (EoL) strategies are one of the main aspects of the sustainable design process. EoL strategy is often referred to as the steps for process in which one manages the product after its lifetime, meaning that after the user discarded the product. In EoL activities, certain processes develop waste, energy, and pollution along with cost if it is introduced in the manufacturing stages. Hence the major focus on the EoL should be given in the design stage of the product.

However, the objectives for the EoL strategies are to maximize the values of the product after the end of the lifetime. In general EoL strategies are of two types (Bauer et al. 2017). Two types and these methodologies allow the product to be in the market with the same quality, guarantee, and application:

1. Recycling strategy – It destroys the added value of the product and recovers only the material.
2. Reuse – This method recovers the material along with its value (to certain amount). The reuse strategies can be of three types as follows (Bauer et al. 2017);

(a) Direct reuse – It is a process in which the quality, physical and aesthetic aspects of the product are kept as same as original. So the product can be used in the same way as before by a different customer.
(b) Remanufacturing – Where the product is allowed to pass through a set of the manufacturing process to retain its original market value in terms of quality aesthetic and cost.
(c) Repurposing – Repurposing is a third reuse end-of-life strategy that complements the two previous ones. Similar to the other reuse strategies, repurposing allows for retention of added value in used products.

Direct Reuse Strategy for EoL

The important parameter of the reuse strategy is that the raw material should be as good as the new one in order to solve the original functional requirements of the product. The products that are reused are often considered as a secondhand product in the market. This implies that the product is new to the new customer. The very important aspect of the direct reuse strategy is the collection of the product. A set of predefined process parameters were listed by previous research workers to collect the used product. Other researchers reported that the collection from the waste stream, professional cleaning, sorting, and testing of the product are the important steps to be done to ensure the functionality of the secondhand product so that they can be reused for a similar application (Feng 2014; Ulrike and Hammerl 2015; Go et al. 2015). In the fashion industry, the concept of reuse recently emerges as a sustainable concept. There are many secondhand stores from normal clothing to vintage wears that were sold as an initiative to sustainability.

Remanufacturing Strategy for EoL

The European Commission defines the remanufacturing process as (European Commission 2015):

"a series of manufacturing steps acting on an end-of-life part or product in order to return it to like-new or better performance, with corresponding warranty."

In contrast to the reuse method, the remanufacturing method required manufacturing steps to bring the product to its original state or to a better state. It is also reported that the remanufacturing process is generally used to recover as much added value from the original manufacturer as possible (Zwolinski et al. 2006). So, the remanufacturing process is a bit complicated than the reuse process as it undergoes some energy consumption and carbon imprint generation at the manufacturing stage. A researcher reported a seven-step activity for the remanufacturing process, namely, inspection, storage, cleaning, disassembly, reassembly, repair, and testing. These steps in part or in full are found in any remanufacturing activity whatever its sector of activity (Sundin and Bert 2005).

Repurposing Strategy for EoL

Repurposing is a process which is very close to the remanufacturing process as it requires few similar steps like that. In general, the repurposing process helps in maintaining the added value products on the market as long as possible. Thus the repurposing of the product delays the disposal or recycling activity (European Commission 2008). The fundamental requirements of the repurposing are to fill the situation where the reuse and remanufacturing cannot be used. However, the main advantage of the repurposing process is its cost effectiveness than the remanufacturing and manufacturing process. This strategy also holds great potential for personalizing new products (Peters 2012). The importance of the processes will be significant if it stimulates designers to generate ideas to improve these steps and to apply different design strategies, such as design for the prevention of waste, for (active) disassembly, and for modularity to facilitate these processes. Detailed guidelines can be found in the previous literature for the end-of-life strategy implementation (Armstrong et al. 2015). Figure 7 represents the various end-of-life strategies from the literature and their routes to achieve the same (Peters 2012).

Challenges in Implementing Sustainable Design Strategies

The challenges in implementing sustainable design strategies in the apparel and fashion designing process are listed in Fig. 8 (Hur and Cassidy 2019).

Existing Sustainable Design Practices in the Fashion Industry: Case Studies

Table 3 details the various existing fashion firms who adapted sustainable fashion practices in their business. Table 3 represents a very few popular examples for sustainable product manufacturer. Out of these companies it is noted that most of the firm holds sustainable practices and follows product-service system, recycling –

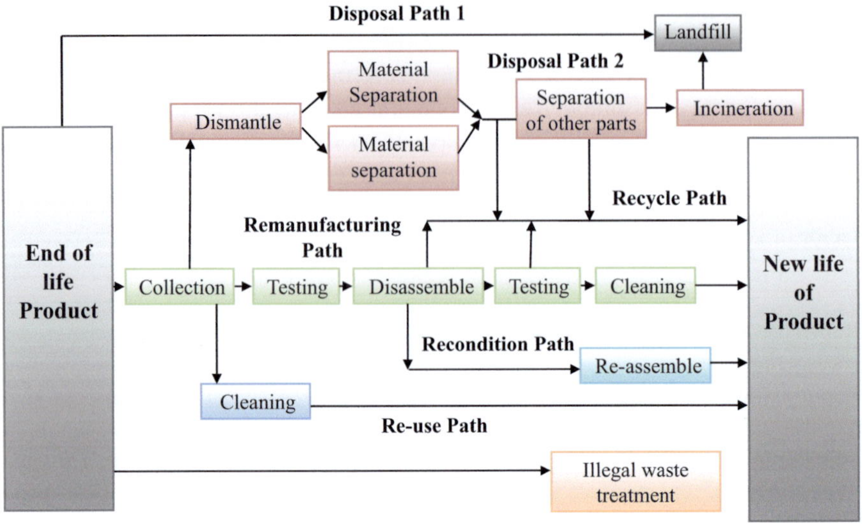

Fig. 7 Various sustainable design strategies for fashion product

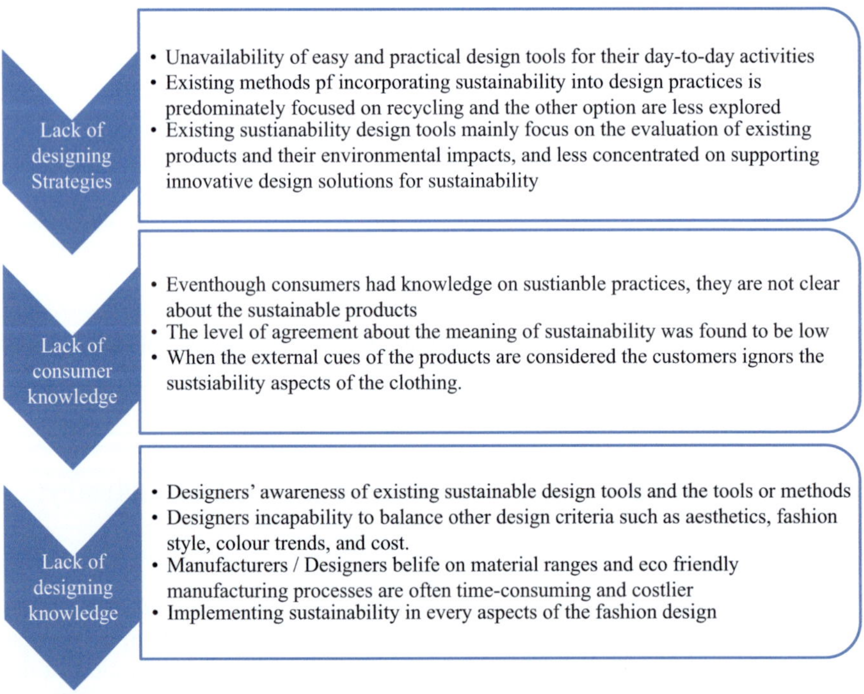

Fig. 8 Challenges in implementing sustainable design strategies

Table 3 Brands and their sustainable practices

S. no	Brand name	Sustainable practices	Strategy adapted	Ref.
1.	MUD jeans	Rental service – In order to retain control of the materials, MUD jeans leases jeans. Although customers can purchase the jeans outright, they can also opt to lease them from MUD jeans/month. After 1 year, the user has three options. They can exchange their jeans for a new pair and continue leasing the new one for another year or end the relationship by returning the jeans to MUD. Free repairs are included in the contract	Product-service system	Lease a jeans, Mud Jeans (2019)
2.	Rent the runway	Monthly rental of high-end designer garments in both the limited and unlimited schemes. Rent the runway has grown to become the leader in the global fashion rental space and is known as the "Netflix for fashion"	Product-service system	How it works. Rent the run way (2019)
3.	Nike	Nike allows customers to design their own shoes	Design for participation	Nike, Design by You (2019)
4.	Girl Meets Dress	Provides rental service for girls and women on a monthly basis with limited and unlimited garments	Product-service system	Girl Meets Dress (2019)
5.	ECONYL®	ECONYL® regenerated nylon in high-end designer handbags, stylish clothes, hosiery, and lingerie. And in sports, you can find it in high-performance sportswear, swimwear, and outdoor apparel	Recycle – End-of-life strategy	Econyl (2019)
6.	Redress R collective	The R collective rescues and upcycles surplus luxury materials that are destined to be wasted and transforms them into beautiful, enduring affordable luxury designer pieces through various product drops with distinguished retailers using socially respectful and brave business practices	Design for zero waste	The R Collective (2019)
7.	Depop	A mobile-based online store where the customer can sell or buy secondhand and vintage clothing	Recycle – End-of-life strategy Product-service system	The Depop Store (2019)
8.	Thredup	Largest secondhand luxury cloth selling online platform	Recycle – End-of-life strategy Product-service system	Thredup (2019)

(continued)

Table 3 (continued)

S. no	Brand name	Sustainable practices	Strategy adapted	Ref.
9.	The clothes library	It is a store that loans out clothes. They source secondhand garments in which the customer can either buy or rent	Recycle – End-of-life strategy Product-service system	The Cloths Library (2019)
10.	Nudie jeans	They use only 100% organic cotton for their production. The company offers free repair service to their customer, and also they collect secondhand material for sale and recycling	Design for longevity Product-service system	The Nudie Jeans (2019)
11.	Houdini	The company initiated the entire sustainability model in their business. They have started the repair facility in all their retail showrooms to extend the lifetime of the product. They had introduced the rental model from 2013	Design for longevity Product-service system	Houdini (2019)
12.	Filippa K	They claim that they have full traceability of raw material from cutting waste to garments. They also mentioned that 10 years of care is a kind of warranty system they offer. The company will help the customers to care for their products for 10 years. As sustainability imitative, the brand also allows their customer to rent their clothing for 4 days at 20% of the full price	Design for longevity Product-service system	Flippa (2019)
13.	Worn again	The company had developed a process in which the cotton and polyester blends can be successfully separated and recycles the cellulose into viscose and polyester. From the low-value clothing, the company produces a virgin-equivalent, cost-competitive *polyester* and c*ellulosic* raw materials to go back into the supply chain as part of a continual process	Design for longevity	Worn Again (2018)
14.	Patagonia	They use certified organic cotton for their product. A high proportion of their materials are made from recycled fabrics, including their polyester, nylon, and wool. They reject fast fashion by creating high-quality, long-lasting products and offer a repair and reuse program. They even go so far as to discourage customers from purchasing too many of their products	Design for longevity Product-service system	Paragonia (2019)

(continued)

Table 3 (continued)

S. no	Brand name	Sustainable practices	Strategy adapted	Ref.
15.	Reformation	It uses eco-friendly materials and reuses offcuts created during the manufacturing process, and it reduces its carbon footprint by manufacturing much of its range close to where it is sold and uses a reputable carbon offset program. They also provide a *RefScale* for each of their garments breaking down the item's impact on the environment	Design for longevity	The Reformation Store (2019)

end-of-life strategy. Concerning to the product longivity, very few companies adapted this stratergy in thier product. However, it is also noted in individual brands websites that the brands also offering longevity in terms of purity of the raw material and quality produced in the manufacturing and not by design. Only one firm, Redress, motivates their designers to follow the zero-waste design strategy to adapt in their designing, where they also do some upcycling kinds of stuff as a waste management strategy.

Conclusions

Even though various brands and fashion brands are following different sustainable activities in their manufacturing process, most of the time, they are all just end-of-life management strategies like recycling and reuse. Except very few, all other fashion brands have extended their service to repair and maintain the product quality and usability for a longer time. However, as mentioned in the literature, the fashion industry still lacks in the application of sustainable designs in their product. There are very minimal efforts made in the design stage to impart the sustainable aspects to the product. In the meantime, few academic researchers had taken this theme in hand, and they are working hard to implement it in the industry. Lack of knowledge or education of the designer about the sustainability is also one of the main barriers in implementing sustainable design practices in the fashion industry.

References

Agri J, Andersson E, Ashkin A, Söderström J (1999) Selling Services: A study of environmental and economic effects of selling functions. CPM report No.6

Armstrong CM, Niinimäki K, Kujala S, Karell E, Lang C (2015) Sustainable product-service systems for clothing: exploring consumer perceptions of consumption alternatives in Finland. J Clean Prod 97(15):30–39

Bauer T, Brissaud D, Zwolinski P (2017) Design for High Added-Value End-of-Life Strategies. In: Stark R, Seliger G, Bonvoisin J (eds) Sustainable manufacturing. Sustainable production, life cycle engineering and management. Springer, Cham

Bell, B. (2004) Good Deeds, Good Design: community service through architecture. Princeton Architectural Press, New York

Black S (2002) Knitwear in fashion. Thames & Hudson, London

Black S (2011) Eco-Chick: the fashion paradox. Black Dog, London

Bogue R (2007) Design for disassembly: a critical twenty-first century discipline. Assem Autom 27(4):285–289

Borden RJ, Schettino AP (1979) Determinants of environmentally responsible behavior. J Environ Educ 10:35–39

Bowen DE, Siehl C, Schneider B (1989) A framework for analyzing customer service orientations in manufacturing. Acad Manag Rev 14(1):75–95

Ellie Violet Bramley (2019) Is buying vintage clothing the most eco way to shop? https://www.theguardian.com/fashion/2019/apr/25/the-vintage-comeback-is-it-the-solution-to-sustainable-shopping. Accessed 16 July 2019

Brey P (2015) Design for the Value of human Well-being. In: Van den Hoven J, Vermaas P, van de Poel I (eds) Handbook of ethics, values, and technological design. Sources, theory, values and application domains. Springer, pp 365–382

Brooks J (1979) A friendly product. New Yorker. 58.

Business for Social Responsibility (2009) Apparel Industry Life Cycle Carbon Mapping. https://www.bsr.org/reports/BSR_Apparel_Supply_Chain_Carbon_Report.pdf. Accessed 24 July 2019

Sarah Butler (2018) Everyone I know buys vintage: the Depop sellers shaking up fashion. https://www.theguardian.com/business/2018/oct/20/everyone-i-know-buys-vintage-the-depop-sellers-shaking-up-fashion. Accessed 16 July 2019

Byun SE, Sternquist B (2008) The antecedents of in-store hoarding: measurement and application in the fast fashion retail environment. Int Rev Retail Distrib Consum Res 18(2):133–147

Carey L, Cervellon MC (2014) Ethical fashion dimensions: pictorial and auditory depictions through three cultural perspectives. J Fash Mark Manag 18(4):483–506

Carrico M, Kim V (2014) Expanding zero-waste design practices: a discussion paper. Int J Fashion Dgn Technol Educ 7(1):58–64

Carroll JM (2006) Dimensions of participation in Simon's design. Des Issues 22(2):3–18

Centre for Sustainable Fashion (2008) Fashion and sustainability: a snapshot analysis. London College of Fashion, London

Chapman J (2009) Design for (emotional) durability. Des Issues 25(4):29–35

Chapman J (2015) Emotionally durable design. Objects, experiences & empathy. Second edition. Routledge, New York

Chapman J, Gant N (2007) Designers, visionaries and other stories. Earthscan, London

Chen C-W (2018) Guidance on the conceptual Design of Sustainable Product–Service Systems. Sustainability 10:2452

Chouinard Y (2008) Foreword. In: Hethorn J & Ulasewicz C (eds) Sustainable fashion: why now? A conversation about issues, practices, and possibilities, Fairchild, New York, ix–x

Vikki Chowney (2011) Burberry attempts 'mass customisation' with Burberry Bespoke. https://econsultancy.com/burberry-attempts-mass-customisation-with-burberry-bespoke/. Accessed 24 July 2019

Clark H (2008) Slow + fashion - an oxymoron - or a promise for the future…? Fashion Theory 12(4):427–446

Anja Claire Crabb (2014). The future shape of design: fashion and longevity, Global fashion 2014. http://gfc-conference.eu/files_download/GFC2014/CRABB_The_future_shape_of_design_fashion_and_longevity.pdf. Accessed 20 July 2019

Alexander Daamen (2015) Co-Creation vs. Mass Customization in Fashion Industry Understanding New Shoppers' Differential Perceived Value and Preferences, Thesis - Master of Science in

Business Administration, UniversidadeCatólica Portuguesa. https://repositorio.ucp.pt/bitstream/10400.14/18753/1/20150910_Mastersthesis_AlexanderDaamen.pdf, Accessed 24 July 2019
Darshitamodi (2013) Upcycling fabric waste in design studio, thesis submitted to National Institute of Fashion Technology, Mumbai. http://14.139.111.26/jspui/bitstream/1/71/1/upcyclingfabricwasteindesignstudio.pdf. Accessed 16 July 2019
Desmet PMA, Pohlmeyer AE (2013) Positive design: an introduction to design for subjective Well-being. Int J Des 7(3):5–19
Dickson M, et al. (2012) The slow fashion movement, reversing environmental damage, not just a label. http://www.notjustalabel.com/editorial/. Accessed 20 July 2019
Alex Diener (2010) Afterlife: an essential guide to design for disassembly. https://www.core77.com/posts/15799/afterlife-an-essential-guide-to-design-for-disassembly-by-alex-diener-15799. Accessed 20 July 2019
Donato J (2014) Design products for sustainability. British Plastic Foundation. http://bpf.co.uk/Sustainable_Manufacturing/Design/Designing_Sustainability.aspx. Accessed 20 July 2019
Dowie T, Simon M (1994) Guidelines for designing for disassembly and recycling. Manchester Metropolitan University
Easey M (1995) Fashion marketing. Blackwell Science, Oxford
Econyl (2019). https://www.econyl.com/apparel/.
Ehrenfeld J (2008) Sustainability by design. Yale University Press, London
Eid M, Diener E (2004) Global judgments of subjective Well-being: situational variability and long-term stability. Soc Indic Res 65(3):245–277
Elkington J (1997) Cannibals with forks: the triple bottom line of 21st-century business. Capstone, Oxford
Ellen MacArthur Foundation (2017) A new textiles economy: redesigning fashion's future, http://www.ellenmacarthurfoundation.org/publications. Accessed 15 July 2019
ENDS Report (2006) Retailers to train suppliers on chemical risks, 273:23
Esty D, Winston A (2009) Green to gold. How smart companies use environmental strategy to innovate, create value, and build competitive advantage. Wiley, New Jersey
European Commission (2008) Directive 2008/98/EC of the European Parliament and of the Council of 19 November 2008 on Waste and Repealing Certain Directives. http://eur-lex.europa.eu/legal-content/EN/TXT/?uri=celex:32008L0098.
European Commission (2015) Communication from the commission to the European Parliament, the Council, the European Economic and Social Committee and the Committee of the Regions. Closing the Loop—An EU Action Plan for the Circular Economy. Bruxelles, Belgique. http://ec.europa.eu/environment/circular-economy/. Accessed 27 July 2019
Farrer J (2011) Remediation: Discussing Fashion Textiles Sustainability. In Gwilt A, Rissanen T (eds) Shaping sustainable fashion. Changing the way we make and use clothes. Earthscan, London. 19–33
Feng Z (2014) Intégration Des Considérations Environnementales En Entreprise: Une Approche Systémique Pour La Mise En Place de Feuilles de Routes. Université de Grenoble, Grenoble, France. http://www.theses.fr/s95811.
Fletcher K (2008) Sustainable fashion & textiles: design journeys. Earthscan, Oxford
Fletcher K (2012) Durability, fashion, sustainability: the processes and practices of use. Fash Pract 4(2):221–238
Fletcher K, Grose L (2011) Fashion & sustainability: design for change. Laurence King, London
Flippa K (2019). https://www.filippa-k.com/en/sustainability-at-filippa-k.
Friedman V (2010) Sustainable fashion: what does green mean?, Financial Times. http://www.ft.com/cms/s/2/2b27447e-11e4-11df-b6e300144feab49a.html#axzz1z1DRM8Y9. Accessed 16 July 2019
Fuad-Luke A (2002) 'Slow Design': A paradigm shift in design philosophy? https://fluido.files.wordpress.com/2006/07/slow-design.pdf. Accessed 13 May 2013
Fuad-Luke A (2009) Design activism. Beautiful strangeness for a sustainable world. Earthscan, London

Galantucci LM, Percoco G, Spina R (2004) Assembly and disassembly by using Fuzzy logic & genetic algorithms. Int J Adv Robot Syst 1(2):67–74

Gam HJ, Cao H, Bennett J, kamp CH, Farr C (2011) Application of design for disassembly in men's jacket. Int J Cloth Sci Technol 23(2/3):83–94

Gay K (2011) Zero waste line of sewing patterns. http://www.mysomapatterns.com/why-zero-waste-fashion.html. Accessed 20 July 2019

Girl Meets Dress (2019). https://hire.girlmeetsdress.com/pages/how-it-works. Accessed 27 July 2019

Go TF, Wahab DA, Hishamuddin H (2015) Multiple generation life-cycles for product sustainability: the way forward. J Clean Prod 95(May):16–29

Goedkoop MJ, van Halen CJG, te Riele HRM, Rommens PJM (1999) Product service systems, Ecological and Economic Basics (Ministry of Housing, Spatial Planning and the Environment Communications Directorate

Gwilt A (2009) Generating sustainable fashion: opportunities, innovation and the creative fashion designer, in E rouse. Fashion and Well-Being CLTAD/IFFTI, London, pp 439–452

Gwilt A (2014) What prevents people repairing clothes? An investigation into community-based approaches to sustainable product Service Systems for Clothing Repair. Making Futures Journal, 3

Gwozdz W, Nielsen KS, Gupta S, Gentry J (2017) The relationship between fashion and style orientation and Well-being. A Mistra Future Fashion Report 5

Henion KE (1972) The effect of ecologically relevant information on detergent sales. J Mark Res 9:10–14

Hofstede G, Hofstede GJ, Minkov M (1991) Cultures and organizations: software of the mind. McGraw-Hill, London

Houdini (2019). https://houdinisportswear.com/en-eu/sustainability/design-philosophy.

How it works. Rent the run way (2019). https://www.renttherunway.com/plans#comparison-table. Accessed 27 July 2019

Hur E, Cassidy T (2019) Perceptions and attitudes towards sustainable fashion design: challenges and opportunities for implementing sustainability in fashion. Int J Fash Dsgn Technol Educ 12:208. https://doi.org/10.1080/17543266.2019.1572789

https://www.nike.com/nike-by-you. Accessed 24 July 2019

IISD, International Institute for Sustainable Development (2015) Definitions & concepts. http://www.iisd.org/susprod/principles.htm.

Ilgin MA, Gupta SM (2012) Remanufacturing modeling and analysis. CRC Press, Hoboken

Johnston A (2012) The first steps towards considerate design incorporating cradle to cradle principles. London: College of Fashion. http://innovatingsustainablefashion.files.wordpress.com/2012/07/cradle-tocradle_copenhagen_final_small.pdf. Accessed 16 July 2019

Joore P, Brezet H (2015) A multilevel design model: the mutual relationship between product-service system development and societal change processes. J Clean Prod 97:92–105

Joy A, Sherry JF Jr, Venkatesh A, Wang J, Chan R (2012) Fast fashion, sustainability, and the ethical appeal of luxury brands. Fashion Theory 16(3):273–296

Kerr W, Ryan C (2001) Eco-efficiency gains from remanufacturing: a case study of photocopier remanufacturing at Fuji Xerox Australia. J Clean Prod 9(1):75–81

Khar SS, Ayachit SM (2013) Looking backwards to go forward – use of traditional Indian pattern making to develop contemporary methods for global fashion. Int J Fashion Dgn Technol Educ 6:181–189

Kim W (2013) we are more than a meetup - we are a movement. http://www.meetup.com/frugal-fashionista-clothing-swap-alexandria/. Accessed 17 July 2019

Langley E, Durkacz S, Tanase S (2013) Clothing longevity and measuring active use. London. www.wrap.org.uk/system/…/Clothing%20longevity%20SUMMARY%20REPORT.pdf. Accessed 20 July 2019

Lawson B (2006) How designers think: the design process demystified (4th ed.) Architectural Press, Oxford

Lease a jeans, Mud Jeans (2019). https://mudjeans.eu/lease-a-jeans/. Accessed 27 July 2019
Lee Y (2008) Design participation tactics: the challenges and new roles for designers in the co-design process. CoDesign 4(1):31–50
Maussang N, Zwolinski P, Brissaud D (2006) Which representation for sets of products and associated services during the design process? International conference on integrated design and manufacturing in mechanical engineering, Grenoble, France
Maxwell D, van der Vorst R (2003) Developing sustainable products and services. J Clean Prod 11(8):883–895
McAloone TC, Bey N (2009) Environmental improvement through product development: a guide. Danish Environmental Protection Agency, Copenhagen
McDonough W, Braungart M (2002) Cradle to cradle: remaking the way we make things. North Point Press, New York, NY
McLaren A, Oxborrow L, Cooper T, Hill H, Goworek H (2015) Clothing longevity perspectives: exploring consumer expectations, consumption and use, PLATE conference - Nottingham Trent University. https://www.plateconference.org/clothing-longevity-perspectives-exploring-consumer-expectations-consumption-use/.
McLennan J (2004) The philosophy of sustainable design. Ecotone Publishing, Washington, DC
McNeill L, Moore R (2015) Sustainable fashion consumption and the fast fashion conundrum: fashionable consumers and attitudes to sustainability in clothing choice. Int J Consum Stud 39(3):212–222
McQuillan H, TimoRissanen JR (2013) The cutting circle: how making challenges design. Res J Textiles and Apparel 17(1):39–49
Morgan LR, Birtwistle G (2009) An investigation of young fashion consumers' disposal habits. Int J Consum Stud 33:190–198
Mule JY (2012) Design for Disassembly Approaches on product development. Int J Sci Eng Res 3(6)
Niinimäki K (2013) Sustainable consumer satisfaction in the context of clothing. In: Vezzoli C, Kohtala C, Srinivasan A (eds) Product-service system design for sustainability. Greenleaf, Sheffield
Nike, Design by You (2019). https://www.nike.com/in/nike-by-you. Accessed 27 July 2019
Paragonia (2019). https://www.patagonia.com/resource-use.html
Peters, JR Kristeldewulf (2012) Design for end of life: a design methodology for the early stages of an innovation process, International conference on engineering and product design education 6 & 7 September 2012, Artesis University College, Antwerp, Belgium
Prahalad CK, Ramaswamy V (2004) Co-creation experiences: the next practice in value creation. J Interact Mark 18(3):5–14
Rahman M, Haque M (2016) Investigation of fabric wastages in knit t-shirt manufacturing industry in Bangladesh. Int J Res EngTechnol 5(10):212–215
Ramirez, Stefanie Ann (2014) Slow fashion: developing a conceptual apparel design process. LSU Doctoral Dissertations 2053. https://digitalcommons.lsu.edu/gradschool_dissertations/2053.
Rathinamoorthy R (2018a) Consumer awareness on sustainable fashion. In: Muthu S (ed) Sustainable fashion: consumer awareness and education. Textile science and clothing technology. Springer, Singapore, pp 1–36
Rathinamoorthy R (2018b) Sustainable apparel production from recycled fabric waste. In: Muthu SS (ed) Sustainable innovations in recycled textiles. Springer, Singapore, 19–52
Rathinamoorthy R, Surjit R, Karthik T (2019) Clothing Swap: Gateway to Sustainable Eco-friendly Fashion, In: Leticia Myriam Torres Martínez, Oxana Vasilievna Kharissova, Boris Ildusovich Kharisov (eds), Handbook of ecomaterials. Springer International Publishing, Singapore, 1599–1622
Nathalie Remy, Eveline Speelman, and Steven Swartz (2017) Style that's sustainable: a new fast-fashion formula. http://www.mckinsey.com/business-functions/sustainability-and-resource-productivity/our-insights/style-thats-sustainable-a-new-fast-fashion-formula. Accessed 16 July 2019

Rissanen T (2008) Creating fashion without the creation of fabric waste. In: Hethorn J, Ulasewicz C (eds) Sustainable fashion: why now? A conversation about issues, practices, and possibilities. Fairchild, New York, pp 184–206

Rissanen T (2013) Zero-waste fashion design: a study at the intersection of cloth, fashion design and pattern cutting. Doctoral dissertation. Sydrey. University of technology. https://opus.lib.uts.edu.au/handle/10453/23384. Accessed 20 July 2019

SEC (2009) Design as a driver of user-centred innovation. Commission staff working document. Brussels, Commission of the European Communities. http://ec.europa.eu/enterprise/policies/innovation/files/design_swd_sec501_en.pdf. Accessed 20 July 2019

Seligman Martin EP, Mihaly Csikszentmihalyi (2014) Positive psychology: an introduction. Flow and the foundations of positive psychology Springer, 279–298

Akriti Shakya (2016) Integrated waste minimization techniques in apparel design: a sustainable perspective, doctor of philosophy, dayalbagh educational institute Dayalbagh, India

Shedroff N (2009) Design is the problem, the future of design must be sustainable, Rosenfeld media. Brooklyn, New York

Smith R, Yates J (2018) Flourishing fashion: an interpretive phenomenological analysis of the experience of wearing a happy outfit. Fash Theory 1(1):5. https://static1.squarespace.com/static/591218e0f7e0abcf6ce40add/t/5b312b30aa4a9930e2760271/1529956381568/Flourishing_Fashion_Rebecca_Smith_and_Julia_Yates_Fashion_Studies_Issue_1.pdf.

Soh SL, Ong SK, Nee AYC (2014) Design for Disassembly for remanufacturing: methodology and technology, 21st CIRP conference on life cycle engineering. Procedia CIRP 15:407–412

Strähle J (2017) Green fashion retail. In: Green fashion retail. Springer, Singapore, pp 1–6

Carolyn F. Strauss, Alastair Fuad-Luke (2008) The Slow Design Principles A new interrogative and reflexive tool for design research and practice. http://raaf.org/pdfs/Slow_Design_Principles.pdf. Accessed 20 July 2019

Sundin E, Bert B (2005) Making functional sales environmentally and economically beneficial through product remanufacturing. J Clean Prod 13(9):913–925

Tanvir SI, Mahmood T (2014) Solid waste for knit fabric: quantification and ratio analysis. J Environ Earth Sci 4(12):68–80

Thackara J (2006) In the bubble: designing in a complex world. The MIT press, Boston, MA

The Cloths Library (2019). https://theclotheslibrary.com/.

The Depop Store (2019). https://www.depop.com/.

The Nudie Jeans (2019). https://www.nudiejeans.com/blog/re-use-is-now-online.

The R Collective, redress (2019). https://www.redressdesignaward.com/the-r-collective.

The Reformation Store (2019). https://www.thereformation.com/pages/2018-year-in-review.

Thøgersen J (2000) Knowledge barriers to sustainable consumption. In: Bone PF, France KR, Wiener J (eds) Marketing and public policy conference proceedings. American Marketing Association, Chicago, pp 29–39

Thorpe A (2007) The designer's atlas of sustainability. Island Press, Washington, DC

Thredup (2019). www.thredup.com.

Tischner U, Charter M (2001) Sustainable product design. In: Charter M, Tischner U (eds) Sustainable solutions: developing products and services for the future. Greenleaf, Sheffield, pp 118–138

Tyler D, Heeley J, Bhamra T (2006) Supply chain influences on new product development in fashion clothing. J Fash Mark Manag 10(3):316–328

Ulrike G, Hammerl B (2015) Integrative re-use systems as innovative business models for devising sustainable product–service-systems. J Cleaner Prod 97(June):50–60

UNEP (2002) Product-service systems and sustainability: opportunities for sustainable solutions', Kenya: United Nation Environmental Programme Division of Technology Industry and Economics Production and Consumption Branch

Veenhoven R (2011) Greater happiness for a greater number: is that possible? If so, how? In: Sheldon KN, Kashdan TB, Steger MF (eds) Designing positive psychology. Oxford University Press, Oxford, pp 392–409

Walker S (2007) Design Redux. In: Chapman J, Gant N (eds) Designers, visionaries and other stories. A collection of sustainable design essays. Earthscan, London, pp 57–73

Sasha Walton (2016) Wearing Wellbeing: Positive Perspectives on the Ethics of Fashion, The fashion conversation. http://www.thefashionconversation.com/2017/12/01/wearing-wellbeing-positive-perspectives-on-the-ethics-of-fashion/. Accessed 24 July 2019

Waste & Resources Action Programme [WRAP] (2017) Valuing our clothes: The cost of UK fashion. http://www.wrap.org.uk/sites/files/wrap/valuing-ourclothes-the-cost-of-uk-fashion_WRAP.pdf, Accessed 15 July 2019

Waste and Resources Action Programme [WRAP] (2016) Guidance on Design for Extending Clothing Life, http://www.wrap.org.uk/sustainable-textiles/scap/extending-clothing-life/guidance-extending-clothing-life

Watson CA (2004) The sartorial self: William James's philosophy of dress. Hist Psychol 7(3):211

World Commission on Environment and Development (1987) Our common future. Oxford University Press, Oxford

Worn Again (2018). www.wornagain.info,

www.Threadless.com. Accessed 24 July 2019

Zwolinski P, Miguel-Angel LO, Daniel B (2006) Integrated design of remanufacturable products based on product profiles. J Clean Prod 14(15–16):1333–1345

Contribution of UV Technology to Sustainable Textile Production and Design

Feristah Unal, Ozan Avinc, Arzu Yavas, Hüseyin Aksel Eren, and Semiha Eren

Abstract The textile and fashion industry is one of the oldest and largest sectors in the World. In the textile industry, it is possible to create both functional and artistic designs by using different materials and different techniques together in harmony. However, it is very important that the applied process types should not harm the environment and living things. Therefore, one of the issues that have come up recently is sustainable textile design. As a result of the developments in today's world, companies invest in technology in order to provide competitive advantage in many sectors and to increase environmental awareness. Textile industry inevitably should re-design their processes in order to reduce the potential environmental damage leading to more sustainable planet for future generations. Therfore, the usage of sustainable, renewable materials and sustainable eco-friendly production methods should be increased in textile industry more and more in order to overcome possible increasing environmental problems arising from the processes of textile sector. In this respect, for instance, ultraviolet (UV) technology offers many alternative innovative usage types not only in the textile industry but also in many other different industrial areas. The usage of UV technology offers so much positive attributes to the textile production and design and also provides sustainable and eco-friendly opportunities for many different areas of the textile industry. The new usage types of UV technology in the textile sector has still been explored today and developed day by day. In this chapter, the information about Ultraviolet (UV) and UV technology, the usage areas of UV technology, the contribution and the production efficiency advantages of UV technology to sustainable textile production and design (such as the use of UV technology in decolorization and purification of textile wastewaters, as pre-treatment and surface modification processes prior to coloration processes, for UV-curing process and for pilling problem prevention etc.) were given in detail.

F. Unal · O. Avinc · A. Yavas (✉)
Textile Engineering Department, Engineering Faculty, Pamukkale University, Denizli, Turkey
e-mail: aozerdem@pau.edu.tr

H. A. Eren · S. Eren
Textile Engineering Department, Uludag University, Bursa, Turkey

S. S. Muthu, M. A. Gardetti (eds.), *Sustainability in the Textile and Apparel Industries*, Sustainable Textiles: Production, Processing, Manufacturing & Chemistry, https://doi.org/10.1007/978-3-030-37929-2_8

Keywords Sustainable · Design · UV · Ultraviolet · Textile · Environment-friendly

Introduction

With the acceleration of industrialization in all sectors of the world, the consumption of the resources and resultant waste production accumulation is unfortunately rapidly and steadily increasing. As in all modern industries, substances left and released to the environment by textile manufacturers are also generally not harmless. For this reason, recently, reducing the amount of damage to the environment and reducing the consumption of the resources for ecological protection, the utilization of environmentally friendly fibers or other materials, reducing the amount of pollution generated, and the development of methods to remove the pollution and pollutants have gained more and more importance. The textile and fashion industry is one of the oldest and largest industries in the world. Today, the textile and fashion industry has been one of the most environmentally damaging sectors to the ecology. In order to solve this ever-increasing problem in the sector, sustainable materials and production methods should be utilized. As is known, textile design fulfills many different purposes in our lives. But proper and more sustainable textile process design is also important for the sustainability of our planet. For this purpose, replacing harmful chemicals with environmentally friendly substances and reducing waste accumulation and resource consumption are of great importance. Again for this purpose, the utilization of sustainable new technologies in the production of textile materials is being investigated. In textile finishing processes, many scientific studies have been carried out on energy-, water-, and time-saving technologies such as plasma, ultrasound, enzyme, ozone, microwave, nanotechnology, and ultraviolet (UV) technology in pretreatment, dyeing, printing, and finishing processes. Moreover, textile industry should re-design their processes in order to reduce the potential environmental damage leading to more sustainable planet.

Developments in the textile industry lead to changes in design and production styles. Changing materials and manufacturing processes continue to influence design from the past to the present. Design is an area where science, technique, and technology combine together in a harmony via creative thinking and aesthetics. If a good design contains both artistic values and is supported by technical equipment and scientific methods, the resulting product shows the ability to combine art with industrial production. It also allows the use of environmentally friendly technologies such as UV technology for clean production in textile. UV rays used in different fields of the textile industry are a kind of electromagnetic wave.

In 1865, the British physicist James Maxwell and the German scientist Hertz discovered electromagnetic waves through his experimental work (Glaze et al. 1987). Radio waves, microwaves, infrared rays, visible light, ultraviolet rays, X-rays, and gamma rays are the types of electromagnetic waves. Ultraviolet is known as a part of the non-ionizing region of the electromagnetic spectrum (United

Nations Environment Programme, the International Labour Organisation, and the World Health Organization 1994). It was first discovered by the German physicist Johann Wilhelm Ritter. Firstly it was expressed as chemical rays by Ritter and then called as ultraviolet (Glaze et al. 1987). The word meaning is derived from Latin and means that it is "beyond violet" (Ansiklopedi 2016). The position of ultraviolet rays in the electromagnetic spectrum is characterized by a wavelength (United Nations Environment Programme, the International Labour Organisation, and the World Health Organization 1994; Bintsis et al. 2000). Wavelengths vary between 10 and 400 nm (nanometer). According to the energy emitted by ultraviolet, they are classified as UV-A, UV-B, UV-C, and vacuum UV (Perincek et al. 2007; Karahan et al. 2007; Saravanan 2007; Mutlu et al. 2003; Perkins 2000; Reinert et al. 1997; Özkütük 2007).

Although most of the UV rays are held by gases in the atmosphere, some of them reach to the earth. In case of excessive exposure to UV rays, it is known to result in human health damage. The usage areas and purposes of UV are changing with the developing technology. It is used in many industries with the advantage of being an environmentally friendly technology. UV technology, which is widely used in disinfection of liquids (Özkütük 2007; Keyser et al. 2008; Koutchma 2009; Pala and Toklucu 2010; Teknolojisi 2019; Fredericks et al. 2011), is used in the pasteurization process of liquid foods in the food industry (Keyser et al. 2008; Koutchma 2009; Pala and Toklucu 2010). It is also used for disinfection of hemodialysis systems in medicine (Sezdi and Benli 2016; Şimşek and Sultan 2013) and disinfection process of textile wastewater in textile sector (Colonna et al. 1999; Georgiou et al. 2002; Konstantinou and Albanis 2004; Robinson et al. 2001). Especially in textile enterprises, massive amount of water is generally used during the finishing processes leading to high wastewater load. The use of environmentally friendly new methods in the treatment of textile wastewater provides the recovery of purified water (Bahtiyari 2009; Kitiş et al. 2009). The logic of the disinfection process with UV technology is almost the same for every sector. UV rays impair the DNA structure of harmful microorganisms formed in the liquids. It breaks the chains between two spirals in the DNA and causes the death of microorganisms (Pest Products 2011; Aydın 2009). In the purification of industrial wastewater, chemical degradation processes are provided with UV/O_3 or UV/H_2O_2. In addition, TiO_2/UV is widely used in the photocatalytic oxidation process, which enables the treatment of wastewater (Glaze et al. 1987; Colonna et al. 1999; Robinson et al. 2001; Güneş and Talınlı 2007; Zhang et al. 1998; Arslan and Balcioğlu 1999). Since no chemicals are used during these processes, there is no change in the chemical structure of liquids (Fredericks et al. 2011; Aydın 2009). Many studies in this area are included in the literature (Glaze et al. 1987; Colonna et al. 1999; Robinson et al. 2001; Güneş and Talınlı 2007; Zhang et al. 1998; Arslan and Balcioğlu 1999). Besides, UV technology is used also for anti-pilling process of wool and cotton fabrics (Perincek et al. 2007; El-Sayed and El-Khatib 2005). Although chemical processes such as chlorination can be widely used for this purpose, the formation of adsorbable organohalogens (AOX) during the process has led to the development of alternative methods such as UV technology (El-Sayed and El-Khatib 2005).

In 1998, Siroflash anti-pilling process was introduced by Millington. The Siroflash anti-pilling process for knitted fabrics is a process involving exposure to short-wavelength ultraviolet radiation (UV-C) of the fabric or garment surface. A light wet oxidation process is applied to the fabric exposed to ultraviolet radiation using, for instance, hydrogen peroxide or permonosulfuric acid salts (El-Sayed and El-Khatib 2005; Millington 1998a). Thus, the weak fibers which cause pilling on the surface of the textile materials are removed from the fabric surface, and the materials are prepared for the next processing step (such as dye and printing) (Perincek et al. 2007).Textile products may need to be modified before dyeing to increase dyestuff intake. Some studies in the literature suggest that UV technology provides significant advantages in dyeing (Millington 2000; Iqbal et al. 2008). By using ecologically acceptable UV-assisted treatments, modification of textile materials, in particular wool, is provided (El-Sayed and El-Khatib 2005). It was reported that more dyestuff intake and good fixation at low temperatures can be provided with the modification of the fiber surfaces. As an outcome of oxidation of surface fibers by UV radiation, treated textile materials exhibited higher color strength (Millington 2000; Iqbal et al. 2008). The influence of UV radiation on natural and synthetic dyes also gives quite important results (Bhatti et al. 2011). UV/O_3 is generally used as a surface modification process. Short UV radiation combined with ozone disintegrates covalent bonds in many organic materials. For example, this process may result in photooxidation of poly(trimethylene terephthalate) (PTT) film. PET and PTT fabrics treated with UV/O_3 radiation are dyed with black disperse dyes, and UV radiation-treated fabrics displayed deep dyeing effect (Karahan et al. 2007; Jang and Jeong 2006). In addition, UV technology can also be used for curing processes. UV curing has found different application areas in various industrial sectors. It offers advantages such as providing ultrafast polymerization of solvent-free formulations. UV curing that is used in textile printing process is used to produce prepress printing plates. In addition, the use of UV-curable inks during printing provides fast drying. This situation significantly increases the processing speed and reduces energy consumption (Decker 2002, 1996). Today, the ability of the countries to achieve superiority in the international market depends on technological innovations. For this purpose, advantages of new technologies such as UV technology in textile field are very valuable (Perincek et al. 2007). The usage of UV technology offers so much positive attributes to the textile production and design and also provides sustainable and eco-friendly opportunities for many different areas of the textile industry. Textile industry should re-design their processes in order to reduce the potential environmental damage leading to more sustainable planet. Moreover, increasing competition between the companies enhances their interest in the utilization of novel technologies such as ultraviolet technology in their textile processes. Thus, while the quality of their products and services may increase, their potential hazardous damage to the environment could be diminished and/or eliminated.

It is known that UV technology can be used for various purposes in different areas of industry. In this chapter, the information about ultraviolet (UV) and UV technology, the usage areas of UV technology, and the contribution and the production efficiency advantages of UV technology to sustainable textile production and design (the use of UV technology in decolorization and purification of textile wastewaters, as pretreatment and surface modification processes prior to coloration processes, for UV curing process and for pilling problem prevention, etc.) were given in detail.

In this chapter, the contribution of UV technology to sustainable textile production and design is reviewed. Firstly, brief information about electromagnetic field and electromagnetic spectrum, ultraviolet, ultraviolet rays, and history of ultraviolet is given. Then, the different contributions of ultraviolet technology to sustainable process design in the textile sector were introduced. Finally, different application areas of ultraviolet technology in the textile production such as decolorization and purification of textile wastewaters, pretreatment and surface modification process prior to coloration processes, curing, and the pilling problem prevention are examined in detail.

Electromagnetic Waves and Electromagnetic Spectrum

Electromagnetic waves are a kind of vibration that can spread very quickly in space. The invention of electromagnetic waves and basic works were carried out by James Clerk Maxwell and Heinrich Hertz. Electromagnetic waves are classified according to the wavelength. Physical events we experience on a daily basis show us the presence of an electromagnetic wave, for example, sunrise, cooking your meal with microwave, etc. The types of electromagnetic waves are radio waves, microwaves, infrared rays, visible light, ultraviolet rays, X-rays and gamma rays. These are separated from each other by only wavelength (Sengupta and Sarkar 2003). Electromagnetic waves spread over a very wide area according to wavelength and energy. Electromagnetic spectrum is obtained by classifying electromagnetic waves according to frequency and wavelength. The light waves we can see are also part of the electromagnetic spectrum. Electromagnetic waves are classified from the longest wavelength to the shortest wavelength in the electromagnetic spectrum. The smallest frequency or the largest wavelength of electromagnetic waves is radio waves. The highest frequency or the smallest wavelength of electromagnetic waves is gamma rays. The frequency increases, and the wavelength decreases from left to right in the electromagnetic spectrum (Aksoy 2019).

Ultraviolet

Ultraviolet is called "beyond violet," and the color of the uppermost frequencies of seeable light is violet. Ultraviolet possesses a higher frequency than violet light, and ultraviolet rays are electromagnetic radiation of between visible rays and X-rays. Ultraviolet rays have longer wavelengths than X-rays and shorter wavelengths than visible rays (Ansiklopedi 2016; Attwood and Sakdinawat 2017). They are not visible to the human eye because of the shortness of their wavelength (Karahan et al. 2007; Kaizenisg 2012; Diffey 2002). Ultraviolet (UV) radiation is a form of light energy from the sun (Mutlu et al. 2003). They have wavelengths between 10 and 400 nanometers (Ansiklopedi 2016; Özkütük 2007). It consists of frequencies greater than 789 THz. A single light cluster has energy of 3.26 eV (Diffey 2002). Ultraviolet rays, which are harmful to the skin, cause sunburn and skin cancer (Ansiklopedi 2016). Ultraviolet also provides the formation of vitamin D which provides strengthening of the bones of all living creatures living on land. The UV spectrum thus possesses effects both beneficial and harmful to human health (Ansiklopedi 2016). Scientists had separated the ultraviolet part of the spectrum into three as UV, far UV, and far-off UV. This distinction made by scientists is based on the energy of UV radiation. Wavelength of UV light is expressed by energy. The sun emits radiation from all wavelengths in the electromagnetic spectrum and is known as the greatest source of ultraviolet radiation (Kaizenisg 2012; Diffey 2002). About 9% of the energy emitted from the sun is ultraviolet radiation. Fourteen percent of the ultraviolet radiation emitted from the sun belongs to the wavelength region smaller than 3000 Angstrom (A°) (Kaizenisg 2012). Ultraviolet rays form a very low fraction in the solar spectrum but affect all living organisms and their metabolism (Saravanan 2007). More than half of the ultraviolet (UV) rays from the sun are kept in the atmosphere (Kaizenisg 2012; Diffey 2002). Some ultraviolet waves pass the earth's atmosphere, but are kept by some gases such as the ozone layer (Perincek et al. 2007; Karahan et al. 2007).

The electromagnetic spectrum of ultraviolet radiation (UVR), which can be described the most broadly as 10–400 nanometers, could be subdivided into a number of ranges advised by the ISO standard ISO-21348 (Ansiklopedi 2016; Özkütük 2007). Ultraviolet rays are grouped among themselves according to different wavelengths. Physicists make this classification as close UV (320–380 nm), medium UV (200–320 nm), and vacuum UV (10–200 nm) (Perincek et al. 2007; Karahan et al. 2007; Saravanan 2007). The term UV-A represents a region of 320–400 nm, while the term UV-B represents a region of 290–320 nm, and the UV-C region represents the region below 290 m. The order of impact has been determined as UV-C > UV-B > UV-A (Saravanan 2007; Mutlu et al. 2003; Perkins 2000; Reinert et al. 1997; Özkütük 2007). An example of ultraviolet rays is shown in Fig. 1.

Long-Wave UV (UV-A)

Wavelengths of UV-A rays are between 320 and 400 nm. At the same time, UV-A rays have the highest wavelength and least energy compared to other UV rays. UV-A rays emitted from the sun are held by the atmosphere. Most UV-A rays are not affected by ozone and pass directly through the ozone layer. In other words, 95–98% of UV-A rays reach to the earth (Ansiklopedi 2016; Perincek et al. 2007; Karahan et al. 2007; Mutlu et al. 2003; Madronich et al. 1998). Excessive exposure to UV-A radiation from the sun causes skin problems and diseases. In particular,

Fig. 1 Ultraviolet rays (Wikimedia Commons 2019)

it reaches and spreads to the deep layers of the skin (dermis). This type of UV is widely used in lighting systems in industry (Karahan et al. 2007).

Middle-Wave UV (UV-B)

The wavelengths of UV-B rays are between 280 and 320 nm. The wavelengths and energy of the UV-B rays are in the middle of the UV band. It is stronger than UV-A rays with longer wavelength rays. Seventy to eighty percent of the UV-B rays are kept by ozone, water vapor, oxygen, and carbon dioxide. UV-A rays form the largest part of UV rays reaching the earth through the atmosphere, and a small portion of it also forms UV-B rays. Two to five percent of the UV-B rays from the sun reach the earth, and UV-B radiation is potentially very harmful (Perincek et al. 2007; Karahan et al. 2007; Mutlu et al. 2003; Madronich et al. 1998). The UV-B rays, which significantly affect human health, reach the upper layer of the skin and cause skin cancers. In addition, it also causes health problems such as temporary blindness and cataract. In industry, these rays are generally known to be used in lighting systems (Perincek et al. 2007; Karahan et al. 2007).

Short-Wave UV (UV-C)

UV-C rays are known as rays that have the shortest wavelengths and the highest energy at the ultraviolet band. The wavelength of UV-C rays varies between 200 and 280 nm (Perincek et al. 2007; Mutlu et al. 2003). Some sources evaluate vacuum UV as UV-C. Therefore, according to these sources, the wavelengths of UV-C rays indicate 100–280 nm (Karahan et al. 2007; Madronich et al. 1998). UV-C rays emitted from the sun are filtered by the ozone layer or absorbed by other gases in the atmosphere (Perincek et al. 2007; Mutlu et al. 2003). It causes serious health problems when exposed to UV-C rays. UV-C rays can be produced by using electrical energy as a result of electronic industrial processes. UV-C rays, which have lost their energy when they come into contact with any surface, are recently used in surface modifications (Perincek et al. 2007; Karahan et al. 2007).

Vacuum UV

Vacuum UV wavelength is between 100 and 200 nm (Diffey 2002). Vacuum UV, its wavelengths shorter than 200 nm, are strongly absorbed by molecular oxygen in the air, although the longer wavelengths of about 150–200 nm could propagate through nitrogen (Ansiklopedi 2016; McPherson et al. 1987). Vacuum UV with very short wavelength produces ozone by separating the oxygen in the air. Ozone (O_3) can only be measured in vacuum. Vacuum UV radiation is used in surface cleaning/purification, photooxidation, surface activation, and ozone production (Diffey 2002).

History of Ultraviolet

The existence of ultraviolet rays was first discovered by German physicist Johann Wilhelm Ritter in 1801. Ritter has studied the effect of X-rays on chemical substances and observed that invisible rays just beyond the violet end of the visible spectrum darkened silver chloride-soaked paper more quickly than violet light itself. In other words, he noticed that there was an energy output in the dark band beyond the violet light (Ansiklopedi 2016). He called these rays as oxidizing rays to accentuate its chemical reactivity. Thus it is easier to distinguish them from the heat rays. The term of chemical rays which is the simpler expression was adopted shortly afterwards. During the nineteenth century, there were people who thought that ultraviolet rays, known as chemical rays, were a completely different kind of radiation from light. In time, the terms chemical rays and heat rays were called ultraviolet and infrared rays, respectively. In 1878, it was observed that short-wavelength light killed bacteria. Thus, sterilizing effect was first discovered. In 1903, it was learned that the most effective wavelength was around 250 nm, and in the 1960s, ultraviolet radiation has been found to be effective on DNA (Glaze et al. 1987). In 1893, the German physicist Victor Schumann discovered that ultraviolet radiation at wavelengths below 200 nm was absorbed by the air. For this reason, ultraviolet radiation at wavelengths below 200 nm is called "vacuum ultraviolet" by Victor Schumann.

Table 1 Usage areas and its related wavelengths of UV technology in different sectors (Ansiklopedi 2016; Özkütük 2007; Keyser et al. 2008; Koutchma 2009; Pala and Toklucu 2010; Teknolojisi 2019; Fredericks et al. 2011; Pest Products 2011; Decker 2002, 1996; Gaillard and Gilbert 1997; Feyler 2004; Apollonio et al. 2006; Smith 2004; Mager et al. 1999; Elman and Lebzelter 2004)

Wavelength (nm)	Application area	Reference
13.5	Excessive ultraviolet lithography	Ansiklopedi (2016)
30–200	Photoionization, ultraviolet photoelectron spectroscopy	Ansiklopedi (2016)
200–400	Judicial test, medicament detection	Ansiklopedi (2016), Gaillard and Gilbert (1997)
200–270	Disinfection of liquids Disinfection of fluid food products (antimicrobial effect)	Ansiklopedi (2016), Pest Products (2011)
230–365	UV-ID, sticker chasing, barcodes	Ansiklopedi (2016), Feyler (2004)
230–400	Optical sensors, several instrumentation	Ansiklopedi (2016)
240–280	Cleansing of surfaces and water (DNA absorption has a peak at 260 nm)	Ansiklopedi (2016), Özkütük (2007), Keyser et al. (2008), Koutchma (2009), Pala and Toklucu (2010), Teknolojisi (2019), Fredericks et al. (2011)
254	Protein analysis, DNA sequencing	Ansiklopedi (2016); Apollonio et al. (2006)
270–360	Protein analysis, DNA sequencing	Ansiklopedi (2016), Gaillard and Gilbert (1997), Smith (2004)
280–400	Medical imaging of cells	Ansiklopedi (2016)
300–320	Light therapy in medicine	Ansiklopedi (2016), Mager et al. (1999)
300–365	Curing and printer inks	Ansiklopedi (2016), Decker (2002), Decker (1996)
407–420	Light therapy in treatment	Ansiklopedi (2016), Elman and Lebzelter (2004)

Usage Areas of Ultraviolet Technology

Ultraviolet technology is now being used in almost all sectors. There is a usage area spreading in food, cosmetic, medical, electrical, and textile sectors. It is effective in damaging the living cells and killing bacteria in sectors such as medicine and food. Therefore, it is a type of radiation used in phototherapy or water sterilization. Moreover, UV technology can also be used for forensic analysis and drug detection (Ansiklopedi 2016; Gaillard and Gilbert 1997). Ultraviolet rays make some chemical reactions easier (Ansiklopedi 2016; Karahan et al. 2007). Therefore, the use of aqueous foods in pasteurization is also increasing (Keyser et al. 2008; Koutchma

2009; Pala and Toklucu 2010). In addition, the use of UV technology in disinfection of wastewater also prevents the environmental pollution caused by textile wastewater (Georgiou et al. 2002; Konstantinou and Albanis 2004; Robinson et al. 2001). It is known that in the textile sector, UV technology can be used not only to disinfect wastewater but also to increase dye intake via its application prior to coloration processes such as dyeing and printing. It can also be used as an alternative to the chlorination process in the production of black and navy blue fabrics especially used in Japan as an official costume (Perincek et al. 2007). It is also used for UV curing process of polymers and printing inks (Decker 2002, 1996). It also provides removal of weak fibers that cause the formation of pilling on the surface of woolen knitted materials. Moreover, UV technology is also used for surface modifications of textile materials (Perincek et al. 2007; Karahan et al. 2007). The different application areas of UV technology in different sectors and its related wavelengths are shown in Table 1 (Ansiklopedi 2016).

Uses of Ultraviolet Technology in Textile Sector

When it comes to the relation of UV and textile, one may think, at first thought, that textile materials (apparel, garment, clothing, curtains, etc.) are generally used to protect us against the harmful effects of UV rays which were generally emitted from the sun (Saravanan 2007). However, the relation between the textile materials and UV rays of the sun is not that limited. In the recent years, physicochemical methods performed with UV radiation are being more preferred instead of chemical methods in the surface modifications of textile materials. In UV surface treatments, generally photoinitiators are used, and thus the yield obtained is increased (Ansiklopedi 2016; Karahan et al. 2007). The use of UV radiation to change the surface properties of fabric has a commercial importance, and it is considered as an alternative to conventional methods. While the fabric surface is modified by UV radiation, the main properties of the textile materials do not change. During this process, high-density surface fibers serve as a layer to prevent UV rays from penetrating into the interior. Thus, the main fibers providing the fabric strength are protected. In order for this process to be carried out correctly, the surface fibers must be directly absorbed by UV rays, or when exposed to UV rays, suitable photoinitiators should be used to produce large amounts of reactive free radicals. The photoinitiators used in textile applications are odorless, non-toxic, and cheap and can be easily removed by washing. For this reason, aromatic photoinitiators are generally used in most UV applications (Perincek et al. 2007).

UV treatment also benefits the coloration process (such as dyeing and printing) of textile products. Owing to the modification of the fabric surface with UV rays, more dyestuffs can be fixed more rapidly leading to darker shades and colors. It also increases the wettability of hydrophobic fibers in the printing process and prevents the pilling problem formation especially for the knitted fabrics (Perincek et al. 2007).

Ozone is formed when UV rays react with air oxygen. The UV rays then react with the ozone formed, resulting in the formation of monoatomic oxygen, which is highly reactive and suitable for the oxidation of the fibers. In addition, UV photons ensure that the bonds on the fiber surface are broken and change the structure, so as to react with ozone and monoatomic oxygen. This double effect is important for rapid oxidation of the fiber surface and that the fiber exhibits the desired interactions with the matrix in the composite materials (Karahan et al. 2007; Qin et al. 2015). Seventekin et al. injected acrylic acid on polyester using a high-pressure UV lamp. They used benzophenone as the initiator and made the polyester fibers hydrophilic. And then the researchers measured the resistance to washing of the dirt-repellent properties of the polyester fibers which became hydrophilic (Karahan et al. 2007; Kurbus et al. 2003).

UV Technology Usage in Sustainable Process Design

Design in textile means the combination of color, texture, and lines with creativity (Gürcüm 2016). However, the process for the formation of a product is also included in the design. Process design includes the works that determine how the product or parts of the product will be produced in which kind of production process, through which hardware tools. Design has been interpreted with various definitions by designers and researchers as "decision-making in uncertainty," "finding the right physical components of a physical structure," "imaginary leap from the realities of the present to the probabilities of future," "problem-solving process," "cognitive process," "reflection in action," and "knowledge-based activity" (Gürcüm 2016). The design is described by Elizabeth Adams Hurwitz succinctly as the search for what is necessary (Ansiklopedi 2019).

When all the comments made for the design are evaluated, the design can be seen as a process for solving a specific problem (Gürcüm 2016). The goal of this process is to identify the problem and to find sustainable and creative solutions to meet the needs. One of the most important elements of design is functionality. Product functionality is provided by process designs and current production technologies. It is very important that these processes are environmentally friendly, sustainable, and clean energy (Gürcüm 2016). Sustainable process design looks for the answer to the question of how to produce the product in a sustainable way. Today, a new aesthetic understanding is formed with process designs that will make a difference in textile materials created with innovative technologies (Gürcüm 2016; Erbiyikli 2012). With the interpretation of traditional or known applications with creative and innovative developments, new opportunities for unusual materials are developed, and new concepts are developed, and new insights are formed (Erbiyikli 2012).

For instance, Semiha Eren's work in 2019 is an example of the use of UV technology in sustainable process design (Eren 2019). In this study, photocatalytic clearing process of disperse dyed polyester fabric with UV light was emphasized. Dyed polyester fabrics treated with different concentrations of hydrogen peroxide were exposed to 254 nm of UV radiation. The fabrics were treated with UV radiation for 5 min, 10 min, 20 min, and 30 min, respectively. The photocatalytic clearing of disperse dyed polyesters was compared with conventional reduction clearing. It was concluded that the use of UV in the photocatalytic clearing process of dyed polyester fabrics causes less damage to the environment than the traditional method. They also obtained comparable washing fastness results. In addition, the use of UV radiation in photocatalytic clearing is stated to be a different, novel, and sustainable process design (Eren 2019).

In the below sections, other than the use of UV technology for the clearing of dyed polyester, other different contributions of ultraviolet technology to sustainable process design in the textile sector (decolorization and purification of textile wastewaters, pretreatment and surface modification process prior to coloration processes, curing, and the pilling problem prevention) were examined in more detail.

The Use of UV Technology in the Decolorization and Purification of Textile Wastewaters

UV treatment systems provide an environmentally friendly and highly efficient disinfection for a sustainable environment and healthy living (Teknolojisi 2019). Textile industries manufacture large amounts of wastewater because of their high consumption of water primarily in the dyeing and finishing operations (Georgiou et al. 2002). Wastewater containing a great variety of organic contaminants in a wide range of concentrations may cause disturbance to the ecological system of the receiving water (Colonna et al. 1999; Georgiou et al. 2002; Konstantinou and Albanis 2004; Robinson et al. 2001; Kurbus et al. 2003). The release of this wastewater into the environment may lead to serious environmental pollution. The chemical reactions in the wastewater phase (such as oxidation and hydrolysis) result in the formation of dangerous byproducts (Konstantinou and Albanis 2004; Kurbus et al. 2003). This poses a danger to the health of living creatures in the environment (Konstantinou and Albanis 2004). Generally, biological purification systems and methods are utilized in the treatment of such industrial wastewater (Colonna et al. 1999; Güneş and Talınlı 2007). But, because of the large degree of aromatics present in dye molecules and the stability of modern dyes, conventional biological treatment methods are insufficient for decolorization and degradation (Konstantinou and Albanis 2004; Robinson et al. 2001; Arslan and Balcioğlu 1999). It is now possible to oxidize and break down wastewaters by using oxidation technologies. Many oxidation processes are used for this aim. These include chemical disintegration processes (UV/O_3, UV/H_2O_2), photocatalysts (TiO_2/UV, photophone reagents), and chemical oxidation processes (O_3, O_3/H_2O_2, Fenton) (Glaze et al. 1987; Colonna et al. 1999; Robinson et al. 2001; Güneş and Talınlı 2007; Zhang et al. 1998; Arslan and Balcioğlu 1999; Kurbus et al. 2003). Advanced oxidation processes such as TiO_2/UV have been studied in detail in the literature (Arslan and Balcioğlu 1999). In the photocatalytic oxidation, TiO_2 should be irradiated and excited in a near-UV energy to induce charge separation. It is hard to conclude whether the photocatalytic oxidation is superior to the photosensitizing oxidation mechanism; however the photosensitizing mechanism will aid to improve the overall efficacy and make the photobleaching of dyes utilizing solar light more feasible (Arslan and Balcioğlu 1999; Epling and Lin 2002). Effective destruction of azo dyes belonging to different chemical groups seems possible by TiO_2/UV (Arslan and Balcioğlu 1999).

Although the combination of ozone and UV technology is widely used for the treatment of textile wastewater, it can take on a task that catalyzes the formation of ultraviolet-hydrogen peroxide or chlorinated compounds (Perincek et al. 2007). UV wavelengths of 200–280 nm result in disassociation of H_2O_2 with mercury lamps emitting at 254 nm which is the most commonly utilized one. It is known that UV/H_2O_2 systems generate hydroxyl radicals (OH) that are highly powerful oxidizing species. Hydroxyl radicals could oxidize organic compounds (RH) fabricating organic radicals (R) that are highly reactive and could be further oxidized (Glaze et al. 1987; Georgiou et al. 2002; Kurbus et al. 2003). Thus, the dyestuffs in textile wastewater are disintegrated, and many organic impurities are destroyed (Perincek et al. 2007; Kurbus et al. 2003).

Many research studies have been done on the purification of textile wastewater with UV technology and are still being carried out. Some studies in the literature have tested several azo-reactive dyes and cotton textile wastewater under UV irradiation in the existence of hydrogen peroxide (H_2O_2). Researchers used a 120 W UV lamp, protected by a quartz tube, emitting at 253.7 nm as a radiation source. Complete destruction of the color of the dye solutions was achieved in the first 20–30 min of irradiation. UV/H_2O_2 proved that it was capable of the complete degradation and mineralization of the azo-reactive dyes (Georgiou et al. 2002). UV radiation in the existence of hydrogen peroxide results in complete decolorization and mineralization of sulfonated azo and anthraquinone dyes in a relatively short period. But, anthraquinone dyes generally seem to be a little more resistant to the treatment than azo dyes. UV/H_2O_2 treatment seems a very encouraging technique for reduction of organic pollutants in the industrial wastewaters. UV/H_2O_2 treatment seems to be a quite successful method for reduction of organic pollutants in the industrial wastewaters (Colonna et al. 1999; Robinson et al. 2001). Semiconductor photocatalysis usually results in partial or complete mineralization of organic pollutants (Chakrabarti and Dutta 2004). In the process of UV radiation treatment, semiconductors catalyze redox reactions with air and water (Chakrabarti and Dutta 2004).

After the biological treatment of wastewater, UV disinfection is carried out before being given to the natural environment. These waters are used for recycling or irrigation purposes (Aydın 2009). The greatest advantage of UV technology is its environmentally friendly nature and its chemical disinfection procuration. It does not change the chemical properties of water (such as conductivity and pH) since the chemical is not given into the water. When chlorine or chlorinated disinfectants are used in the water, carcinogenic compounds called THM (trihalomethanes) can be formed. Carcinogenic byproducts are not formed in the disinfection processing with UV technology. In addition to all these positive attributes, disinfection with UV technology is also very fast (Aydın 2009).

The Use of UV Technology as a Pretreatment and Surface Modification Process Prior to Coloration Processes

UV technology can be utilized for bleaching and surface modification processes prior to coloration processes such as dyeing and printing. In textile finishing, bleaching of the product is a quite important process since it affects the brightness and color shade of the dyeing process. In particular, the removal of the natural yellowish color of cotton and some impurities is possible by bleaching. The most common bleaching agent utilized in textile is hydrogen peroxide (H_2O_2) (Eren 2018). However, in the bleaching process with hydrogen peroxide, caustic soda is required for activation of hydrogen peroxide, and moreover stabilizer is needed to prevent catalytic damage. What is more, wetting agents for better penetration and high temperature are also required. Therefore, methods with less environmental impacts for bleaching of cotton continue to be explored. In a study by Eren S. (2018), UV technology is used as a cleaner bleaching method (Eren 2018). The hydrogen peroxide treatment carried out under UV radiation is considered an advanced oxidation treatment. The main oxidizing agent in the H_2O_2/UV advanced oxidation process is the hydroxyl radical (-OH) which is created by the application of UV radiation on H_2O_2. It was reported that the H_2O_2/UV advanced oxidation process, which also can be used in bleaching soybean fabrics, is also very suitable for cotton bleaching (Eren 2018). In here, Eren S. examined the photocatalytic bleaching of cotton by H_2O_2/UV. She used 254 nm UV light and a specially designed 470 Watt cabinet with 18 UV lamps in this study. Whiteness of cotton samples bleached by H_2O_2/UV treatment with the whiteness of traditionally bleached (pad-steam method) cotton fabric samples was compared. It was concluded that the best whiteness levels were obtained for 40 and 60 min treatment times with H_2O_2/UV treatment (Eren 2018). Thus, the bleaching process carried out with the help of UV technology proves that this technology type can be an alternative to traditional methods both by being environment friendly and providing sufficient whiteness (Eren 2018).

Another application type of the UV technology in the textile industry is a pre-modification process prior to the coloration process of the textile material. In fact, color is the most important element of design and ultimate purchase decision. Precursor radiation treatment on fabrics and garments is a value added process for coloration process. In here, the modification of the fiber surface allows more dye uptake and higher fixation rates at low application temperatures leading to increased wettability (Bhatti et al. 2011; Kim et al. 2005). This also facilitates the applicability of different pattern designs.

The most important advantages of UV technology over other methods applied are energy saving, low-temperature process applicability, low environmental effect, simple, economical, and high treatment speed (Bhatti et al. 2011). Nevertheless, the number of radiation curing applications for nonwoven fabric bonding, fabric coating, and pigment printing is low in the textile industry (Ferrero and Periolatto 2011). It was stated that the influence of UV radiation on natural and synthetic dye results in important outcomes (Bhatti et al. 2011). Some studies in the literature prove that UV irradiation adds value to coloration and also enhances the dye uptake capability of the cotton fabrics through oxidation of the surface fibers of cellulose and besides resulted in even shade with higher color strength (Millington 2000; Iqbal et al. 2008).

It was stated that the UV irradiation of fabric is also another factor that influences the color strength of the fabric (Millington 2000; Iqbal et al. 2008; Bhatti et al. 2011). The effect of UV irradiation on color strength was tested using appropriate dyes and fabrics.

According to the test results, the UV-irradiated fabric displayed better color strength than the UV-non-irradiated fabric. It has been observed that the fabric exposed to UV radiation for 120 min exhibited maximum color strength value (Bhatti et al. 2011). The reason of this may be the result of the cellulose oxidation due to UV light exposure. Oxidation of cellulose on UV radiation expressively upsurges the dye intake into the substrate (Micheal and El-Zaher 2005).

The research carried out by Michael and EL-Zaher in 2005 proved that the UV treatment of cellulose fiber created spaces between the fibers which take more dye, and as an outcome, the interaction between dye and cellulose fabric becomes more substantial. It was reported that the dye particles rush fast onto the fabric, and as an outcome, darker shades were obtained (Bhatti et al. 2011; Micheal and El-Zaher 2005; Adeel et al. 2012). In the studies conducted by Millington, it has been observed that as a result of the UV processes called photo-modification, better hydrophilicity can be imparted to the textile material. Also the color productivity obtained as a result of dyeing and printing processes is increased, and even the coloration takes place in a shorter time and at a lower application temperature. They have reported that the tendency of pilling was also reduced in cotton and wool knitted fabrics. If the number of cross-links due to the effect of UV rays is high, it loses its fabric durability and acquires a rigid structure, which has a negative effect on the comfort of the garment. In the case of excessive number of cross-links due to UV rays' effect, it loses its fabric casting and acquires a rigid structure. For this reason, it has a negative effect on the comfort of the garment (Karahan et al. 2007). As with all high-energy sources, extreme UV radiation results in photo-degradation and degradation of the physical properties of the product (Karahan et al. 2007; Waddell et al. 1992). Previous study by K. R. Millington suggested that photo-modification of the surface fiber could lead to more dye uptake to produce deeper shades (Millington 1998b). It is concluded that UV radiation provides faster fixation of dyes and enhances the wettability of hydrophobic fibers to improve depth and shade in printing (Millington 1998b).

Jang and Jeong studied the surface modification of PET and PTT fibers using UV/O_3 combination. At the end of the treatment, the surface which gains anionic character has increased dyeability due to electrostatic interaction with cationic dyes (Karahan et al. 2007; Jang and Jeong 2006). Lately, UV/O_3 treatment has become more and more popular and practical since its process type has several advantages in comparison with the other well-known surface modification methods. The UV light, which is less than 340 nm, breaks the covalent bonds of C-C, C-O, and C-H in the structure of many organic materials together with the ozone, and this causes photooxidation. Molecular oxygen absorbs UV light emission and manufactures ozone with a reaction of the molecular oxygen and ground-state oxygen, and ozone absorption of 253.7 nm light leads to singlet atomic oxygen manufacture. It was stated that UV/O_3 radiation can be used to obtain deep tones when PET and PTT fabrics are dyed with black disperse dyes (Jang and Jeong 2006).

In another study, UV/ozone irradiation, as a surface modification process, and fluorocarbon finishing combinations were applied to polyester fabrics (Rahmatinejad et al. 2016). In here, the enhancement in polyester substrates hydrophobicity was done by surface modification via chemical pretreatment, UV-ozone irradiation, and fluorocarbon finishing combinations. The outcomes of the study exhibited the usefulness of UV-ozone treatment for creating proper surface roughness to improve the hydrophobicity of polyester fabrics after fluorocarbon finishing, especially when the fabric was pretreated with NaOH and H_2O_2 solutions. Moreover, it was also stated that the obtained highest water repellency levels and the best air permeability characteristics that resulted in noteworthy rise in the substrate hydrophobicity did not display any adverse effect on tensile properties and strength deterioration (Rahmatinejad et al. 2016).

The effects of UV radiation on wool fabrics before the printing process were studied (Millington 2000). UV radiation pretreatment helps to achieve good color yield and brightness. Wool pretreatment is traditionally carried out by an oxidative chlorination process. The gaseous chlorine or dichloroisocyanuric acid is used in this process. However, this process may damage both the fabric and the environment due to the release of the adsorbable organohalogen compounds in the wastewater. For these reasons, the International Wool Secretariat (IWS) prioritizes the development of alternative methods. Alternative chlorine-free oxidative processes for printing preparation may also have commercial application in the pulling and anti-hairing processes of the wool fabric. A special technique, Siroflash, for the preparation before printing for woolen fabric is based on the use of conventional wet oxidation utilizing hydrogen peroxide or permonosulfuric acid (PMS) after continuous ultraviolet (UV) irradiation of the dry wool fabric. In other words, the Siroflash process utilizes this technology following a conventional wet oxidation utilizing hydrogen peroxide or permonosulfuric acid (PMS) after continuous UV irradiation. The main purpose of the Siroflash process is to examine the basic methods of disrupting disulfide bonds of keratin and the possible methods of increasing the degree of fiber damage on the surface of a wool fabric to an equal level with conventional process. UV radiation (especially UV-C, run 200–280 nm) breaks both peptide and disulfide bonds in wool keratin and cannot penetrate the cortex of the fiber cuticle during short-term operations. The Siroflash process comprises the exposure of the wool fabric to UV radiation followed by wet oxidation utilizing chlorine-free oxidants. UV-treated wool fabric results in deeper black shades than non-UV-treated wool counterpart when dyed with 1:1 metal complex dyes. Because hydrogen peroxide is used during the process, the fabric prepared by the Siroflash method is much whiter than the fabric treated with dichloroisocyanuric acid (Millington 1998c).

Chemical changes induced by short-term UV radiation are confined to the fiber surfaces. It was reported that the color changes in wool keratin because of UV radiation have also been observed (Millington 2000). When the wool fabric is exposed to

UV radiation, there are some physical and chemical changes on the wool fiber surface. This process changes the texture of the wool as well as improves gray and black tones. Finally it was also stated that UV curing technology is utilized for the modification of the surface of the wool to help its finishing process (Ferrero and Periolatto 2011).

In the study of Khoddami et al. (Rahmatinejad et al. 2015), the utilization of UV/ozone as an environmentally friendly fast treatment for the surface modification of the wool substrate was investigated. It was reported that both-sided UV/ozone irradiation, in comparison with only one-sided UV/ozone irradiation, led to the slightly better performance as judged for both pad-dry-cure and print-dry-cure fluorocarbon finishing treatments. It was concluded that after the application of this novel method (both-sided UV/ozone modified and then fluorochemical finish treated via the print-dry-cure process) on wool fabric, the inner side of the wool fabric, which was next to the skin, can ensure desirable comforting moisture absorption for human body and protect the human skin from the negative hydrophobic effects of fluorocarbon chains; furthermore, the outer side of the wool fabric displayed highly durable (to repeated washings and abrasion) water and oil repellency characteristics to the outside world, aiding the wool fabric exhibits better dimensional stability with less shrinkage performance after the repeated washings and higher air permeability performance without detrimental effects on the tensile properties (Rahmatinejad et al. 2015).

There are many commercial applications starting from the curing, finishing, color enhancement, and characterization of dyed fabrics with radiation processes. The best known advantages of this technology are improvement of color shades, increasing color fastness and color strength, low cost, and reduction of dye concentration to achieve similar depth of shades (Bhatti et al. 2011). In addition, curing of silk, wool, and cotton can be achieved with this technology. Pilling problem in cotton and wool fabrics can be eliminated with this technology. Besides, the UV technology also can improve the finishing and mercerization processes. Thus, UV radiation has developed the textile material according to ISO, EPA, and FAO standards (Bhatti et al. 2011).

UV Curing

UV light curing process can be applied to the light-dependent polymerization of multifunctional monomers or oligomers. The UV curing process has become a well-accepted technology used in different industrial applications due to its various advantages (Decker 2002; Fouassier and Rabek 1993; Pappas 1992). It is believed that UV curing method is the most effective way of rapidly converting a solvent-free liquid resin to a solid polymer at ambient temperature. Under the intensive light, cross-linking polymerization of acrylate-based resins proceeds in a fraction of a second in order to form a three-dimensional polymer network. The three-dimensional network of polymers exhibits excellent resistance to organic solvents, chemicals, and heat. The cross-linking in this process can easily be obtained by UV (Decker

2002, 1996, 2001). The main benefit of using UV radiation to initiate the chain reaction lies in the very high polymerization rates that can be achieved under intense light. This allows the liquid-to-solid phase change to occur within 1 second (Decker 2002, 1996, 2001).

Photo-initiated polymerization of multifunctional monomers or UV radiation curing was applied in various industrial sectors previously. This technology is now used for ultrafast drying of protective coatings, varnishes, printing inks, and adhesives to produce high-resolution images required for the production of microcircuits and printing plates. Besides, solvent-free formulations provide many advantages such as low-energy consumption (Decker 2002, 1996). Most studies concerning photo-curing have been carried out on the development of highly efficient photoinitiators and highly reactive monomers or polymers. A wide variety of high-performance compounds are now commercially available. Acrylate-based resins, which are polymerized by a radical mechanism, are the most commonly used systems hardening by UV due to their large reactivity (Decker 1996; Siegel 2006). UV curing is typically a process which converts a multifunctional monomer into a cross-linked polymer through chain reaction initiated by the reactive species (ions or free radicals) produced by UV radiation (Decker 1996). Most monomers that are exposed to UV light cannot produce efficient initiators high enough. Therefore, a photoinitiator must be added to the formulation.

There are two main classes of UV-curable resins which vary according to polymerization mechanisms. The first one is the photo-initiated radical polymerization of multifunctional monomers, especially acrylates or unsaturated polyesters. The second one is known as the photo-initiated cationic polymerization of multifunctional epoxides and vinyl ethers (Decker 1996). Radical-type systems are widely utilized in today's UV curing applications due to their higher reactivity (Decker 1996). One of the first UV-curable resins to be utilized in large-scale applications consists of a mixture of unsaturated polyester with styrene and fumaric or maleic structures (Decker 1996). The usage of UV curing process, which is used in different application areas, in textile printing processes is common. The use of UV-curable inks ensures fast drying and increases the speed of processing. For instance, to obtain a high-quality glossy image, UV technology can be used in the last step in which the UV-curable printing varnish is applied (Decker 1996; Siegel 2006; Caiger and Cockett 2000). Because of its many advantages such as low-energy consumption, short startup time, fast and reliable curing process, environment-friendly nature, room temperature curing possibility, and space saving, UV curing technology is used as an alternative to conventional thermal hardening techniques in textile finishing processes like pigment dyeing, coating, and pigment printing (Neral et al. 2006). In an EU project, inkjet printing and pigment printing were applied to textile fabric by UV curing method. Within the scope of this project, the use of a prototype UV scanner was also investigated (Neral et al. 2006). Some of the studies also investigated the suitability of curing of screen printing pigment prints with UV scanner, which is a multifunctional photoinitiator, and the evaluation of UV-treated prints on cotton fabric (Neral et al. 2006). In order to obtain an efficient cure during the UV curing processes, the best possible overlap between the absorption spectrum of the photoinitiator, the transmission spectrum of the pigment, and the emission spectrum of the light is required (Neral et al. 2006).

In the study of Karim et al. (Karim et al. 2014), UV-curable inkjet printing was carried out on poly(lactic acid) fabrics. The wavelength of UV radiation during application was 200 nm. Thermally hardened inkjet-printed PLA fabrics were compared with UV-curable inkjet-printed PLA fabrics. It was concluded that UV-curable inkjet PLA fabrics provided better color fastness and show better behavior (Karim et al. 2014).

In another study, radical and cationic UV curing methods were applied to impart water and oil repellency properties to cotton, polyester, and polyamide fabrics. After the application, the polymerization was controlled by the weight increase, while the water and oil repellency properties are determined by the contact angle, moisture absorption, and water vapor permeability angle (El-Molla 2007). UV curing is used not only during printing but also in prepress printing plates. UV technology also reduces the total energy consumption of the process while providing an alternative adaptation strategy to the problem of volatile organic compounds (Decker 1996; Siegel 2006; Caiger and Cockett 2000).

UV curing process is particularly suitable for printing on plastic packages requiring high durability. Therefore, it is applicable not only to conventional cellulosic substrates such as paper and cardboard but also to synthetic polymeric substrates (Caiger and Cockett 2000). UV-curable varnishes are widely utilized to obtain highly durable and resistant coatings for the protection of all kinds of substrates such as wood, plastics, metals, glasses, optical fibers, leather, paper, fabrics, etc. So, the viscoelastic properties of the cured coating are particularly designed to suit the characteristics of the substrate (Decker 1996). Hard and scratch-resistant, aromatic polyether and polyester-based coatings are utilized for surface protection of hard substrates such as metal and organic glass. Low-modulus polyurethane elastomers, which exhibit good corrosion and impact resistance, are required to maintain flexible supports. UV curing adhesives are used for laminating the two parts. Thus, a rapid pressure-sensitive adhesion is achieved. UV-curable systems provide pollution control, economic gain, and high performance (Decker 1996).

The Use of UV Technology to Prevent Pilling Problem

Surface phenomena such as shrinkage or puckering occur during the use and washing of garments. In addition to dyeing processes, shrinkproof processes are the most common chemical processes applied to textile materials made from wool fibers. Pilling also occurs on the surface of the fabric during use. Pilling is a small surface fiber ball and is a physical problem sometimes consisting from the fibers containing contaminants. Chlorination is the most commonly used process to prevent felting and beading of the wool fabric. Recently, concerns about the release of adsorbable organohalogens (AOX) into the environment during this process have been increasing. Alternatively, these environmental concerns lead to the development of AOX-free processes. In addition, different radiation techniques such as ultraviolet and plasma can be used as an alternative to chlorination in the wool processing (El-Sayed and El-Khatib 2005). There are many processes to decrease pilling, but no process could guarantee zero pilling in wear (Millington 1997, 1998a, b).

In a study conducted by K. R. Millington, it was suggested that UV radiation could remove the pilling problem for knitted wool and cotton fabrics (Millington 1998b). In 1998, Millington developed a method of exposing the fabric to ultraviolet radiation, followed by a mild oxidation process utilizing hydrogen peroxide or permonosulfuric acid. This method developed for knitted fabrics is called Siroflash anti-pilling (El-Sayed and El-Khatib 2005; Millington 1997, 1998a, b). The fibers on the fabric surface are exposed to UV-C radiation, and then the wet oxidation process applied sensitizes surface fibers. As the surface fibers are much weaker after the applied process, the beads on the fabric surface cannot hold onto the fibers leading to a decrease on pilling problem. This process is a very effective and chlorine-free method to prevent pilling in wool knitted fabrics.

A variation of the Siroflash processes is also applied to cotton and cotton/wool blend fabrics. Hydrogen peroxide is applied to the fabrics before the fabrics were exposed to the UV radiation. It is then treated with continuous UV irradiation with a medium-pressure arc system akin to those utilized commercially for UV curing of polymer films. This process, which is a variation of the Siroflash process, is ineffective on pure wool knitted fabrics. Low-power antiseptic UV tubes are used to provide an effective treatment on wool knitted fabrics. Irradiation time is also very important in this process. The irradiation times vary depending on the fabric construction. Stronger UV sources providing high levels of UV-C radiation can provide fast, effective, anti-pilling treatment for wool. Also, a protease enzyme with UV/H_2O_2 can also be used to attenuate surface fibers (Millington 1997, 1998a, b). After this process, wool and cotton knitwear pilling problem had eliminated from the surface of the fabric by UV radiation treatment without affecting the strength of the fiber (Bhatti et al. 2011; Kim et al. 2005). Also, the wool fabric characterization conforms to the standards established by ISO. For this reason, the continuous UV irradiation of the fabric followed by batch oxidation is of a great commercial value (Millington 1997, 1998a, b).

Conclusions

It is possible to create both functional and artistic designs by using different materials and different techniques together in the textile industry. But it is very important that applied process types should not harm the environment and living things. One of the issues that have come up recently is sustainable textile design. As a result of the developments in today's world, companies invest in technology in order to provide competitive advantage in many sectors and to increase environmental awareness. Textile industry should re-design their processes in order to reduce the potential environmental damage leading to more sustainable planet. Moreover, increasing competition between the companies enhances their interest in the utilization of novel technologies such as ultraviolet technology in their textile processes. Thus, while the quality of their products and services may increase, their potential hazardous damage to the environment could be diminished and/or eliminated. Therefore, in textile finishing processes, many scientific studies have been carried out on energy-, water-, and time-saving technologies such as plasma, ultrasound, enzyme, ozone, microwave, nanotechnology, and ultraviolet (UV) technology in pretreatment, dyeing, printing, and finishing processes. For instance, the usage of UV technology offers so much positive attributes to the textile production and design and also provides sustainable and eco-friendly opportunities for many different areas of the textile industry. UV technology, which is one of these technologies, is applied in medicine, food, cosmetics, and other sectors. Some studies in the literature emphasized that the applications of UV technology in textile sector can increase the productivity of production. The environment-friendly nature of ultraviolet technology is also very important in terms of sustainability. It also provides an advantage in the process of creating the desired pattern and design. The use of UV technology in decolorization and purification of textile wastewaters, as pretreatment and surface modification processes prior to coloration processes, for UV curing process, and for pilling problem prevention are some of the examples of the successful contribution of UV technology to sustainable textile production and design. UV technology, which is especially important for textile sector, can widely be used in curing applications. In addition, the functions of UV technology on different textile materials (e.g., the prevention of shrinkage and pilling in wool fabrics) are also supported by experimental studies. We understand that the use of UV technology in textile processes has been developed day by day. It was shown that UV technology increases the production efficiency in textile sector. Bleaching processing can be performed in a shorter time with the help of UV technology. For instance, the best whiteness levels could be obtained with H_2O_2/UV treatment. Short-wavelength UV rays provide the surface modification of woolen textile materials in particular and thus give advantage to the dyeing process of textile material (as a premodification process prior to dyeing and printing processes). In addition, the development of both aesthetic and functional effects can be enhanced through UV technology. It allows creating more vivid areas in coloration. Thus, it is easier and cleaner to create different patterns and designs on the textile material. In the future, it is predicted that many dry and clean technologies such as UV technology will be used more and more frequently for design purposes since the sustainability in design and fashion is gaining more and more importance day by day.

Current studies show that UV technology will be used more actively in sustainable textile process design in the near future. The only way to bring functionality to design products will be better understood by incorporating new technologies into process designs. It is thought that designs inspired by artistic textiles can be brought into production and industry by adding different values with sustainable UV technology in production processes. It is very important to use UV technology in premodification processes in this field. Since UV technology affects dye uptake of textile materials, the usage of UV technology may create differences in coloration. Thanks to the use of UV technology in textile process design, it is envisaged that the variety of textile products developed for completely different purposes will increase and their usage areas will be increased leading to more sustainable future.

References

Adeel S, Bhatti IA, Kausar A, Osman E (2012) Influence of UV radiations on the extraction and dyeing of cotton fabric with Curcuma longa L. Ind J Fibre Textile Res 37(1):87–90

Aksoy E (2019) Elektromanyetik Spektrum. https://prosafety.com.tr/elektromanyetik-spektrum-radyoaktivite/.

Wikipedia Özgür Ansiklopedi (2016) Ultraviolet. https://en.wikipedia.org/wiki/Ultraviolet.

Wikipedia Özgür Ansiklopedi (2019) Tasarım. https://tr.wikipedia.org/wiki/Tasar%C4%B1m.

Apollonio LG, Pianca DJ, Whittall IR et al (2006) A demonstration of the use of ultra-performance liquid chromatography–mass spectrometry [UPLC/MS] in the determination of amphetamine-type substances and ketamine for forensic and toxicological analysis. J Chromatogr B 836(1–2):111–115

Arslan I, Balcioğlu IA (1999) Degradation of commercial reactive dyestuffs by heterogenous and homogenous advanced oxidation processes: a comparative study. Dyes Pigments 43(2):95–108

Attwood D, Sakdinawat A (2017) X-rays and extreme ultraviolet radiation: principles and applications. Cambridge University Press, United Kingdom

Aydın K (2009) Ultraviyole Işınları İle Suların Dezenfeksiyonu. IX. Ulusal Tesisat Mühendisliği Kongresi, Tepekule Kongre ve Sergi Merkezi, İzmir, 6–9 Mayıs 2009

Bahtiyari M İ (2009) Çevre dostu yeni yöntemlerin tekstil ön terbiyesindeki bazı kullanım alanlarının araştırılması. Dissertation, Ege Üniversitesi

Bhatti IA, Adeel S, Abbas M (2011) Effect of radiation on textile dyeing. In: Hauser PJ (ed) Textile dyeing. InTech, New York, p 1

Bintsis T, Litopoulou-Tzanetaki E, Robinson RK (2000) Existing and potential applications of ultraviolet light in the food industry–a critical review. J Sci Food Agric 80(6):637–645

Caiger N A, Cockett M A (2000) Ink jet printer with apparatus for curing ink and method. US Patent 6,145,979, 14 Nov 2000

Chakrabarti S, Dutta BK (2004) Photocatalytic degradation of model textile dyes in wastewater using ZnO as semiconductor catalyst. J Hazard Mater 112(3):269–278

Colonna GM, Caronna T, Marcandalli B (1999) Oxidative degradation of dyes by ultraviolet radiation in the presence of hydrogen peroxide. Dyes Pigments 41(3):211–220. https://doi.org/10.1016/S0143-7208(98)00085-0

Decker C (1996) Photoinitiated crosslinking polymerisation. Prog Polym Sci 21(4):593–650. https://doi.org/10.1016/0079-6700(95)00027-5

Decker C (2001) UV-radiation curing chemistry. Pigment and resin technology 30(5):278–286

Decker C (2002) Kinetic study and new applications of UV radiation curing. Macromol Rapid Commun 23(18):1067–1093. https://doi.org/10.1002/marc.200290014

Diffey BL (2002) Sources and measurement of ultraviolet radiation. Methods 28(1):4–13

Elman M, Lebzelter J (2004) Light therapy in the treatment of acne vulgaris. Dermatol Surg 30(2):139–146. https://doi.org/10.1111/j.1524-4725.2004.30053.x

El-Molla MM (2007) Synthesis of polyurethane acrylate oligomers as aqueous UV-curable binder for inks of ink jet in textile printing and pigment dyeing. Dyes Pigments 74(2):371–379

El-Sayed H, El-Khatib E (2005) Modification of wool fabric using ecologically acceptable UV-assisted treatments. J Chem Technol Biotechnol 80(10):1111–1117

Epling GA, Lin C (2002) Photoassisted bleaching of dyes utilizing TiO_2 and visible light. Chemosphere 46(4):561–570

Erbiyikli N (2012) Tekstil ve Moda Tasarımı Açısından Sanat ve Bilim. Akdeniz Sanat Dergisi 4(7)

Eren S (2018) Photocatalytic hydrogen peroxide bleaching of cotton. Cellulose 25(6):3679–3689

Eren S (2019) Photocatalytic clearing of disperse dyed polyester. AATCC Journal of Research 6(5):10–15

Ferrero F, Periolatto M (2011) Ultraviolet curing for surface modification of textile fabrics. J Nanosci Nanotechnol 11(10):8663–8669. https://doi.org/10.1166/jnn.2011.3447

Feyler D (2004) UV2D reader, age verification and license validation system. US Patent 10/809,325, 26 March 2004

Fouassier JP, Rabek JF (eds) (1993) Radiation curing in polymer science and technology: practical aspects and applications. Springer Science & Business Media, London
Fredericks IN, Toit MD, Krügel M (2011) Efficacy of ultraviolet radiation as an alternative technology to inactivate microorganisms in grape juices and wines. Food Microbiol 28(3):510–517. https://doi.org/10.1016/j.fm.2010.10.018
Gaillard Y, Gilbert P (1997) Use of high-performance liquid chromatography with photodiode-array UV detection for the creation of a 600-compound library application to forensic toxicology. J Chromatogr A 763(1–2):149–163
Georgiou D, Melidis P, Aivasidis A, Gimouhopoulos K (2002) Degradation of azo-reactive dyes by ultraviolet radiation in the presence of hydrogen peroxide. Dyes Pigments 52(2):69–78. https://doi.org/10.1016/S0143-7208(01)00078-X
Glaze WH, Kang JW, Chapin DH (1987) The chemistry of water treatment processes involving ozone, hydrogen peroxide and ultraviolet radiation. J Int Ozone Assoc 9(4):335–352. https://doi.org/10.1080/01919518708552148
Güneş Y, Talınlı İ (2007) Pestisit endüstrisi zararlı atıklarının arıtılmasında inhibisyon itüdergisi/e. 17(2):79–86
Gürcüm BH (2016) Geleneksel tekstil tasarimi için tasarim algoritmasi önerisi. J Int Soc Res 9(47):383–395
Iqbal J, Bhatti IA, Adeel S (2008) Effect of UV radiation on dyeing of cotton fabric with extracts of henna leaves. Ind J Fibre Textile Res 33(2):157–162
Jang J, Jeong Y (2006) Nano roughening of PET and PTT fabrics via continuous UV/O3 irradiation. Dyes Pigments 69(3):137–143
Kaizenisg (2012) Ultraviyole ışınlar. http://www.kaizen-isg.com/egitimler/ultraviyole-isinlar/. Accessed 22 May 2019
Karahan HA, Demir A, Özdoğan E, Öktem T, Seventekin N (2007) Some methods used for the surface modification of textiles. Tekstil ve Konfeksiyon 17(4):248–255
Karim NM, Rigout M, Yeates SG, Carr C (2014) Analysis of UV-cured and thermally-cured inkjet printed poly (lactic acid) fabrics. NIP & Digital Fabrication Conference. Soc Img Sci Technol 14:92–95
Keyser M, Műller IA, Cilliers FP, Nel W, Gouws PA (2008) Ultraviolet radiation as a non-thermal treatment for the inactivation of microorganisms in fruit juice. Innovative Food Sci Emerg Technol 9(3):348–354. https://doi.org/10.1016/j.ifset.2007.09.002
Kim TK, Kim MK, Lim YJ, Son YA (2005) Degradation of the disazo acid dye by the sulfur-containing amino acids of wool fibers. Dyes Pigments 67(2):127–132
Kitiş M, Yiğit N Ö, Köseoğlu H, Bekaroğlu Ş Ş (2009) Su Ve Atıksu Arıtımında İleri Arıtma Teknolojileri-Arıtılmış Atıksularin Geri Kullanımı. https://xzenon34.files.wordpress.com/2012/03/5-ileri-aritma-teknolojileri_ders-notu_cevre-gorevlisi-egitimi_aralik-09_m-kitis.pdf.
Konstantinou IK, Albanis TA (2004) TiO2-assisted photocatalytic degradation of azo dyes in aqueous solution: kinetic and mechanistic investigations: a review. Appl Catal B Environ 49(1):1–14
Koutchma T (2009) Advances in ultraviolet light technology for non-thermal processing of liquid foods. Food Bioprocess Technol 2(2):138–155
Kurbus T, Slokar YM, Le Marechal AM, Vončina DB (2003) The use of experimental design for the evaluation of the influence of variables on the H_2O_2/UV treatment of model textile waste water. Dyes Pigments 58(2):171–178
Madronich S, Mckenzie RL, Björn LO, Caldwell MM (1998) Changes in biologically active ultraviolet radiation reaching the Earth's surface. J Photochem Photobiol B Biol 46(1–3):5–19. https://doi.org/10.1016/S1011-1344(98)00182-1
Mager D, Pass HI, Tecotzky M (1999) Photodynamic therapy apparatus and methods. US Patent 5,944,748 31 Aug 1999
McPherson A, Gibson G, Jara H, Johann U et al (1987) Studies of multiphoton production of vacuum-ultraviolet radiation in the rare gases. J Opt Soc Am B 4(4):595–601. https://doi.org/10.1364/JOSAB.4.000595

Micheal MN, El-Zaher NA (2005) Investigation into the effect of UV/ozone treatments on the dyeing properties of natural dyes on natural fabrics. Colourage 52(1):83–88
Millington KR (1997) Wool and wool-blend fabric treatment. US Patent 5,595,572, 21 Jan 1997
Millington KR (1998a) Using ultraviolet radiation to reduce pilling of knitted wool and cotton. Text Res J 68(6):413–421
Millington K R (1998b) Potential applications of UV surface treatments in the textile industry. In: Radtech '98 North America UV/EB Conference, Chicago, 19–22 April 1998
Millington KR (1998c) The use of ultraviolet radiation in an adsorbable organohalogen-free print preparation for wool and in wool dyeing: the Siroflash process. J Soc Dye Colour 114(10):286–292
Millington KR (2000) Comparison of the effects of gamma and ultraviolet radiation on wool keratin. Color Technol 116(9):266–272
Mutlu B, Şen O, Toros H (2003) Ultraviole radyasyonun insan sağlığı üzerine etkileri. III. Atmosfer Bilimleri Sempozyumu, İstanbul Teknik Üniversitesi, İstanbul, 19–21 Mar 2003
Neral B, Šostar-Turk S, Vončina B (2006) Properties of UV-cured pigment prints on textile fabric. Dyes Pigments 68(2–3):143–150
Özkütük N (2007) Ultraviyole lambalarının kullanımı. 5. Ulusal Sterilizasyon Dezenfeksiyon Kongresi, Celal Bayar Üniversitesi, Manisa, 4–8 Nisan 2007
Pala ÇU, Toklucu AK (2010) Ultraviyole Işın (UV) Teknolojisinin Meyve Sularına Uygulanması. Academic. Food J 8(1):17–22
Pappas SP (ed) (1992) Radiation curing: science and technology. Springer Science- Business Media, New York
Perincek SD, Duran K, Körlü AE, Bahtiyari Mİ (2007) Ultraviolet Technology. Tekstil ve Konfeksiyon 17(4):219–223
Perkins WS (2000) Functional finishes and high performance textiles. Textile chemists and colorists and American dyestuff. Reporter 32(4):24–27
Pest Products (2011) Ultraviolet Light, UV Rays, What is Ultraviolet, UV Light Bulbs, Fly Trap. http://www.pestproducts.com/uv_light.htm. Accessed 16 May 2019
Qin W, Vautard F, Drzal LT, Yu J (2015) Mechanical and electrical properties of carbon fiber composites with incorporation of graphene nanoplatelets at the fiber–matrix interphase. Compos Part B 69:335–341. https://doi.org/10.1016/j.compositesb.2014.10.014
Rahmatinejad J, Khoddami A, Avinc O (2015) Innovative hybrid fluorocarbon coating on UV/ozone surface modified wool substrate. Fibers and Polymers 16(11):2416–2425
Rahmatinejad J, Khoddami A, Mazrouei-Sebdani Z, Avinc O (2016) Polyester hydrophobicity enhancement via UV-ozone irradiation, chemical pre-treatment and fluorocarbon finishing combination. Prog Org Coat 101:51–58
Reinert G, Fuso F, Hilfiker R, Schmidt E (1997) UV protecting properties of textile fabrics and their improvement. Textile Chemist & Colorist 29(12):31–43
Robinson T, McMullan G, Marchant R, Nigam P (2001) Remediation of dyes in textile effluent: a critical review on current treatment technologies with a proposed alternative. Bioresour Technol 77(3):247–255
Saravanan D (2007) UV protection textile materials. AUTEX Res J 7(1):53–62
Sengupta DL, Sarkar TK (2003) Maxwell, hertz, the Maxwellians, and the early history of electromagnetic waves. IEEE Antennas and Propagation Magazine 45(2):13–19. https://doi.org/10.1109/MAP.2003.1203114
Sezdi M, Benli İ (2016) Disinfection in hemodialysis systems. Medical Technologies National Congress (TIPTEKNO'16), IEEE, Antalya, 27–29 Oct. 2016. doi: 10.1109/TIPTEKNO.2016.7863132
Siegel S B (2006) Ink jet UV curing. US Patent 7,137,696, 21 Nov 2006
Şimşek M, Sultan N (2013) Disinfection methods and applications for hepatitis viruses. Viral Hepatitis J 19(2):37–42. https://doi.org/10.4274/Vhd.58966

Smith F (2004) Handbook of forensic drug analysis. Elsevier, New York
UV Teknolojisi (2019) AO Smith Innovation has a name. https://aosmith.com.tr/teknolojiler/uv-teknolojisi/.
United Nations Environment Programme, the International Labour Organisation, and the World Health Organization (1994). ISSN 0250-863X. https://www.who.int/uv/publications/EHC160/en/.
Waddell WH, Evans LR, Gillick JG, Shuttleworth D (1992) Polymer surface modification. Rubber Chem Technol 65(3):687–696. https://doi.org/10.5254/1.3538634
Wikimedia Commons (2019) https://commons.wikimedia.org/wiki/File:Ultraviolet_LEDs.jpg
Zhang F, Zhao J, Shen T et al (1998) TiO2-assisted photodegradation of dye pollutants II. Adsorption and degradation kinetics of eosin in TiO_2 dispersions under visible light irradiation. Appl Catal B Environ 15(1–2):147–156. https://doi.org/10.1016/S0926-3373(97)00043-X

Repurposing Design Process

Rachel Eike, Erin Irick, Ellen McKinney, Ling Zhang, and Eulanda Sanders

> *"No matter what you choose to buy, just use it – for a long time."*
> *Carlee Green-Goff, founder of repurposing company Elizabeth Green.*

Abstract The fashion industry has innumerable damaging impacts to the environment (Zaffalon, Text World 16:34, 2010). Presently, the vast majority of all textile-based products, including clothing and home good fashions, end up in landfills (Kozlowski et al. J Clean Prod 183:197-207, 2018). Consumers often purchase new clothing because the style is outdated rather than because of lack of functionality. In other words, what consumers discard can still be functional and valuable in another form. The current phenomenon of fast fashion and increased turnover of merchandise has led to an abundant quantity of functional production-level textile waste and secondhand clothing (Fletcher, Sustainable fashion and textiles design journeys. Earthscan, London, 2008). It has been suggested that the greatest opportunity for reclaimed fashion goods is to repurpose them into new products (Hawley, Recycling in textiles. Woodhead Publishing Limited, Cambridge, 2006a; Hawley, Cloth Text Res J 24: 262, 2006b). Design efforts that employ reuse, repurposing, or upcycling techniques could assist in assigning renewed value from unwanted yet still functional, discarded clothing.

In this chapter, the term "repurposing" is used to describe the process that utilizes discarded textiles to create new fashion (textile-based) products. Textile

R. Eike (✉) · E. McKinney · L. Zhang · E. Sanders
Iowa State University, Ames, IA, USA
e-mail: rjeike@iastate.edu; emckinne@iastate.edu; lzhang1@iastate.edu; sanderse@iastate.edu

E. Irick
University of Wyoming, Laramie, WY, USA
e-mail: eirick@uwyo.edu

S. S. Muthu, M. A. Gardetti (eds.), *Sustainability in the Textile and Apparel Industries*, Sustainable Textiles: Production, Processing, Manufacturing & Chemistry, https://doi.org/10.1007/978-3-030-37929-2_9

"recycling," described by Lewis et al. (Int J Fash Des Technol Educ 10:353-362, 2017), is the process of returning a textile product back into its original fiber form and is not covered in this chapter.

Repurposing researchers (Irick, Examination of the design process of repurposed apparel and accessories: An application of diffusion of innovations theory. Oklahoma State University, Stillwater, 2013); (Irick & Eike, Teaching the repurposing mindset: The introduction of a repurposing project into an advanced apparel construction course. International Federation for Home Economics Conference. Sligo, 2017) have identified four levels of repurposing: (1) re-style to repurpose, (2) subtractive repurposing, (3) additive repurposing, and (4) intentional patternmaking to repurpose. This chapter provides analysis of the four repurposing levels through case study application to detail the creative design process employed by select designers for the purpose of repeatability and advancing research connected to repurposing. Case studies walk the reader through research and discovery, sampling of techniques, descriptions of full-scale design, and reflection to share learned experiences alongside detailed images of completed repurposed fashion designs. The chapter concludes with a cross-case analysis of all repurposed designs and suggests future directions to advance "repurposing" endeavors for industry and/or academic design scholars.

Keywords Repurposing · Design process · Textile waste diversion · Sustainable design

Chapter banner image: Femme du Pee collection by Carlee Green-Goff (collection features level 2 and level 3 repurposing approaches)

Introduction

The fashion industry significantly contributes to negative environmental impacts and excess waste generation (Zaffalon 2010) and has been studied by a multitude of authors and organizations. Fashion is something that always changes; however, its meaning remains intact. Fashion is an expression of identity and communicates personal values, interests, and group connection. Fashion suggests a passing trend that reaches temporary widespread adoption and then cycles out of popularity – making room for a new trend for consumer acceptance. Fashion can been described as the discarding of clothes that are fully functional for purely semiotic or symbolic reasons. As a result of fast fashion (fashion cycles entering and exiting at amplified speeds), the rate of apparel consumption and disposal has increased over recent years, resulting in an abundance of still functional, unwanted clothing, much of which ends in landfills (Kozlowski et al. 2018). It has been suggested that through the repurposing of unwanted items, renewed value can be given to these products and is one solution to overconsumption-disposal issues that are connected to overflowing secondhand clothing markets (Fletcher 2008; Young et al. 2004).

While researchers have suggested the use of secondhand textile-based items as a resource for new product development and have referenced examples of products produced via repurposing to explain differences in approaches, there is still an unclear design process to follow that provides in-depth description and analysis that would allow for repeatability. This chapter provides case study analysis of the creative processes of eight repurposing designers, organized by the four repurposing levels put forth by Irick and Eike (2017). Case studies walk the reader through each designer's creative process and reflection on learned experiences while highlighting multiple images of the featured design.

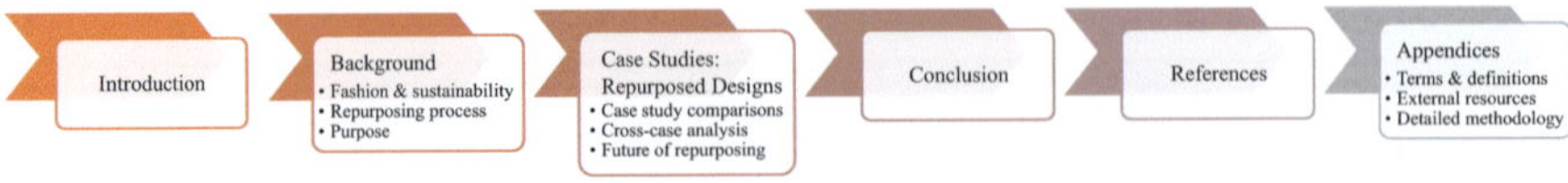

Fig. 1 Chapter outline

The chapter begins with a brief literature review to provide support for repurposing as a sustainable fashion effort. Repurposing and the repurposing process are explained in detail through each case study and analyzed within each level. The chapter concludes with a cross-case analysis of techniques used, challenges identified, and suggestions for industry integration and future research. Appendices include (a) Key Terms and Definitions, (b) External Resources to Learn More About Repurposing, (c) Research Methods Applied on Case Studies, and (d) Case Study Interview Protocol. Please review Fig. 1 for illustration of chapter outline. Audience members for this repurposing chapter may include academic professionals, students, designers, and industry collaborators.

Background and Literature Review

Fashion and Sustainability

The critical need for sustainability in apparel stems from the myriad of environmental impacts caused by clothing production and consumption processes (Zaffalon 2010). Presently, only about 10% of discarded clothing goes into the secondhand clothing market. The vast majority of all other textile-based products, approximately 2.35 million tons, including clothing and home goods, end up in landfills per year (EPA 2019; Kozlowski et al. 2018; Vennström 2012). When disposed of in a landfill, textile products release ozone-depleting methane gases and asthma-causing airborne particulates and are unable to decompose properly (Blackburn 2009). The excessive amount of textile products, namely, clothing, that ends up in landfills each year is suggested to be caused from the rise of fast fashion culture (Birtwistle and Moore 2007; Fletcher 2008; Joung and Park-Poaps 2013). As a result, consumers often purchase new clothing because the style is outdated rather than because of a lack of functionality (Joung and Park-Poaps 2013). Therefore, what consumers discard can still be functional and valuable in another form.

Sustainable apparel, often referred to as "sustainable fashion," is one response to these pressing environmental concerns (Gordan and Hill 2015) and has rapidly gained popularity with apparel companies and consumers over the past several decades. Sustainable fashion is subjective, difficult to define, and often interchangeable with "green," "eco," or "organic" fashion (Brown 2010; Gordan and Hill 2015: 1). Fashion that is typically labeled as "sustainable" may address issues connected to community and fair trade, ecological and organic sourcing, slow design/traditional craft, recycling, repurposing, and corporate social responsibility, for example. At the most basic level, sustainable fashion refers to ethically and environmentally conscious production and consumption that does not disrupt ecological balance (Gordan and Hill 2015).

The emergence of sustainable fashion has led many designers to explore creative design processes and techniques that could lessen the environmental burden of the textile industries. Sustainably driven brands, designers, and academic scholars have increasingly integrated forms of sustainable design approaches into their design practices to mitigate both pre-consumer and post-consumer textile waste. Results of these design practices have led to further research and development to advance technologies in the fashion industries. Such works have manifested in the forms of biodegradable and renewable fibers (Fletcher 2008); upcycling and downcycling previously used materials (Brown 2010; Chanin et al. 2008); creation of adaptable, trans-seasonal, and multifunctional clothing (Brown 2010; Fletcher and Grose 2011); exploration of low-chemical, no-chemical, and natural dying processes (Fletcher and Grose 2011); participation in co-design partnerships (Fletcher and Grose 2011); ethical cultivation (organic farming) of fibers (Fletcher and Grose 2011); mending and altering to prolong a garment's life (Eike et al. 2018; Fletcher and Grose 2011); and creative pattern cutting such as zero- and minimal waste design (Carrico and Kim 2014; McQuillan 2011).

The central focus of this chapter, repurposing via reuse and redesign – approaches that utilize pre-consumer textiles and post-consumer fashion goods – are among these identified sustainable fashion practices that have been investigated and employed by contemporary designers and scholars (Chanin et al. 2008; Hawley 2006a, b; Janigo and Wu 2015). Repurposing has been suggested to be the most widely sought-after method of sustainable design (Hawley 2006a, b). Regardless of the sustainable design process employed, the underlying goal remains the same: reduce environmental burden and restore balance.

Repurposing It has been suggested that the greatest opportunity for reclaimed fashion goods is to repurpose them into new products (Hawley 2006a). Design efforts that employ repurposing techniques could assist in assigning renewed value to unwanted yet still functional, discarded clothing. Janigo and Wu (2015) identified five approaches to sustainable apparel disposal: (1) repair and alteration, which changes details and fit of clothing; (2) upcycling, which is turning lower-value item(s) into a higher-value item; (3) downcycling, which entails changing higher-value item(s) into a lower-value product; (4) wearing/using secondhand fashion items; and (5) recycled/redesigned clothing from used material goods. Note: The term recycling in this chapter embodies the definition from Lewis et al. (2017) where recycling is a process of returning a textile product back into its original fiber form; Janigo and Wu's description of recycling is more similar to "repurposing." This list of sustainable apparel disposal options has led to more in-depth research within each option to advance knowledge and understanding. For example, motivations, attitudes, and behaviors associated with secondhand fashion items have been researched by scholars (Fratzke 1976; Park and Choo 2012; Kim and Kim 2013). This chapter focuses to advance understanding of the design process associated with repurposing as minimal research has been conducted on this process topic in the detail that would allow for repeatability by others.

Currently, many repurposed garments/products are individually produced where each item is completely unique, considered "one of a kind," and sold at a premium price (Fletcher 2008). Dunn (2008) suggested a design shift is needed in order to transition repurposed apparel from wearable art (limited production) to a repeatable process for standardized (repeatable) production.

Repurposing Levels

Irick (2013) surveyed repurposing designers whose products varied from one-of-a-kind items to small-scale production to form an initial understanding of processes unique to repurposing. Organization of the repurposing process and development of the different repurposing levels were outcomes of this study. A design process model for repurposing textile and apparel goods was proposed. Initially, Irick (2013) identified three levels of repurposing: (1) redesigning and repurposing (later termed "re-style to repurpose"), (2) subtractive repurposing, and (3) additive repurposing. Further work to test the repurposing design process led Irick and Eike (2017) to study the process in a longitudinal repurposing study within an Advanced Apparel Construction course. As a result of this research, a fourth level of repurposing was added and labeled as "intentional patternmaking." The steps designers engage in as they move through the repurposing design process differ, depending on the level of repurposing employed. Figure 2 outlines the present version of the "process for repurposing." The repurposing model referenced in this chapter was adapted from the C2CAD model put forth by Gam et al. (2009) which was structured using the cradle-to-cradle design framework (McDonough and Braungart 2002) as the basis for their C2CAD sustainable product development model.

In level 1, designers may make minor alterations in fit or style, such as removing sleeves or re-sizing along a garment's side seams and/or adding embellishments, like beadwork or adhesive appliques, to the original garment. This level is considered the most simplistic of repurposing, as novices to the area of sewing/construction may be able to achieve a renewed style or design from a discarded product. This level involves little or no deconstruction. Re-style to repurpose may involve some sewing construction, but this would occur along existing seamlines to address fit/shape.

In level 2, subtractive repurposing, designers cut the entire pattern shapes/pieces from a larger-sized clothing or non-clothing (e.g., interior textiles, beddings, etc.) product – treating this larger item as the source of textile yardage. This level may involve deconstruction but is not required for achieving an end product. Level 2 repurposing requires introductory to moderate sewing/construction knowledge and skills. These skills are necessary as the designer is cutting pattern pieces from the larger textile product for design development. These patterns may be drafted by the designer or purchased commercially.

In level 3, additive repurposing, pieces of fabric, either deconstructed from an original garment/product or pre-consumer scraps, are pieced together to create a new textile that is then made into a full-scale product. Level 3 repurposing requires moderate to advanced sewing/construction knowledge and skills. Knowledge and skills

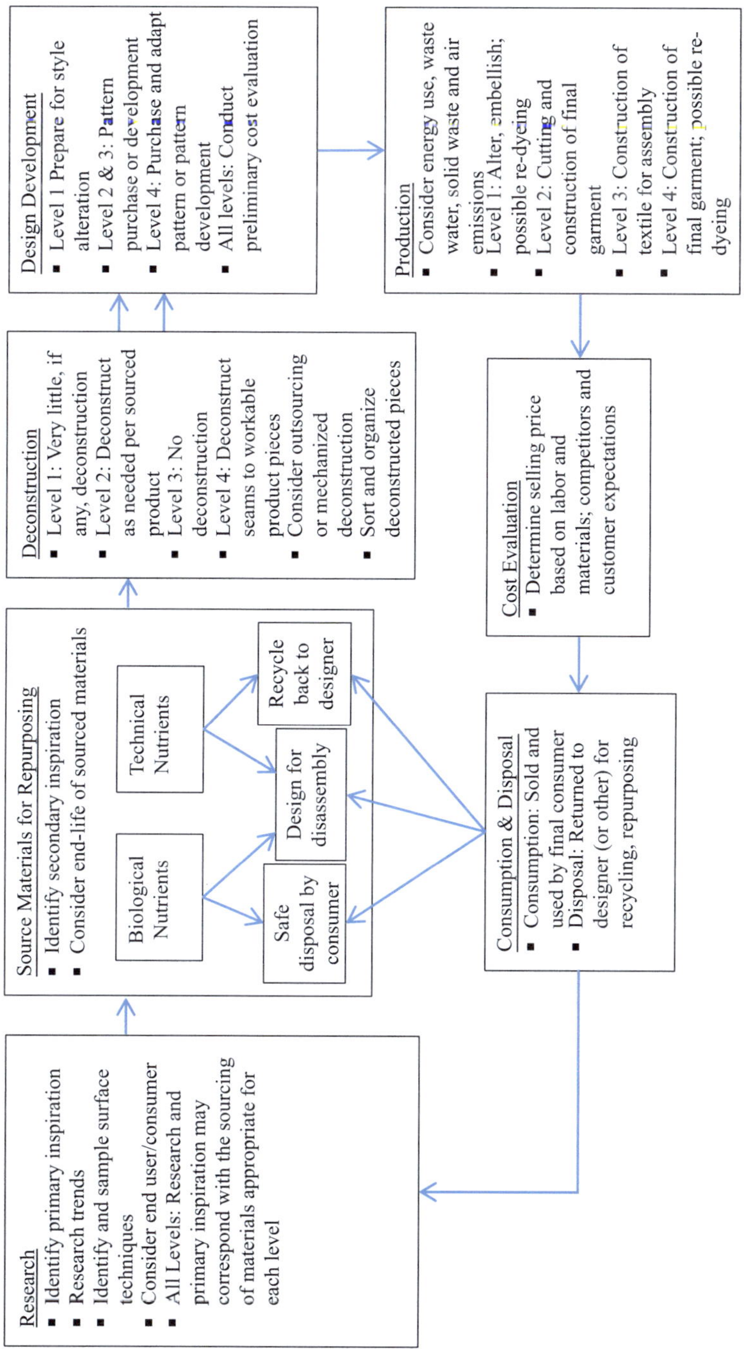

Fig. 2 Process for repurposing

related to apparel construction are necessary in this level because the designer will need to construct the pieces of textiles together to create larger pieces of fabric that will then be assembled into a full product. Patterns used may be drafted or purchased, and pieced textiles may be constructed together in a way to take shape of the patterns.

In level 4, intentional patternmaking, designers deconstruct discarded garments and create patterns based on the sizes and shapes of fabric available. New product patterns are purposefully designed and cut to utilize available fabric, working within existing shapes and area. Patternmaking techniques focus around effective use of the unwanted/discarded fashion products in order to produce a new wearable garment or product that results in minimal or zero waste. Level 4 repurposing requires advanced construction and patternmaking skills. These advanced skills are necessary in this repurposing level because the designer will need a strong understanding of pattern development in order to work within existing shapes. It may be possible for a designer to use a purchased pattern, but alterations would be needed to adjust the pattern to accommodate the material while addressing patternmaking requirements, such as accurate seam allowances.

Further work is needed to more deeply understand the usefulness and challenges of repurposing techniques for repeatability and potential industry adoption, as well as future research directions. While the proposed model is useful for understanding levels of repurposing, as well as the steps of the repurposing design process, a more in-depth understanding of individual designers' creative processes is needed. Further, understanding how repurposing processes employed by designers compare within the repurposing levels may help industry focus on specific adoption levels and practices that best suit their business. Finally, the repurposing process [outlined by Irick and Eike 2017] should continue to be tested and evaluated in light of repurposing processes employed by more designers. In doing so, the model may be refined.

Purpose

The purpose of this chapter is to explain repurposing design processes executed through select cases at each of the repurposing levels and answer the following research questions:

1. What is the creative process employed by a designer when developing a repurposed garment?
2. How do repurposing processes employed by designers compare within each repurposing level?
3. How do repurposing processes employed by designers compare between the four repurposing levels?
4. How do repurposing processes employed by designers compare to the *repurposing process* [outlined by Irick and Eike (2017)]?

The goal of this research is to expand knowledge in the domain of apparel repurposing and to "illuminate a process or set of processes: why they were taken, how they were implemented, and with what result" (Schramm 1971; Yin 2009: 17). Each case walks the reader through the process employed by the designer.

Repurposing Design Process Case Studies

The case studies of creative repurposing design processes outlined in this chapter answer *Research Question 1: What is the creative process employed by a designer when developing a repurposed garment?* The case studies include the four levels of repurposing: (1) re-style to repurpose, (2) subtractive repurposing, (3) additive repurposing, and (4) intentional patternmaking to repurpose. At the end of each section, a comparative analysis identifies a key "take-away" concept for the specific level and answers *Research Question 2: How do repurposing processes employed by designers compare within each repurposing level?* For details pertaining to methods and analysis of case studies, please review Appendices C and D.

Section 1: Re-Style to Repurpose

Case Study 1: On the Fringe

On the Fringe is a children's wear repurposed design created by Leslie Browning-Samoni completed in August 2018. This design was re-styled from an adult T-shirt ($2.00) and features a detachable fringe necklace made from embroidery thread and beads ($8.00) with a separate butterfly wing accessory (repurposed from a damaged costume). *On the Fringe* targets parents who are seeking an interesting kid's product for their child. To these parents, a sustainable product is a bonus, but the item itself needs to be special or unique on its own. Browning-Samoni estimates a resale price point of $40.00 for the dress only and $55.00 with wing accessory (based upon materials and labor expenses).

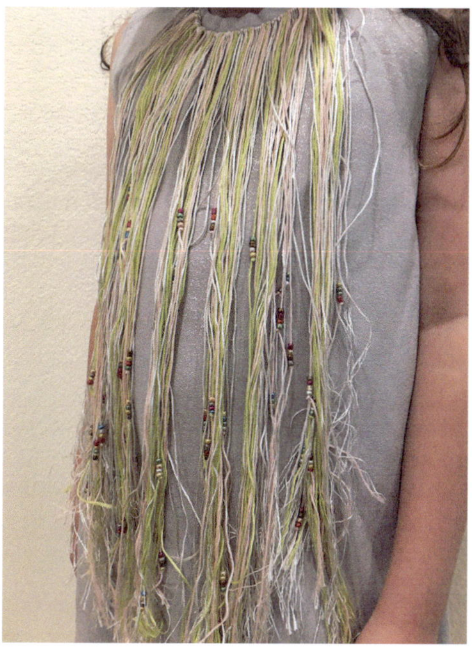

Browning-Samoni began her design process by conducting research on sustainable design strategies used in children's clothing and then sourced materials. "The fabric of the original t-shirt is a lovely silky, sparkly silver that I thought kids would like the look and of, therefore the underlying dress is the original fabric manipulated with a seam down the center back and pleats under the arms to tailor it to fit a smaller body," said Browning-Samoni. She chose to make the fringe necklace a detachable accessory to increase versatility while adding an element of fun. She pulled inspiration from her daughters who love to dance. "The fringe adds to the movement of their bodies as they dance and play around the room," explained Browning-Samoni.

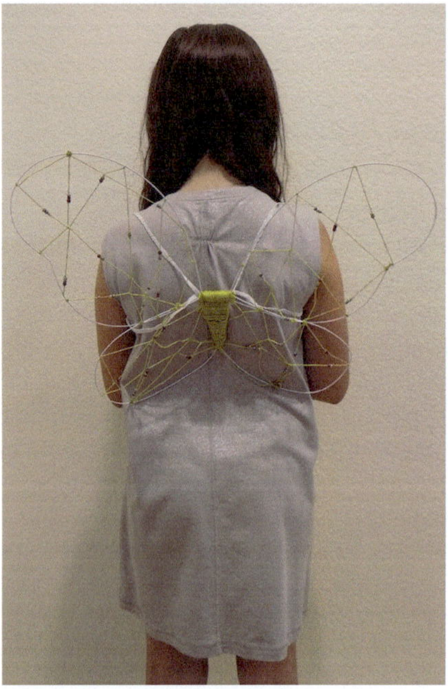

Browning-Samoni shared some challenges she encountered while completing *On the Fringe* that others may find helpful when considering this level of repurposing. One of the challenges she focused on involved the scaling down in size to accommodate a child's body and shape. She explained that one of the most difficult areas to adjust was the large armhole opening. She chose to control the excess fabric with an inverted box pleat. Browning-Samoni also expressed the challenge of thrifting a garment that would meet her expectations for fiber content (wear and launderability), design outcome, and cost.

When prompted about disposal or discarding plan for *On the Fringe*, Browning-Samoni suggested the inclusion of a note (in some form) to the consumer that says, "Be kind to the planet and to others - When you are done with this garment, please pass it along through a resale store, an online resale group, or give to a friend. Use the beads and fringe to create a necklace or other fun item of your own." Browning-Samoni continued to explain how the mix of fibers of the original T-shirt (57% cot-

ton, 38% modal, 5% spandex) would be challenging to recycle, so the best option would be to keep it as a garment for as long as possible and then use pieces of it to make doll clothing, a small interior item, or cleaning rags.

Case Study 2: Upside-Down Jean Crop Top

Upside-Down Jean Crop Top is a women's apparel top, targeting juniors' and misses' categories, created by Sophia Luu in September 2018. This design was restyled from a pair of jeans ($8.00) and features a center-back zipper and bateau/off-the-shoulder neckline with flared sleeves. "*Upside-Down Jean Crop Top* targets females 14–27 years old who are bold and daring with their clothing, desire sustainable items, and prefer a streetwear aesthetic," describes Luu. The suggested retail price point for *Upside-Down Jean Crop Top* is $85.00.

Luu's design process began with her desire to repurpose a denim product while maintaining the integrity of the original design. She wanted to keep the original existing shape of the pants and components (such a fly zipper and pockets) but repurpose the

pants to be a different clothing area altogether while challenging herself to complete a (nearly) zero-waste design. Her research began by searching the Internet for design inspiration and educational videos that involved deconstructing and reconstructing flat-felled seams, which is a garment seam type typically used in the construction of denim jeans. Luu preserved as much of the original fabric as possible by deconstructing with a seam ripper and then adding tucks and darts on the front and back to control excess fabric.

When draping the cropped top, Luu commented on how challenging it was to control the sleeve shape so that fit would be comfortable and the design would not be too revealing (around the neck/chest and shoulders). Luu also shared a surprising challenge she encountered when developing *Upside-Down Jean Crop Top*, which was connected to concern for hygiene. Luu recalled a question posed by a peer, "How can customers be more comfortable with repurposed products and reassured of hygiene?" Luu agreed that this was an important point that may cause hesitation for some consumers to purchase repurposed designs as well as designers to use secondhand products as material resources.

When prompted about ideal consumption and disposal consumer practices for *Upside-Down Jean Crop Top*, Luu said that overtime, she hopes that the consumer would feel comfortable interacting with this garment to extend its life – manipulating it as desired (re-styling and embellishing with fabric paint or embroidery) to continue using and wearing it instead of just "sentencing it to the life of 'just another object'." Luu continued on to reference Kate Fletcher's (2016) philosophy of the "Craft of use," where the item owner should acknowledge all of the people that the fabric (i.e., jeans) has touched – from cotton field farmers, to designers and factory workers, to consumers of initial product, and then future consumers of the material. Luu hopes that the pair of jeans repurposed through the creation of *Upside-Down Jean Crop Top* will continually function and be repurposed into something useful, and then when it is no longer able to serve a purpose, the trims such as the zippers and button tack will be removed and reused (if possible), and the denim fabric will be mechanically recycled.

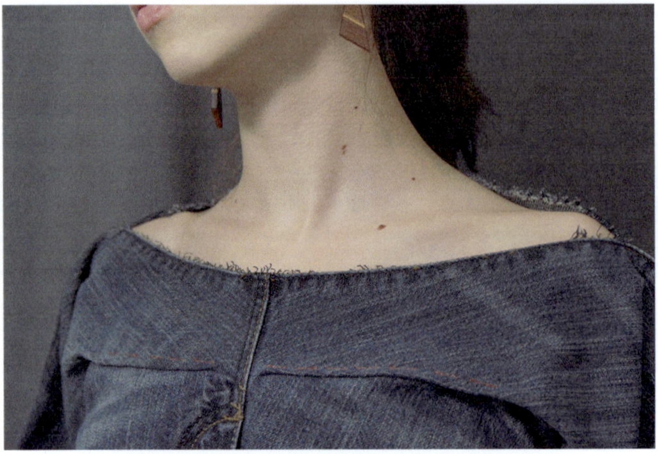

Re-Style to Repurpose Case Study Comparative Analysis

Reviewing the design processes of both Browning-Samoni (*On the Fringe*) and Luu (*Upside-Down Jean Crop Top*), similarities and differences both exist. Both designers began by conducting research on sustainable design approaches and selecting a main focus for their piece. Browning-Samoni sought to expand sustainable children's wear offerings, while Luu chose a specific textile (denim, jeans) to guide ideation of design possibilities. Both of these designers repurposed secondhand clothing items and performed minor alterations to achieve alternative styles compared to the original garments. Browning-Samoni incorporated style lines (seams and pleats) to take an adult-sized shirt and transform it into a child/youth-sized dress. Luu created a neckline for her cropped top by releasing the inseam of the jeans and incorporating darts to control excess fabric fullness. *On the Fringe* also included embellishments (detachable fringe necklace and butterfly wings) that changed the aesthetic of the original shirt while appealing to her target consumer.

When selecting materials/textiles to use in their designs, neither designer considered how their design would be disposed. However, when analyzing the designs of these products, they could fairly easily be deconstructed and reassembled into something different. *On the Fringe's* center-back seam and underarm pleats could be opened/released, and fabric panels could be inserted to increase the shirt size. The detachable fringe necklace could be paired with another garment for an alternative style for the child or worn as an accessory by an adult. The necklace fringe could be separated and created into another accessory or threaded for needlepoint stitching onto the child's dress in a new configuration or onto another product all together. *Upside-Down Jean Crop Top* has the potential to be returned to the original jean form by releasing darts and re-stitching pant inseam. The zipper at center back could be removed, and the seat of the jeans could be re-seamed to form a pair of pants. This design could also be converted into a skirt through the further releasing of the inseam and then top-stitching of the leg pieces together to form a large circular opening. There is also potential to change the arm/leg shape to be more fitted.

Level 1 take-away: Browning-Samoni and Luu emphasized the importance of sustainability messaging to consumers with a focus on "how to" continue using their designs through the use of embellishment and surface design techniques. Simple shaping techniques, such as dart and pleats, can be used with minimal disassembly to repurpose unwanted apparel.

Section 2: Subtractive Repurposing

Case Study 1: La Courtepointe

La Courtepointe is one ensemble from Carlee Green-Goff's *Femme du Pee Pee* collection (featured at the beginning of chapter) that was completed in April 2017. This upscale women's wear design was created from a quilt that her grandmother made, an unwanted blanket from a family member, and three men's chambray button-up shirts purchased from an estate sale ($6.00). *La Courtepointe* is comprised of a skort, pullover sleeveless top, and reversible coat. "*La Courtepointe* targets a woman who cares for her environment and has a sentimental side for items with a history and is looking for something unique for her wardrobe. The targeted consumer is someone who prefers customized options, oversized/comfortable style and fit, and values high quality craftsmanship," describes Green-Goff. She estimates a retail price point of $600 for the coat, $100 for the top, and $120 for the skort.

Green-Goff was inspired for the direction of *Femme du Pee Pee* by a story told while visiting a "history of dress" museum in Paris, France. The story involved the eighteenth-century dress (extreme hoops and corsets) and how these fashions required bathroom-specific assistants, referred to as the "femme du pee pees" or "the potty women." These assistants were given unwanted clothing of the upper-class women, which they would rework into something new. This reworking process inspired Green-Goff's collection. From this inspiration, she conducted research of trends, including silhouettes, colors, and textures, and crafted a narrative for her ideal consumer. She then developed a concept board which guided her sourcing of textile materials, primarily from estate sales and family member donations. She then deconstructed sourced products to repurpose, sketched designs, and identified where and how pieces of the quilt, blanket, and shirts would be used. She created samples to ensure fit and trialed different ways of constructing seams and finishing edges that would allow for versatility (reversible coat) and desired effects (frayed edging). Green-Goff stated, "*La Courtepointe* and the *Femme du Pee Pee* collection was all about taking something old and forgotten and creating something new and beautiful." When reflecting upon her process, Green-Goff said that once she made the decision that she wasn't going to purchase any new/virgin fabrics, it channeled her creativity.

When prompted about discard plan for *La Courtepointe*, Green-Goff said that she hopes that the consumer will keep it and wear it. Referencing the quilted coat, she said that as the original quilt was made from small pieces of fabric, these could

easily be replaced if damaged. She also thought that since the pattern pieces of *La Courtepointe* are fairly large in shape (due to the boxy silhouette), consumers may be able to deconstruct and reuse materials. Another suggestion, made by Green-Goff, was to frame the garment or section of the fabric as a decorative item for the home. This would be a good option (after the garment is no longer wearable or mendable) to keep the memory behind the quilt, for example.

Case Study 2: NEXT

NEXT is a women's wear repurposed design created by Rebecca Nelson (brand name BEX) that was completed in May 2019. This contemporary pant suit was

repurposed from two sets of curtains purchased from a local secondhand retailer ($40.00). *NEXT* features an asymmetrical vest with waistline ruffle, structured mandarin collar, and exposed center front zipper. The armhole area of the vest was created from the top edges of curtains and features metal grommets, through which rope was strung between to close the side area to create an interesting view.

BEX began her design process by learning more about repurposing through research and then developed a vision of the consumer to whom she aimed to sell this ensemble. "*NEXT* targets the female consumer living in a major metropolitan area who wants non-traditional, experimental clothing that is made to last," says BEX. She estimates a resale price point of $300 for this ensemble and envisions *NEXT* will be worn in a professional yet modern business environment. The subtle geometric print of the tan curtain reminded BEX of a modernized cheetah animal print. She imagined a young professional walking to/from work along a busy city street in the summertime and wanted to create a garment that would allow for the breeze to keep her cool in the wild concrete jungle. BEX stated "let the fabric inspire you and let it be what it wants to be." This statement reflects her repurposing process compared to the pro-

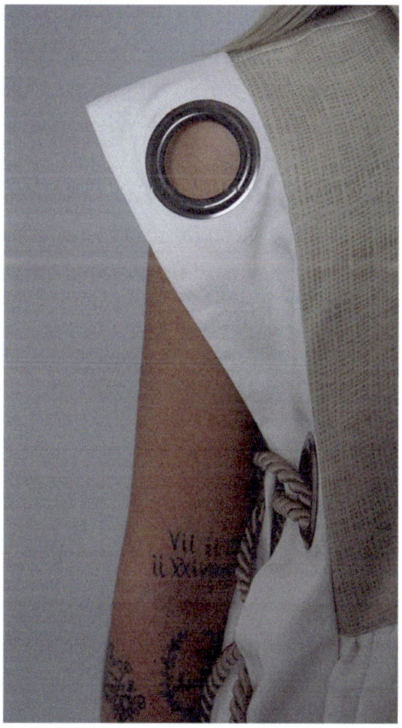

cess she employs when designing from new/virgin materials. BEX expressed how sourcing took more time than buying new yardage from a fabric vendor but in the end created a more interesting garment. After sourcing textile materials, she sketched a few designs, reflecting upon her target consumer's interest in experimental clothing.

From here, BEX developed patterns, created a prototype for proper fit, made adjustments where needed, and completed *NEXT* from the curtains as seen in images. BEX used patternmaking software (Lectra Modaris) to create *NEXT*. While challenging at times, the software versus hand drafting sped up the patternmaking process.

When prompted about how she envisions the consumer will discard of *NEXT*, BEX imagines that this design will be resold to another consumer via secondhand clothing stream. She also noted that *NEXT* has the potential to be repurposed again, but the individual would need to have knowledge and skills in patternmaking and/or construction.

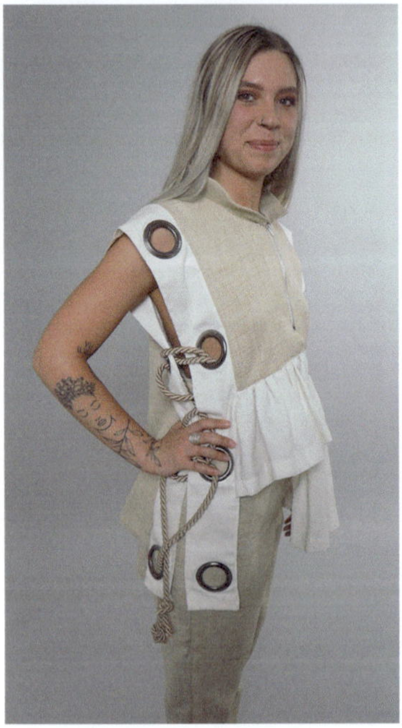

Subtractive Repurposing Case Study Comparative Analysis

Upon review of the design processes by Green-Goff (*La Courtepointe*) and BEX (*NEXT*), it is apparent how garments can drastically differ through the use of interior textiles. Both designers primarily utilized textiles from home textiles including bedding/blankets and curtains. These fabrications allowed the designers to work from a fairly large piece of fabric and treat the product (blanket/curtain) similarly to yardage of material for laying and cutting pattern pieces while incorporating design elements that are identifiable to the original product, such as curtain rod grommets and blanket edging (binding and fringe). In these design examples, both designers found it helpful to identify a target consumer and create a narrative/description of this individual. Crafting the narrative helped designers to source and select materials before creating design sketches for product development. This process assisted the designers in establishing an appropriate retail price point based upon production expenses and targeted retail location.

When selecting materials to use in their designs, neither Green-Goff nor BEX considered how their design would be disposed, initially, but were able to provide some suggestions to prolong product life. Green-Goff suggested the mend-ability of *La Courtepointe*'s quilt-based coat. Repurposing quilts could be advantageous as they are originally made from small fabric pieces. If a garment was soiled or damaged, replacement of a small portion of the quilt may be possible. When analyzing the designs of both *La Courtepointe* and *NEXT*, these products both contain unique components that could be integrated into new products. For example, *NEXT* utilized large grommets found on the upper edge of a curtain. The grommet section of the vest could be deconstructed and sewn into another product, such as a hammock or sunshade (with the addition of fabric in between grommet sections) so that rope or clips could be attached for adjustable use. Applying grommets of large size is a challenging process as it typically requires special equipment/tools for secure clamping of grommet pieces together. Repurposing products that contain pre-existing grommets, for example, into other items is a resourceful way to acquire specialized notions.

Level 2 take-away: Analysis of Green-Goff's and BEX's designs may suggest exploring home textiles for unique textiles and components for repurposing, as they offer large expanses of fabric to cut pieces from.

Section 3: Additive Repurposing

Case Study 1: Altered

Altered is a repurposed design created by Megan Romans that was completed in November 2018. This wearable art design was developed from a discarded bridal

gown purchased from a local secondhand retailer ($10) and donated formal wear alteration scraps from a mass-market bridal retailer. *Altered* was designed and developed as an experimental piece that utilized techniques of textile layering and applique. This design targets a sustainably conscious female teen or young adult

consumer who is seeking a high-quality, unique, and artistic dress to wear for a special formal event. Romans estimates a retail price point of $2500 for the dress and would suggest selling it in a high-end formal wear boutique.

Romans began her design process for *Altered* by conducting research on companies that utilize pre- and post-consumer textile waste in their product offerings as well as examples of designs produced as creative outputs of academic scholarly activity. This research led Romans to experiment with discarded products (such as neckties) to piece materials together and form a new textile fabric, as well as working with available textile scraps from her studio workspace. She was drawn to the gap that exists in sustainable design repurposing works between the pre-consumer waste that is typically connected to apparel industry cutting and manufacturing processes and the post-consumer waste that is associated with apparel products that are discarded because they no longer function or are no longer desired by the original consumer. "Textile waste from the alteration process (i.e., alterations of a formal gown) is an untapped area to explore as a design resource because the waste generated commonly comes from the hem area of a gown – producing similar shaped waste," stated Romans. Once Romans decided to focus her attention on utilizing discarded formal wear hems, particularly bridal gown hems and bridesmaid dress hems, she quickly embraced the textile variations found in 5–7 layered-gowns, such as polyester satin, chiffon, lightweight and heavyweight lining, crinoline, net, tulle, mesh, organza, lace, and taffeta, to separate and intentionally stagger and layer to create hurricane-inspired wave shapes. Before developing the waved shapes, Romans deconstructed the purchased bridal gown into original pattern pieces, re-patterned to fit the needs of *Altered,* and then reassembled to create the dress foundation. She then used larger fabric scraps to created organically shaped textile appliques onto which she stitched the multilayered wave design. After multiple textile appliques were formed, she then secured (by machine and hand stitching) onto the dress foundation.

In reflection on the design process to create *Altered*, Romans thought the applique technique applied atop of the pieced dress foundation was a good way to utilize similarly shaped (hem) waste from the alteration process to create a functional and unique product. However, covering the entire dress with appliques was a very time-intensive process that resulted in a heavy dress. "The technique of using appliques to create an effect which mimics that of a hurricane creates an interesting visual surface design," suggested Romans, "but may be most appropriate and applicable on an individual product-basis (i.e., not at a mass-produced level) due to its time-consuming nature."

When prompted on the disposal or discard plan for *Altered*, Romans reflected about how the original textiles were primarily from bridal dresses and how special event attire may more appropriately be preserved as a memento or family heirloom. Preserving in this manner may encourage reuse of this dress for a different special event by a close friend of family member. Romans also mentioned that deconstruction for redesign of the dress is possible. "Deconstructing the dress along applique lines would allow for the appliques to stay intact as well as the dress foundation – deconstruction for further repurposing would give renewed life to the appliques in an interesting way, such as an interior decorative pillow," stated Romans. She further suggested a more focused textile sourcing process so that all fabrics used would all be of the same fiber content. This would then allow for mechanized textile recycling (hems used in wave appliques were a mix of 100% synthetics (polyester and nylon) and blends of synthetics).

Case Study 2: Kids' Heroes Fight for GREEN

Kids' heroes fight for GREEN

is a repurposed pullover jacket design created by Bingyue Wei in October 2018. This design was developed from 12 bags purchased from a local secondhand retailer

($45.00) and features atypical color combinations. "*Kids' heroes fight for GREEN* targets celebrities who are looking for a bold piece that communicates a message to their social media followers. To these celebrity consumers, price is not as important as the attention they will be able to draw when wearing this eye-catching piece," describes Wei. She estimated a resale price point of approximately $1300.00 for the pullover.

Wei conducted extensive research on sustainable design approaches in creative scholarship. Two areas she identified were underrepresented in creative juried apparel exhibitions included creative works for the male consumer and use of children's items as a material resource. These two factors served as the primary source of inspiration for Wei as she began her creative process. After deciding on the direction of her piece, she develops a color palette and sourced secondhand children's items to include in her design. She was drawn to the bright colors and cartoon graphics found on backpacks, cases (e.g., makeup or pencil), and tote bags and began ideation of silhouettes that would be producible using the structured vinyl/plastic, nylon materials, and notions (webbing) typical to backpacks. Wei sampled production equipment to identify appropriate sewing machine style for stitching structured materials before beginning full-scale work.

One of Wei's biggest challenges she encountered involved equipment that would allow for sewing inflexible materials (such as a long-arm cylinder-bed machine) as compared to a typical straight-stitch machine mounted into a flat table. Wei also identified the challenge of working with used backpacks as they were dirty and peeling/flaking on the interior, due to the fact that many backpacks are treated with a waterproof coating that breaks down over time.

Wei reflected back upon her process and said that she should have sourced truly unusable bags/totes. She explained that the bags she used for the final product were

high quality (unlike the ones she used for her prototype). "Once I cut into these bags, they were no longer usable as a backpack, and a child may have really liked or needed this bag for school – it may have been what his/her family could afford – and now this isn't an option," said Wei. She said that if she were to approach this design again or replicate in the future, she would source bags that were truly unusable or in poor condition.

Additive Repurposing Case Study Comparative Analysis

Analysis of the design processes by Romans (*Altered*) and Wei (*Kids' heroes fight for GREEN*) demonstrates different approaches to "piecing textiles" in order to create a new product. Romans primarily utilized scrap fabric waste from formal wear gown hems to piece a hurricane-inspired surface design into appliques for adhering onto a pieced dress foundation, while Wei deconstructed unwanted bags, cut into desired shapes, and pieced them together to create a lively jacket. Both designers were heavily inspired by textiles. In their processes, both found it helpful to sample piecing techniques prior to full-scale product development. Textiles selected for piecing considerably impact the potential of end product shape in this repurposing level. For example, Wei used structured textiles from backpacks and tote bags that were made of plastic (polypropylene) or vinyl (PVC) which made construction of the jacket challenging, thus leading to the production of an oversized/non-tailored silhouette. In comparison, Romans incorporated a mix of synthetic and blended textiles in varying weights which offered a soft hand and malleable drape to create her formfitting gown design.

Designers provided suggestions for second-life options of their pieces. Romans suggested preservation for reuse or deconstruction along appliques for future product repurposing. With the current popularity of delicate and romantic-inspired fashion (apparel and home décor) products, transforming *Altered* into a resellable alternative product is a viable option. Wei suggested an alternative design approach when working with three-dimensional shapes, such as the backpacks incorporated into *Kids' heroes fight for GREEN*. This alternative approach is called "modular design" which is a design process that allows the consumer to disassemble and change components (reassemble), resulting in a renovated product (Ljungberg 2005). Wei suggested integrating the bags with the jacket body (perhaps with zippers) so that the 3D bags/pockets could be removed and exchanged with different bags or non-3D-shaped pieces to give the jacket a renewed aesthetic. Wei also mentioned the quality of original products that were selected for use in *Kids' heroes fight for GREEN*. Upon analyzing her process and modular design suggestion, along with quality materials for continued use, it would be helpful for the original bag to also function as a backpack. Perhaps the jacket could be sold with the arm straps and padded back section to allow the consumer to utilize all materials from the original item.

Level 3 take-away: Analysis of pieces created by Romans and Wei suggests a focus of "designing for disassembly" in the ideation and development process to reduce unnecessary consumption and waste. Further, selected materials can greatly influence the silhouette of the final garment.

Section 4: Intentional Patternmaking to Repurpose

Case Study 1: Repurposed Bustier

Repurposed Bustier is a repurposed design created by Sunhyung Cho in September 2018. This design was made from a donated size 8 women's suit coat

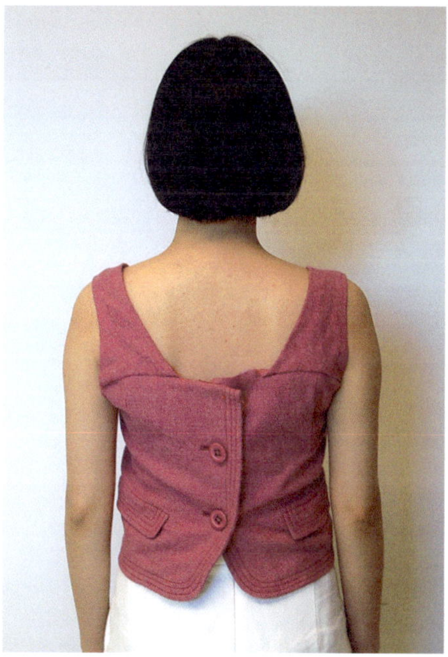

and repurposed into a bustier-style bodice. "*Repurposed Bustier* targets a specialized consumer group who is interested in unique designs and values sustainability," stated Cho. She estimated the price point for *Repurposed Bustier* resale to be around $275 based on the time spent in ideation and development.

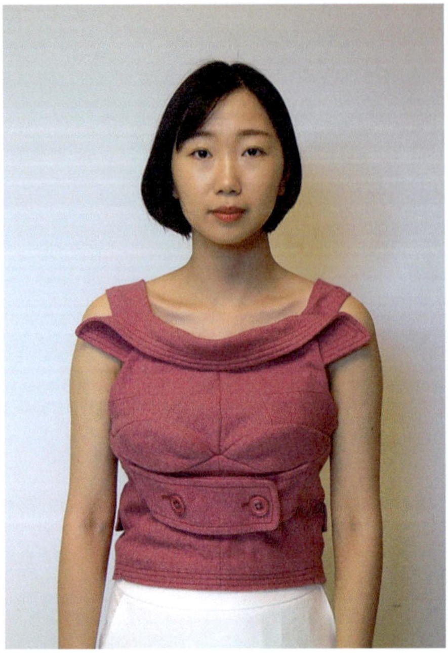

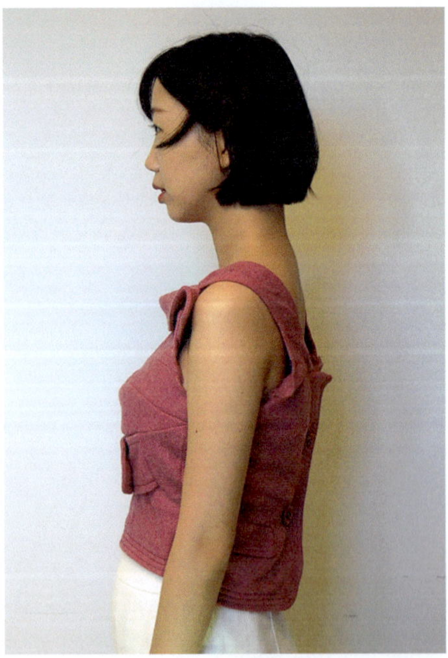

Cho's design process began by conducting research on repurposing from research articles (Paras and Curteza 2018) as well as design inspiration, such as pieces from Maison Margiela, who repurposes RTW pieces into high-end unconventional designs. Once information was gathered, Cho experimented with positioning the original garment's components and design lines in unconventional ways to create a new silhouette. Cho was given a women's magenta suit jacket to repurpose from a friend. Cho spent much time in the ideation phase while working on *Repurposed Bustier* by shifting the different parts of the suit jacket over the dress form to get inspiration for the shape and direction of the design. Once she decided to use the shoulder seam to create a sweetheart design line over the bust area, her design progressed. She used draping techniques to get the initial shape for each pattern piece and then developed more formal pattern pieces to properly cut symmetrical pieces for the right and left of *Repurposed Bustier.* Cho referred to this process as a "jigsaw" approach to patternmaking where she had an idea for placement, draped the garment on the 3D form, restructured as needed, and then flatten to create a "formal" pattern piece. Cho deconstructed many of the seams of this jacket but also utilized original/existing seams (such as the shoulder and center-back seams) to speed construction.

After completion of *Repurposed Bustier*, Cho reflected upon her design process to share some lessons learned. Cho expressed that techniques associated with level 4 repurposing, such as incorporating original features of components of the initial design, are a very creative and challenging process – but pose new possibilities to combat excessive amounts of unwanted garments (i.e., those typically discarded to secondhand retailers or landfills). Cho expressed her challenge with "simply starting" *Repurposed Bustier* by saying how nervous she was about making the initial cut into the fabric – because once it was cut, the available fabric was gone.

When prompted about disposal of *Repurposed Bustier*, Cho initially commented that she didn't think about the next life of the bustier during her process but followed up by commenting on how a consumption and disposal plan should be considered when designing a product, no matter what "life" it is on (original garment or further repurposed). She imagined that when the consumer of *Repurposed Bustier* no longer desired it, the owner would donate or resell for continued use. Cho thought that an online secondhand resale platform would be best suited for *Repurposed Bustier* as it has a unique aesthetic that may be of interest to a specialized consumer group (and those may not be local).

Case Study 2: Suit of the Youth

Suit of the Youth is a repurposed design created by Rachel Eike in May 2019. This design was made from secondhand retail items of one fully lined men's suit jacket (shell = 100% wool; lining = 100% rayon), size 48 L, and a men's pair of slacks (88% wool, 12% silk), size 36 waist. These items were selected for repurposing based upon the abundance of business attire found in secondhand stores (Irick and Eike 2019). "*Suit of the Youth* targets the consumer who values quality materials, high design, and sustainability efforts in their wardrobe and is likely to be of interest to a young professional who desires a casual approach to office attire," describes Eike. *Suit of the Youth* is an active wear-style ensemble, comprised of a jacket, pant, and sleeveless tank. The process of deconstruction through reconstruction of this ensemble took Eike approximately 40 hours to complete. She estimates

approximately $600 as the resale price for *Suit of the Youth* to account for materials, labor, and retail markup.

Eike conducted research on repurposing and decided to make a level 4 repurposed design as she previously created other (level 3) repurposed pieces. After conducting research, she deconstructed major seams of the suit and slacks via separation and pressed seam allowances open to maximize available fabrics. Separation of construction components, such as pockets, was not deconstructed due to small size and distinct shaping of pattern pieces (e.g., welt pockets and handkerchief pocket). The jacket lining was also removed and deconstructed for inclusion in repurposed design. Facings, such as those found in the collar and lapel, were separated along the shoulder seam, and interfacings/stabilizers were removed where possible (some interfacings were adhered to shell fabric via heat-activated adhesive). Eike took measures of this available "yardage" as well as notes on key components (pockets) to incorporate into the new design. From this deconstruction stage, Eike sketched a design plan after researching patternmaking techniques to control excess material. Patterns were created and placed to feature original components such as pockets (10 total) and belt loops. Eike developed a rough lay plan that would allow her to utilize as much of the fabric as possible. She decided to reuse shapes that were present in the original garments such as the curve of the seat seam in the pants, armscye/armhole of the jacket, and sleeve cap. Eike found it most useful to draft patterns while cutting, which allowed her to continually assess available fabric throughout her experimental process. She was able to utilize much of the original garment – producing a minimal waste design.

One of the greatest challenges that Eike encountered was trying to develop a design from garments that contained numerous pockets. She also explained how higher-quality suit jackets will have a variety of stabilizers (between the shell and the lining) and how these impacted the drape of the repurposed product. She experienced challenges associated with fabric blemishes (holes/tears) from wear and age. Eike stated, "It is necessary to locate these wear spots early on in the design process so that pattern shapes can be developed around these flaws."

When prompted about disposal of *Suit of the Youth*, Eike said that her hope is the owner will keep and wear the ensemble for many years or share/donate to friend or family member. This sharing or swapping with close friends may allow for the recirculation of the look back into their wardrobe. Depending on consumer's knowledge and skills, this garment may be disassembled and used for other products – for example, the suit shell (wool) may be felted into another product.

Intentional Patternmaking to Repurposing Case Study Comparative Analysis

Analysis of the design processes by Cho (*Repurposed Bustier*) and Eike (*Suit of the Youth*) provides different patterning approaches to repurposing garments for maximum textile utilization. Cho used a mixed patternmaking approach by piece shaping through draping over a dress form and then formalizing piece shape for mirrored cutting ("jigsaw" approach). Eike primarily used flat-pattern hand-drafting techniques to create pattern pieces but concurrently cut and constructed these pattern pieces to assemble *Suit of the Youth*. Both designers utilized business wear attire, as there is an abundance in the secondhand market stream (Irick and Eike 2019) possibly due to dress code changes in the workplace (Biecher et al. 1999). Designers identified a common design challenge of working with pre-existing components, such as pockets and design aspects (belt-area feature), that must be taken into account when ideating design options.

When prompted about disposal options for their design, both designers suggested donation, resell, or swapping due to the uniqueness/non-traditional interpretations of their business-inspired designs. Upon further design analysis, their suggestions for donation or resell may be connected to their opinions and feelings from cutting into the cloth. Cho stated her anxiousness about making a cut into the suit jacket because once the cut was made, that specific fabric area for pattern availability no longer existed. However, referencing the other levels of repurposing, there are options to re-style (level 1) these looks through surface design and/or embellishment or deconstruct and piece textiles together (possibly incorporating other textile pieces) to create a new material for new design development (level 3). There is also a strong possibility that these garments, particularly *Suit of the Youth* (as it utilized practically the entire jacket and slacks), may be deconstructed and reconstructed into the original design – if desired.

Level 4 take-away: Analysis of Cho's and Eike's designs points out the need for special attention during the sourcing stage of development to identify design components or features in the original product so that these may be taken into account during ideation. This method may save time when existing design components and seams are able to be reused in the new garment.

Cross-Case Analysis

Comparing these four progressing levels of repurposing, one can see that they are quite different in approach and embody techniques that require different skill levels connected to sewing and/or patternmaking (*Research Questions 1 and 2*). In the cross-case analysis, the following research questions are addressed:

3. How do repurposing processes employed by designers compare between the four repurposing levels?
4. How do repurposing processes employed by designers compare to the *repurposing process* [outlined by Irick and Eike 2017]?

In the cross-case analysis, several key themes emerged regarding repurposing processes employed by designers. These themes are explained within the context of the stages of the *repurposing process.* Themes identified in cross-case analysis are (a) process of conducting research on repurposing, (b) process and importance of sampling, (c) utilization of the secondhand market stream to source materials, and (d) disregard for disposal plan. Some themes were common to all levels, while other themes were specific to only certain levels.

The theme that was common among all case studies was the *process of conducting research*, specifically connected to repurposing and/or surface techniques, as the initial step in the development process. Conducting research before beginning any design project, particularly a design that involves the transformation of a pre-existing product into a different product or form, is of utmost importance to properly prepare oneself for design options or challenges ahead. Taking the time to conduct design research and then advancing onto target consumer research are essential steps to ensure high-quality and appropriate end products that also address designer goals – such as repurposing specific textiles, designing for a specific consumer, or working to create minimal waste designs.

Another theme that emerged through cross-case analysis was the *process and importance of sampling*, either small-scale techniques such as trialing different edge finishes or developing a prototype to test construction steps or equipment to ensure efficient full-scale production. Creating a sample or series of samples allows the designer to build confidence in repurposing techniques, troubleshoot design obstacles, and reconfigure deconstruction, cutting, or assembly plan. Ultimately, producing samples improves organization/efficiency while identifying opportunities for innovation (Bailey 1993).

The third theme observed in case study analyses connects to the "sourcing" stage in the repurposing model: focusing on the *utilization of the secondhand market stream to source materials.* Most of the original products used by designers to repurpose were acquired from local secondhand stores or donated from a friend or family member. Utilizing the secondhand market is a viable option for material/textile sourcing (Hawley 2006a) and prevents unclaimed used textile-based products from ending in landfills (Vennström 2012; EPA 2016; Henninger et al. 2016). Designers identified noteworthy challenges and opportunities connected to the selective sourcing of secondhand goods such as unique features or components to incorporate, wear or wornness of a textile (positive view of distinct surface effect; negative view of low quality or damage), and unexpected creative satisfaction from the sourcing process.

The final theme centers on the consumption stage in the life of a product, which ultimately connects to disposal. Many of the designers *disregarded a disposal plan* for the product they created. Irick (2013) initially outlined the consumption phase to consider "end-of-life" options of a consumed product as safe to dispose by consumer, design for disassembly, or return back to the designer for recycling/repurposing. These options or ideas for disposal were prompted to designers during interviews, which lead to further discussion and future opportunities for industry adoption.

Suggestions to Advance "Repurposing" in the Future

This section suggests future directions to advance "repurposing" endeavors for industry and/or academic design scholars. When conducting the case study interviews, each designer was asked for thoughts or suggestions for large-scale repurposing production to combat consumption and waste issues that impact our society and environment. Synthesis of comments outlined the following topics that warrant further research and consideration for industry embrace and implementation.

Consumer Mindsets:

In order for repurposing to be adopted by the industry at a mass-produced level, it needs to be verified that consumers will purchase goods made from used clothing that will yield profit (Sung 2015; Janigo et al. 2017). One area of concern identified in this chapter, and confirmed by Clark (2005), is connected to cleanliness and hygiene of "used" apparel. One suggestion to combat this discomfort is to establish standardized cleaning solutions and accurate marketing materials to ensure consumers of their health safety.

Another area connected to the consumer mindset is related to a relatively high (premium) resale price point of repurposed goods (Fletcher 2008). Consumers need to be willing to spend more money on customized items, simply because costs associated with production are more for the company/brand. When designing with used/worn items, attention to damaged areas (holes or stains) needs to be taken into account. Textile blemishes can impact pattern shapes and the cutting process which in turn results in time-intensive procedures that ultimately influence resale price. Green-Goff commented on how consumers need to change their mindset associated with value, uniqueness, and quality to realize that these factors will cost more money to produce and purchase, compared to mainstream fast fashions.

Restructuring of Donation Process:

One of the biggest challenges Browning-Samoni identified for large-scale industry adoption of repurposing is the need for a reliable supply chain of (used) materials. Thrift or secondhand stores tend to receive more merchandise than they are able to resell within a timely manner which results in many usable/functional garments ending in landfills (Kozlowski et al. 2018; Spring 2019). A formalized clothing donation program was suggested that channel discarded garments in a different way. The program could sort and organize by style, component or trim features, and/or fiber content – then repurposing designers could source more efficiently, which could allow for a more systematic approach to production where multiple layers could be made in one cut. Technology, such as RFID tags in garments, may be used to automate the sorting process. Research is needed to understand the most useful categories for sorting products for successful repurposing. Luu suggested the inclusion of retail stores' unsold product (dead stock) in the donation stream. Acknowledgment was made that government involvement (city, state, or country level) may be helpful to establish and support this restructuring of donation/discard.

Connected to this idea of restructuring how textile-based items are donated is the formation of partnerships. A major obstacle of getting designs manufactured from repurposed materials is that many factories have minimum orders/quantities they require to produce a product. Finding a factory that will adjust their approach to production to include used materials is a challenge for repurposing brands, so forming a partnership to ensure workflow and financial feasibility is key. Green-Goff declared that factories will need to change their model of business in order to accommodate the sustainability-transitioning apparel industry.

Sustainable Apparel Education:

Education was an overarching theme discussed by numerous designers during their interviews. This idea of education spanned from environmental education, to education of garment laundering and care, to sewing-related education of basic mending skills to patternmaking, to surface design/embellishment techniques. Designers commented on how environmental education and topics of sustainability need to be integrated into primary and secondary education levels, particularly with an emphasis on the role of the consumer.

Citizen education related to care and mending was suggested in order to help form a relationship between the consumer and clothing item by being involved "hands-on." With the decline of Family and Consumer Science (FCS) classes, particularly those connected to clothing and textiles (Danovich 2018), consumers are at a loss – not knowing how to perform basic mending skills, such as sewing on a button. Eike suggested efforts to be made to revitalize "making and mending" to put consumers in the role of repurposing their own unwanted clothing. Eike stated, "Every person should have the knowledge and skills to thoughtfully purchase, care for, and re-create."

Piggybacking off of the need for sewing-focused education is the idea for brands to provide instructions for second-life product ideas. Luu suggested that companies could develop a simplified design (or collection of simple design options) with instructions – similar to a pre-purchased pattern envelope for garment construction – that could be sold as part of an original purchase. This would put the consumer in control for giving new life to their old clothing. Educational opportunities may be possible in the form of workshops (physical or digital).

Conclusions

In closing, this chapter provided an introduction to apparel industry challenges and opportunities to embrace repurposing, from a design standpoint. As the majority of discarded textile goods from the pre-consumer production process, alteration process, and post-consumer garment disposal end in landfills, the act of repurposing poses great potential to fuel creativity and mitigate textile waste. As textile products release ozone-depleted methane gases in landfill settings (Blackburn 2009), utilizing these materials in alternative forms (repurposing into different/new products) has the potential to increase the air, soil, and water quality for those areas in direct proximity to a landfill. Ultimately, this means greater quality of life for earthly inhabitants (human or animal). As around 2.35 million tons of waste comes from the clothing and textile industries each year, which estimates about 40 kilograms (or 88 pounds) per person (Vennström 2012), and earth's population is currently around 7.7 billion people, the act of repurposing a single apparel product by each individual can make a staggering positive impact.

While it may be ambitious to declare this target repurposing goal to mitigate textile waste, this chapter provides direction to guide the design process for repurposing at levels appropriate for each consumer's ability and confidence to carry out technique(s). Thus, the consumer may take on the role of a designer. Research has shown that when individuals are involved in the design process of a product (co-design and slow design), they are more likely to keep it, thus diverting it from disposal route, as they have developed an emotional attachment with the item (Cataldi et al. 2010; Niinimäki and Koskinen 2011; Tullio-Pow et al. 2013). This attachment through self-design, in the situation of individual repurposing, may also decrease consumption of new, unnecessary (fast) fashion goods as consumers' functional and creative needs have been met (Joung and Park-Poaps 2013).

The explanatory findings of this chapter added noteworthy value to the area of sustainable fashion design through repurposing. The four levels of repurposing were described through multiple case study design process analyses to achieve the research purpose and provided suggestions for considerations for designers as they approach repurposing while highlighting a variety of techniques for application. Within case analyses of each repurposing level, key "take-away" messages were identified to assist in future design works, such as utilizing non-traditional textiles (e.g., home décor) for unique components, designing with approaches that allow for disassembly and further use (e.g., use of appliques), selecting textile resources that will allow for one-of-a-kind creations (e.g., integration of notions or pockets for function and/or aesthetic), and messaging on the repurposed item to educate the wearer about the positive environmental impact of their action and provide direction for "how to" dispose – suggestions for repurposing, directions for donation, or opportunities to compost or recycle. As the design stage of the product development process, where critical decisions are made involving material selection, design, and production approach, has a direct influence on the final product's environmental impact (approximately 80%) (Curwen et al. 2013), providing practical guidance on the repurposing design process allows readers to weigh options and design with the considerations outlined in this chapter.

Overall, the design process executed by the designers featured in this chapter followed the development process outlined by Irick and Eike (2017), excluding the disposal stage. However, the disposal stage was identified as a theme of important consideration for future design developments, and this area is suggested for future research to better understand consumer's disposal actions and how these actions are influenced by values, beliefs, and norms (Meyer 2013) as well as capabilities related to sewing and patternmaking to repurpose textile items into high-quality products. It is suggested that the outlined "Process for Repurposing" is an operative process that can be employed by those who design with discarded textiles (individual basis or at a scaled production level) and is projected to evolve with future research.

In conclusion, this chapter outlined many areas for scholars and industry members to consider in their own design practice or research for future sustainable fashion activities. The following listing briefly re-caps some of the opportunities for future research: altering consumer mindsets pertaining to utilization of "used" apparel as a resource for repurposed design (i.e., cleanliness) as well as a changed mindset connected to "value" of a well-designed, high-quality, unique product (i.e., high resale price point), even though it may come from "used" materials. Future research and development opportunities also exist involving the supply chain of pre-consumer textile waste as well as secondhand products to more efficiently get desired materials into the hands of designers who repurpose. Lastly, a vast educational opportunity exists to conduct research, specifically related to the consumer to connect apparel fashion actions to environmental impact. This may include education of garment laundering and care to decrease water and energy usage, education pertaining to information communicated on product labels and what this information means for repurposing or recycling/composting potential, education on pre-purchase evaluation of apparel quality and durability to support more sustainable purchasing practices, and sewing-related education to provide skills for basic mending, to pattern-drafting, as well as application of different surface design or embellishment techniques.

Regardless of the design processes employed, the underlying goal remains the same: reduce environmental burden and restore balance.

Appendix A

Key Terms and Definitions

- Design for disassembly: Design products with multiple components to be separated at the end of their useful life into appropriate components (McDonough and Braungart 2002).
- Disposal: Occurs when the garment leaves the possession of an individual (Winakor 1969). Action of disposing or discarding of may involve trashing, donating, swapping, recycling, repurposing, and/or reselling.
- Downcycling: Refers to repurposing higher-value items such as clothing to a lower-value end use, such as car seat filling or cleaning rags (Birtwistle and Moore 2007).
- Pre-consumer textile waste: Excess material produced during the production of yarn, fabric, and textile products (e.g., fabric scraps from cutting; also referred to as "offcuts").
- Post-consumer textile waste: Result of discarded textile products after consumers have purchased and used the item (e.g., used apparel and home textiles).
- Reclaimed products: Any post-consumer product diverted from the solid waste stream (Hawley 2006b).
- Recycling of textiles: Process of returning a textile product back into its original fiber form (Lewis et al. 2017).
- Remnants: Small, unused textiles, typically end of bolt fabric (yardage).
- Repurposing: Act of re-creating an item for another use. May involve approaches or techniques connected to downcycling, upcycling, or re-design.
- Redesign: Redesign may be considered as a form of upcycling, but value added varies by the extent of the product's change, from adding minor details such as decorative trim, to changes in the garment's silhouette, to complete transformation; tailored process to meet market needs (Janigo and Wu 2015).
- Secondhand: Items of personal property that may be donated or sold to new owners. Typically, secondhand apparel is used/worn at least once but may also be items with tags still attached. Venues through which secondhand items may change hands include person to person, at charities, secondhand retailers, consignment shops, thrift stores, yard or garage sales, or online resale platforms.
- Upcycling: Refers to repurposing lower-value items such as a neck scarf to construct a higher-value end use item, such as a wrap skirt or halter top (Janigo and Wu 2015: 77).

Appendix B

External Resources to Learn More about Repurposing

A Designer's Guide to Reconstruction with Wan & Wong Fashion

- www.redressdesignaward.com/learn
- This guide features a systematic making process of a reconstructed design by Wan & Wong Fashion's Kelvin Wan and Joyce Wong through a series of images.

A Review on Upcycling: Current Body of Literature, Knowledge Gaps and a Way Forward by Kyungeun Sung (2015)

- http://irep.ntu.ac.uk/id/eprint/12706/1/219287_PubSub1825_Sung.pdf
- This paper summarizes the current body of literature on upcycling, focusing on different definitions, trends in practices, benefits, drawbacks, and barriers in a number of subject areas, and illuminates knowledge gaps in the area of upcycling.

Fashion Unraveled

- https://exhibitions.fitnyc.edu/fashion-unraveled/
- This exhibition included mended and altered, repurposed, unfinished, and distressed and deconstructed garments.

Study conducted by Yoon Kyung Lee and Marilyn DeLong (2018) Rebirth Product Development for Sustainable Apparel Design Practice in a Design Studio Class in Fashion Practice, 10:1, 34–52

- https://www.tandfonline.com/doi/full/10.1080/17569370.2017.1413797
- The focus of the study was a process that outlined the basic steps to involve students in a meaningful partnership for product redevelopment.

Redress Design Award Reconstruction Tutorial

- youtu.be/YDkO52LG-U8,
- This 4.5-min video gives an introduction to different approaches to reconstruction design and includes interviews with designers who use the technique.

Reconstruction Design Technique Guide

- www.redressdesignaward.com/learn
- This PDF guide introduces different approaches to reconstruction design and includes brands, designers, project examples, and designer case studies.

Trash to Trend by Reet Aus

- https://issuu.com/runnel/docs/reet-aus
- Doctoral thesis.

Appendix C

Research Methods Applied on Case Studies

In this chapter, the differences in repurposing design processes across the levels of repurposing are explained through case study analysis of each designer's individual development process. The case study method of explanatory research aims to address the "how" and "why" throughout a designer's decision-making process and provides a synthesized analysis of experiences across all cases. Implementation of the case study approach in research offers one way to gather and explain a seemingly intuitive process while forming a framework to evaluate current practices (Yin 2009). Design researchers have stated positive implications to employing case study methodology in design scholarship and research (Breslin and Buchanan 2008; Bye 2010; McKinney et al. 2012).

Research Design

The research design (Table 1) employed in this chapter follows the structure outlined by Yin (2009) for explanatory case studies. Each case is the process carried out by the designer resulted in a completed apparel piece from repurposed textiles. The organization for the cases selected follows a multiple-case version of the classical single case study format where multiple cases (repurposing design processes) are presented and analyzed singly through narrative and then compared across the multiple cases for each repurposing level and then across all cases.

Authors addressed the following four problems when developing the research plan for this chapter: (1) what questions to study, (2) what data are relevant, (3) what data to collect, and (4) how to analyze the results (Yin 2009). This plan was structured to ensure the evidence gathered and analyzed addressed the initial purpose. The goal of this research is to expand knowledge in the domain of apparel repurposing and to "illuminate a process or set of processes: why they were taken, how they were implemented, and with what result" (Schramm 1971; Yin 2009: 17). Each case walks the reader through the process employed by the designer

The following points were established to guide case study development and direction for this chapter:

1. Purpose: To explain repurposing design processes executed through select cases at each of the repurposing levels and answer the following research questions:
   (a) What is the creative process employed by a designer when developing a repurposed garment?
   (b) How do repurposing processes employed by designers compare within each repurposing level?
   (c) How do repurposing processes employed by designers compare between the four repurposing levels?

Table 1 Research design

Level of repurposing	Case 1	Case 2	**Cross-case analysis**
1. Re-style to repurpose	On the Fringe	Upside-Down Jean Crop Top	
	Case Study Comparative Analysis		
2. Subtractive repurposing	La Courtepointe	NEXT	
	Case Study Comparative Analysis		
3. Additive repurposing	Altered	Kids' heroes fight for GREEN	
	Case Study Comparative Analysis		
4. Intentional patternmaking to repurpose	*Repurposed* Bustier	Suit of the Youth	
	Case Study Comparative Analysis		

   (d) How do repurposing processes employed by designers compare to the *repurposing process* [outlined by Irick and Eike 2017]?

2. Analysis of cases: Execute case study interview protocol (see Appendix C).
   (a) Description of process executed by the designer.
      - Questions posed to designer follow a gradual increase in complexity (Yin 2009: 89).
        – Questions asked of the specific interviewee.
        – Questions asked of the individual case.
        – Questions asked of discovered patterns across multiple cases.
        – Questions asked of the entire study.
        – Normative questions about procedural recommendations and conclusions.
   (b) Detailed description of design/artifact.
   (c) Creation of case study database.
3. Logic linking:
   (a) Connecting what is known from literature to process of each designer.
   (b) Criteria for interpreting findings:
      - Analyze reports from case study database.
        – Synthesize interview narratives and descriptions of designs.
      - Comparison of works within and between repurposing levels.

Construct validity, internal validity, external validity, and reliability were all considered during the design of this case study research protocol. Table 2 outlines these criterion items and how each was addressed in this chapter. These items may be helpful to evaluate the quality of research design.

An operational set of measures (interview questions) were developed to match the purpose of this explanatory process and concepts embodied in this chapter (see Appendix C). In addition, a cross-case synthesis was conducted across the four repurposing levels to identify interesting processes, techniques, and considerations

Table 2 Research design criteria for repurposing case study

Criteria	Criteria description	How addressed in case study	Chapter connection
Construct validity	Identifying correct operational measures for the concepts being studied	Definition of specific concepts that are related to purpose and questions of the study; operational measures outlined in protocol match concepts	Section: Case study interview protocol (appendix C); questions posed to interviewees labeled RPQ in protocol are directly connected to phases in the repurposing process diagram; assist in explaining process and comparing between designs and repurposing levels
Internal validity	Seeking to establish a causal relationship, whereby certain conditions are believed to lead to other conditions	Cross-case synthesis in one way to combat independent variable challenges commonly experienced in explanatory case studies; synthesis of findings across multiple cases was performed to develop plausible and fair conclusions supported by data	Section: Cross-case analysis and discussion
External validity	Defining the domain to which a study's findings can be generalized	A theory (whether established or evolving) must be confirmed through finding replication in multiple scenarios; in this chapter, ten repurposed pieces were analyzed using the *repurposing process*	Section: Artifact case studies, subsection – repurposing process analysis within level
Reliability	Demonstrating that the operations of a study can be repeated, resulting in similar outcomes	Tactics to combat documentation challenges that connect to reliability: (1) use a case study protocol and (2) develop a case study database	Section: Case study interview protocol (appendix C) development and execution; database organized (original resources from designers); objectives, issues, and relevant readings about topic included in chapter

identified by the designers interviewed to draw conclusions that may impact future repurposing research, creative scholarship, model developments, or industry-level adoption. Comparison of case study report to the broader concept of repurposing (in apparel) is replicated within each level covered in this chapter. These findings can be generalized to the domain of sustainable apparel design, specifically repurposed apparel

Selection of Cases

Repurposing design process selection for feature in this chapter was performed by the authors. Featured repurposing design processes are works connected to educational courses or personal creative scholarship activities of the authors. Authors gathered and reviewed repurposing design processes and categorized into the four repurposing levels. Following the categorization, each repurposing design process was reviewed for techniques employed and materials used in their process in order to provide the reader with a variety of examples to reference for their own repurposing design exploration.

Data Collection

Data collection was performed through multiple sources which included designer interviews and repurposing design process analyses within each of the repurposing levels and between all levels. An operational set of measures (interview questions) were developed to match the purposes outlined in this chapter. Designers of repurposing design processes were contacted for interview invitation to gather details on their creative process and reflection of repurposing experience. Documentation of case study interview and development of a reference database were developed to assist in the organization and reference of materials during analysis.

Interpretation of Findings

Research Question (1) What is the creative process employed by a designer when developing a repurposed garment? The creative process employed by the designer to develop a repurposed garment is described in detail in each case study. Each step in the *Process for repurposing* (Irick and Eike 2017) is systematically addressed, to build a complete understanding of each designer's creative process.

Research Question (2) How do repurposing processes employed by designers compare within each repurposing level? Within each level (at end of each section), the two cases were compared to the broader concept of repurposing (in apparel). This step was replicated within each level of repurposing covered in this chapter.

Research Questions (3) How do repurposing processes employed by designers compare between repurposing levels and research question and *(4) How do repurposing processes employed by designers compare to the repurposing process?* Answers to these research questions were based on the cross-case synthesis all repurposing design processes across the four repurposing levels. Interviewees' common and unique processes, techniques, and considerations for future repurposing works were analyzed within and between repurposing levels. Specific themes and future research suggestions emerged.

Appendix D

Case Study Interview Protocol

1. Gather contact information from case (artifact) designer – schedule interview.
   (a) Level 1 questions:
   - Name of design.
   - Preferred name as designer to be referenced in chapter.
   - Date in which work was completed.
   - RPQ: Fiber content of repurposed materials used.
   - RPQ: Targeted audience for design (brand, retail price point, etc.)
   - RPQ: Estimated time spent (from start to finish) to complete.
     – Labor, materials, customer expectations (RPQ: cost evaluation).
2. Collect permission to use work(s) in book chapter [copyright form].
3. Ask targeted question pertaining to design process and considerations.
   (b) Level 2 questions:
   - Walk the reader through discovery from implementation research.
   - Sampling of techniques.
     – Fabrication/surface design techniques (RPQ).
   - Description of full-scale design approaches.
   - Detailed images of completed textile-based products.
   - Reflection on personal design/case.
     – Learned experiences.
     – Barriers.
     – Provide suggestions for individual or industry employment of repurposing process.
     – RPQ: Suggestion for consumption and/or disposal of product (reference repurposing process: safe to dispose by consumer, design for disassembly, recycle back to designer for recycling or repurposing).

   (c) Level 3 question:
   - Describe usefulness and challenges of techniques (referencing level of repurposing employed and research conducted by designer).

   (d) Level 4 question:
   - Express potential industry adoption of repurposing process overall (provide examples of different repurposing level works).

(e) Level 5 questions:

- Detail procedural recommendations for large-scale adoption of repurposing (level experienced in designer piece or from any level).
- Identify future research direction for repurposing.

4. Analyze design using the repurposing process model.
5. Write report of each case study design.

Note: *RPQ* repurposing process Question (connected to repurposing process phases).

References

Bailey T (1993) Organizational innovation in the apparel industry. Ind Relations A J Econ Soc 32:30–48

Biecher E, Keaton PN, Pollman AW (1999) Casual dress at work. SAM Adv Manag J 64(17)

Birtwistle G, Moore CM (2007) Fashion clothing – where does it all end up? Int J Retail Distrib Manag 35:210–216. https://doi.org/10.1108/09590550710735068

Blackburn R (2009) Sustainable textiles: Lifecycle and environmental impact. Woodhead Publishing, Oxford

Breslin M, Buchanan R (2008) On the case study method of research and teaching in design. Des Issues 24:36–40

Brown S (2010) Eco fashion. Laurence King Publishing Ltd., London

Bye E (2010) A direction for clothing and textile design research. Cloth Text Res J 28:205–217. https://doi.org/10.1177/0887302X10371505

Carrico M, Kim V (2014) Expanding zero-waste design practices: a discussion paper. International Journal of Fashion Design, Technology and Education 7(1):58–64. https://doi.org/10.1080/17543266.2013.837967

Cataldi C, Dickson M, & Grover C (2010) Slow Fashion. Tailoring a strategic industry approach towards sustainability. Slow fashion forward. Retrieved from http://www.slowfashion.at/en/about-us.html

Chanin N, Stukin S, Rausch R (2008) Alabama stitch book: Projects and stories celebrating hand-sewing, quilting, and embroidery for contemporary sustainable style. Stewart, Tabori & Chang, New York

Clark H (2005) Second hand fashion, culture and identity in Hong Kong. In: Palmer A, Clark H (eds) Old clothes, new looks: second hand fashion. UK, Oxford, pp 155–172

Curwen LG, Park J, Sarkar AK (2013) Challenges and solutions of sustainable apparel product development: a case study of Eileen Fisher. Cloth Text Res J 31(1):32–47. https://doi.org/10.1177/0887302X12472724

Danovich T (2018) Despite a revamped focus on real-life skills, "home ec" classes fade away. In: NPR. https://www.npr.org/sections/thesalt/2018/06/14/618329461/despite-a-revamped-focus-on-real-life-skills-home-ec-classes-fade-away.

Dunn J (2008) ReFashion reDunn. Massey University, Wellington

Eike RJ, Myers B, Sturges D (2018) The impact of service-learning targeting apparel design majors: a qualitative analysis of learning growth. Fam Consum Sci Res J 46(3):267. https://doi.org/10.1111/fcsr.12250

EPA (2016) Advancing sustainable materials management: 2014 fact sheet. United States Environmental Prot. Agency, In. https://www.epa.gov/sites/production/files/2016-11/documents/2014_smmfactsheet_508.pdf

EPA (2019) Facts and figures about materials, waste, and recycling: textiles: material-specific data. United States Environmental Prot. Agency, In. https://www.epa.gov/facts-and-figures-about-materials-waste-and-recycling/textiles-material-specific-data.
Fletcher K (2008) Local and light. In: Sustainable fashion and textiles design journeys. Earthscan, London, pp 137–159
Fletcher K (2016) Craft of use: post-growth fashion, 1st edn. Routledge, New York
Fletcher K, Grose L (2011) Fashion & sustainability: design for change. Laurence King Publishing Ltd., London
Fratzke D (1976) Clothing values as related to clothing inactivity and discard. Iowa State University, Ames, IA
Gam HJ, Cao H, Farr C, Heine L (2009) C2CAD: a sustainable apparel design and production model. Int J Cloth Sci Technol 21:166–179. https://doi.org/10.1108/09556220910959954
Gordan JF, Hill C (2015) Sustainable fashion: past, present, and future. Bloomsbury Academic Publishing Plc, London
Hawley JM (2006a) Textile recycling: a system perspective. In: Wang Y (ed) Recycling in textiles. Woodhead Publishing Limited in association with The Textile Institute, Cambridge, pp 7–24. https://doi.org/10.1533/9781845691424.1.7
Hawley JM (2006b) Digging for diamonds: a conceptual framework for understanding reclaimed textile products. Cloth Text Res J 24(3):262–275. https://doi.org/10.1177/0887302X06294626
Henninger CE, Alevizou PJ, Oates CJ (2016) What is sustainable fashion? J Fash Mark Manag 20:400–416. https://doi.org/10.1108/JFMM-07-2015-0052
Irick E (2013) Examination of the design process of repurposed apparel and accessories: an application of diffusion of innovations theory. Oklahoma State University, Stillwater
Irick E, Eike RJ (2017) Teaching the repurposing mindset: the introduction of a repurposing project into an advanced apparel construction course. In: International Federation for Home Economics Conference. Sligo, Ireland
Irick E, Eike RJ (2019) Analysis of the availability of second-hand clothing as the raw materials for repurposing. In: Regent's and International Textile & Apparel Association Joint Conference. London
Janigo KA, Wu J (2015) Collaborative redesign of used clothes as a sustainable fashion solution and potential business opportunity. Fash Pract 7:75–97. https://doi.org/10.2752/175693815X14182200335736
Janigo KA, Wu J, Delong M (2017) Redesigning fashion: an analysis and categorization of women's clothing upcycling behavior. Fash Pract 9:256–281. https://doi.org/10.1080/17569370.2017.1314114
Joung H-M, Park-Poaps H (2013) Factors motivating and influencing clothing disposal behaviours. Int J Consum Stud 37:105–111. https://doi.org/10.1111/j.1470-6431.2011.01048.x
Kim DH, Kim MS (2013) Perception, purchase behaviors of and the buying motives toward secondhand clothing products. Res J Costume Cult 21:324–337. https://doi.org/10.7741/rjcc.2013.21.3.324
Kozlowski A, Searcy C, Bardecki M (2018) The reDesign canvas: fashion design as a tool for sustainability. J Clean Prod 183:197–207. https://doi.org/10.1016/j.jclepro.2018.02.014
Lewis TL, Park H, Netravali AN, Trejo HX (2017) Closing the loop: a scalable zero-waste model for apparel reuse and recycling. Int J Fash Des Technol Educ 10:353–362. https://doi.org/10.1080/17543266.2016.1263364
Ljungberg LY (2005) Responsible products: selecting design and materials. Des Manag Rev 16(64)
McDonough W, Braungart M (2002) Cradle to cradle: remaking the way we make things. North Point Press, Wellington
McKinney EC, Bye E, Labat K (2012) Building patternmaking theory: a case study of published patternmaking practices for pants. Int J Fash Des Technol Educ 5:153–167. https://doi.org/10.1080/17543266.2012.666269

McQuillan H (2011) Zero-waste garment design: strategies and risk-taking for design practice. In: Gwilt A, Rissanen T (eds) Shaping sustainable fashion: changing the way we make and use clothes. Earthscan, London, pp 83–99

Meyer J (2013) The role of values, beliefs and norms in female consumers' clothing disposal behaviour. University of Pretoria

Niinimäki K, Koskinen I (2011) I love this dress, it makes me feel beautiful! Empathic knowledge in sustainable design. Des J 14(2):165–186. https://doi.org/10.2752/175630611X12984592779962

Paras MK, Curteza A (2018) Revisiting upcycling phenomena: a concept in clothing industry. Res J Text Appar 22:46–58. https://doi.org/10.1108/RJTA-03-2017-0011

Park H-H, Choo T-G (2012) Influence of consumer self-confidence and self-confidence in fashion coordination on attitude toward secondhand fashion goods and purchase intention of secondhand fashion goods. J Korean Soc Cloth Ind 14:544–553. https://doi.org/10.5805/KSCI.2012.14.4.544

Schramm W (1971) Notes on case studies of instructional media projects. Washington, DC.

Spring A (2019) Marie Kondo, you know what would spark joy? Buying less crap, Guard

Sung K (2015) A review on upcycling: current body of literature, knowledge gaps and a way forward. Venice, Italy

Tullio-Pow S, Schaefer K, Rajanayagam K, Navarro H (2013) Community co-design: from magic squares to magic dresses. In: Pedagogy track # 70. International Textile and Apparel Association, New Orleans, pp 53–54

Vennström K (2012) Sustainable fashion consumption: an interactive system between consumers and institutions. Centrum för modevetenskap, Stockholms universitet

Winakor G (1969) The process of clothing consumption. J Home Econ 6:629–634

Yin R (2009) Case study research: design and methods, 4th edn. Sage Publications, Inc., Thousand Oaks, CA

Young C, Jirousek C, & Ashdown S (2004) Undesigned: A study in sustainable design of apparel using post-consumer recycled clothing. Clothing and Textiles Research Journal, 22(1/2), 61–68. https://doi.org/10.1177/0887302X0402200108

Zaffalon V (2010) Climate change, carbon mitigation and textiles. Text World 16:34–35

Doodlage: Reinventing Fashion Via Sustainable Design

Sheetal Jain

Abstract Doodlage, a fashion and lifestyle brand, was originated in 2012 with a simple notion to create exclusive eco-friendly items, with high-style quotient. Kriti Tula, founder of Doodlage, perceived this idea when she noticed massive stack of textile waste while doing her internship at an export house. The brand was born from her love for planet. She thought of reinventing the rejected fabrics and put them to better use. As she was aware of the fact that fashion industry is a major contributor to textile waste that ends up in landfill, she decided to start upcycling this waste to create sustainable fashion. The concept of upcycling provides Doodlage an opportunity to reprocess waste, recover intrinsic value through recycling, and optimize the end-of-life processes toward zero-waste systems. Doodlage tries its best to minimize its production waste, and whatever is left is reused in making bags and home furnishings. The brand emphasizes on developing sustainable business practices at each stage of the fashion supply chain from procurement of raw materials to the disposal of clothes by the consumers.

This case study focuses on the company – *Doodlage* – and *The Four Actions Sustainable Fashion Value* (FASFV) *framework* that emphasizes on elimination, reduction, creation, and rise of specific value proposition factors for transition toward sustainable fashion economy. The FASFV framework can be applied as an audit instrument to help fashion companies to develop sustainable business practices at each stage of the fashion value chain.

Keywords Circular economy · Fashion · India · Sustainable · Upcycle · Zero waste

---

S. Jain (✉)
Founder & CEO Luxe Analytics, Delhi, India

Associate Professor, IILM, Delhi, India

S. S. Muthu, M. A. Gardetti (eds.), *Sustainability in the Textile and Apparel Industries*, Sustainable Textiles: Production, Processing, Manufacturing & Chemistry, https://doi.org/10.1007/978-3-030-37929-2_10

Introduction

The current fashion industry is linked with a linear model, which is based on take-make-dispose rationale (Ellen MacArthur Foundation 2013a, b). Therefore, there is a need for a new economic model, "circular economy," that emphasizes on the notion of restoration and regeneration in place of traditional end-of-life concept (Ellen MacArthur Foundation 2013a, b). In this study, author developed "The Four Actions Sustainable Fashion Value" (FASFV) framework that emphasizes on elimination, reduction, creation, and rise of specific value proposition factors for transition toward sustainable fashion economy.

This paper presents the company, "*Doodlage*," a fashion and lifestyle brand, born with a philosophy to create sustainable upcycled clothing, to protect the mother earth. The FASFV framework is used as a tool to analyze how "Doodlage" works on the principle of building sustainable fashion model throughout its value chain, from sourcing environment-friendly fabrics to innovative designing and from green packaging to offering repair services to customers to prolong end-of-product life.

Methodology

This paper adopts a qualitative exploratory research design to get an in-depth understanding of the relatively unexplored area. Present research uses case study method since it has been found appropriate for sustainability research as argued by Evans (2011: 61) "given the nature of much sustainability research, which looks at cutting-edge ideas, projects and practices, the case study method offers a methodological approach that allows the researcher to make confident claims of potentialities, causality or development."

This case study was conducted in two phases. In the first phase, bibliographic compilation on the given subject was done. In the second phase, data about case company was collected through trade media as well as through semi-structured interviews with founder and designers.

Current State of Fashion Industry

Fashion has been defined as "…a broad term that typically encompasses any product or market where there is an element of style that is likely to be short-lived" (Christopher et al. 2004:367). The fashion industry is the second most polluting industry worldwide. The industry is projected to consume 25% of the world's carbon budget by 2050 (Ellen MacArthur Foundation 2017). With two billion more people ready to join the global middle class by 2030 (World Economic Forum 2018), these challenges will continue to grow unless fundamental changes are made in the operating structure of the fashion industry. Today, people buy much more clothes than they use. Fast fashion business models dominate, and a lot of clothes are either stored in wardrobes or are being dumped (Rydberg 2016). "There is an unhealthy 'throwaway' consumer culture that fosters overconsumption and waste. Consumers are becoming increasingly accustomed to cheap, poor-quality fashion that they can throw in the garbage after a few washes" (Gjerdum Pedersen and Reitan Andersen 2013: 3).

In the last 15 years, production of clothes has almost doubled (Remy et al. 2016), mainly due to the growth of middle-class population and increased per capita sales across the globe. There has been a rise in "fast fashion" phenomenon with rapid turnaround of new styles and increased number of collections offered every year at lesser prices. It is projected that more than half of fast fashion produced is disposed of in less than a year (McKinsey and Company 2016). The present clothing system of manufacturing, distributing, and using clothes functions in linear way which puts pressure on resources, environment, and its ecosystem (Ellen MacArthur Foundation 2017). Each year, less than 1% of materials used to produce clothing is recycled into new clothes, leading to a loss of more than USD 100 billion worth of materials (Ellen MacArthur Foundation 2017). Around 92 million tons of waste is produced by the fashion industry every year and is expected to further rise by around 60% between 2015 and 2030 (Boston Consulting Group (BCG) and Global Fashion Agenda (GFA) report 2017). The major chunk of textile waste ends up in a landfill or is incinerated.

Transition toward Circular Economy: A Sustainable Future

The present socioeconomic system is based on a linear economy, wherein companies create products and consumers use and dispose (Michelini et al. 2017). The linear model suffers unnecessary material losses like excessive energy use, end-of-life waste, and erosion of ecosystems. Therefore, there is a grave call for new economic model – "circular economy" – which brings the notion of restoration and regeneration in place of traditional end-of-life concept (Ellen MacArthur Foundation 2013a, 2013b).

Moving from current linear economic system of "take-make-use-dispose" to closed-loop circular economy (CE) requires embedding the 3R principles (reduce, reuse, and recycle) into production and consumption process (Zhu and Qui 2007). In CE, the life of a product is elongated through reuse, repair, remanufacture, refurbish, redistribution, and recycling, thereby increasing resource efficiency and reducing the need for new products and virgin raw material. It involves incorporation of strategies that stimulate new consumption patterns and establish new business models (Shikaleska et al. 2017). Reformation is required at each phase of value chain starting from product design phase to production, distribution, and consumption. It is vital to reinvent how items are designed and produced, rethink how items are used and consumed, and redefine how items are reused and recycled (Moorhouse and Moorhouse 2017).

As per a recent Accenture Strategy Report (2018), with about two-thirds of consumers across the globe preferring to purchase goods from firms that are purpose driven, it becomes imperative for businesses to relook at their business models and make it more sustainable and eco-friendly in order to survive in long term (Kim 2010). Advanced technology, infrastructure developments, new circular design practices, changing consumption patterns, and growing regulatory pressure are building ecosystem where established fashion brands can pursue circular economy initiatives (Accenture and Fashion for Good Report 2019).

Circular Business Models for Sustainable Fashion Economy

Osterwalder and Pigneur (2010, p.14) described business model as "The rationale of how an organization creates, delivers, and captures value." A sustainable business model is defined as a business model that builds a competitive advantage through superior customer value and contributes to the sustainable development of the organization as well as society (Lüdeke-Freund 2010). Transformation in business models is required to attain systematic change toward sustainability (Gardetti and Muthu 2015).

Osterwalder and Pigneur (2010) pointed nine basic elements of a business model, namely, customer segments, value propositions, channels, customer relationships, revenue streams, key resources, key activities, key partnerships, and cost structure. These nine building blocks can be combined in innovative ways to develop circular business model in fashion industry. The Ellen MacArthur Foundation (2013a, b) and Nguyen et al. (2014) indicated four value-generating principles that firm could adopt when designing their circular fashion business models. First, the "power of the inner circle" is concerned about keeping products alive and using it for as long as possible by original owner through repair and maintenance. Second, the "power of circling longer" refers to keeping items in as many consecutive cycles as possible and prolonging the time of each cycle. Third, the "power of cascaded use" refers to the idea of reusing products and materials within and between industries. For instance, apparel can first be reused in the clothing industry as a secondhand apparel and then can be used in the furniture industry as upholstery. Lastly, the "power of pure circles" highlights the significance of unadulterated material steams, since this is the key to conserving the quality of the materials for many consecutive cycles.

Accenture (2014: 13–14) identified five different types of circular economy models, namely, circular supplies, resource recovery, product life extension, sharing platforms, and product as a service. First, the *circular supplies* business model is about phasing out scarce resources by using fully renewable, recyclable, or biodegradable resources. Second, *resource recovery* is regarding seizing embedded value at the end of one product life cycle to feed into another through innovative recycling and upcycling services. The third business model is product life extension which is concerned with extending the life cycle of products by repair, upgrade, or remanufacture. Fourth, the sharing platform business model promotes collaboration among the product users. Lastly, the *product as a service* business model provides products through lease or pay-for-use arrangements.

Fashion industry can adopt these business models and contribute in the shift toward sustainable fashion economy.

Stubbs and Cocklin (2008: 103) stated that "Understanding of sustainable business models and how sustainable development is operationalized in firms is weak." Schaltegger et al. (2011: 12) further pointed that "Neither theoretical nor empirical research offers sufficient answers to the question what a sustainable business model might be." Therefore, this study aims to bridge this lacuna in the existing literature. It tries to explore the potential for building sustainable business model for fashion industry through the analysis of a fashion startup, "Doodlage," founded by young entrepreneur Kriti Tula, to offer sustainable, upcycled clothing, with a notion to save planet. Doodlage was selected for analysis since it incorporates principles of sustainable design concept throughout its business model, from sourcing environment-friendly fabrics to innovative designing and from green packaging to offering repair services to customers to prolong end-of-product life.

The Four Actions Sustainable Fashion Value (FASFV) Framework

This study dwells on The Four Actions framework developed by Kim and Mauborgne (2005). The study emphasizes on elimination, reduction, creation, and rise of specific value proposition factors for transition toward sustainable fashion economy (Figure 1). The following section will discuss how each action will result in creating value in fashion industry.

Eliminate "Single-Use" Economy

The present system of production, distribution, and consumption of clothing works entirely in a linear way. Huge quantities of nonrenewable resources are extracted to produce garments that are used only for a short period after which the resources are mainly lost to landfill or incineration (Ellen MacArthur Foundation 2017). The environmental footprint is growing due to pollution and waste.

The main concept embraced in circularity is to significantly reduce the production and consumption of raw materials in combination with a strategy to repair, recycle, and reuse resources from waste (Ho and Choi 2012). Designing and producing garments of superior quality and providing access to them via new business

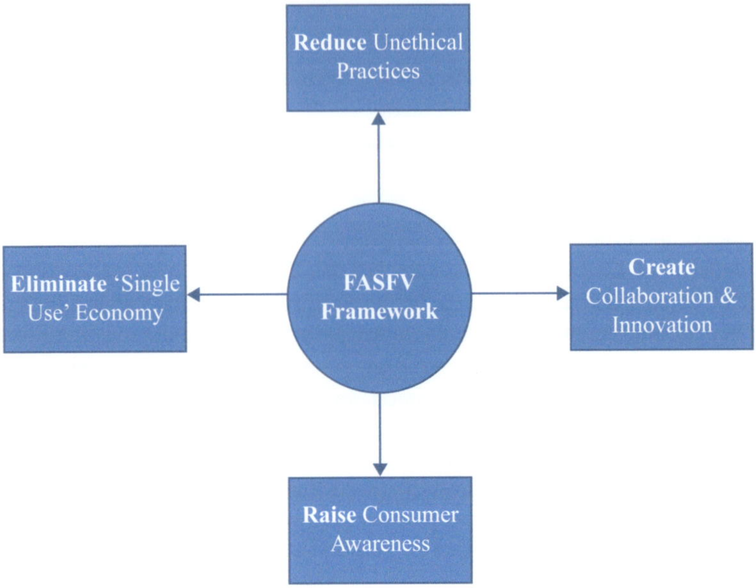

Fig. 1 The Four Actions Sustainable Fashion Value (FASFV) Framework. (Source: Designed by author adapted from Kim and Mauborgne (2005))

models would help shift the perception of clothing from being a disposable product to being a durable item (Ellen MacArthur Foundation 2017).

Sustainable fashion concept has led many designers to rethink innovative ways to design. Material is the starting point to embrace changes in fashion industry. Varied natural fibers, such as soybean, banana, corn, bamboo, and hemp, and biodegradable artificial fibers are required to be adapted by designers to minimize the burden on the environment. Brands should design with the end use in mind. They should think about how a product will be cycled at the end of use (Segran 2019). There is a need for paradigm shift in the way products are procured, designed, manufactured, distributed, consumed, and disposed. Sustainability and closed-loop thinking should be positioned at the heart of each business model (Preston 2012).

Reduce Unethical Practices

Palomo-Lovinski and Hahn (2014, p.87) argued "Sustainable practices in clothing have not, thus far, created a significant impact…and that the fashion industry continues to work in an inefficient manner that creates massive waste and exploits workers." All through the value chain, fashion uses various resources from the planet and society that often result in negative impacts (Hvass 2016). For example, key social concerns linked with fashion production are use of child labor, unfair wages, unhealthy working environment, and exposure to harmful chemicals (Krüger et al. 2012; Allwood et al. 2008). In Bangladesh, majority of garment workers work for 14–16 h shift for each day with wages of USD 68 for a month (Stotz and Kane 2015). Although various initiatives have been taken to address these social issues, still unethical practices remain part of the fashion industry (Hobbes 2015).

Furthermore, there are various environmental issues that occur along the fashion supply chain, like unwarranted use of energy, water, and toxic chemicals (Hvass 2016). For example, to produce a pair of jeans, 3625 liters of water, 3 kg of chemicals, 400 MJ of energy, and 16 m^2 of harvested land are required (Deloitte 2013). To address these unethical practices and achieve sustainable fashion economy, Huber (2000) pointed three strategies: sufficiency, efficiency, and ecological consistency. Sufficiency focuses on reduction in resource use through alteration in consumption patterns (Lüdeke-Freund 2009). It supports business model that enable longevity, repair, resell, lease, and other forms of sustainable fashion consumption (Jain and Mishra 2020; Tukker and Tischner 2006). Efficiency emphasizes on the process of production and use of materials with objective to minimize environmental impact linked with production of each unit of output. Ecological consistency deals with products based on cradle-to-cradle material and design choices so that materials stay in circular material flows (McDonough and Braungart 2013).

Stahel (1994) pointed that reuse and recycling strategies for waste minimization would result in a more sustainable and resource-efficient society. Hethorn and Ulasewicz (2008) stated, "*New* concepts are needed that embrace a rethinking of the process of garment creation, use, and disposal, re-creation, or reuse with the focus on extending the life span of products and the meaning they bring." With growing concern among consumers regarding the influence of their purchases on people and planet, organizations need to realize the benefits and unexploited economic potential of efficient use of waste (Moorhouse and Moorhouse 2017). The focus should be to design with minimal waste. Zero-waste fashion design addresses ineffectiveness in use of fabric by reframing fabric waste and exploring opportunities to discover it in new forms (Rissanen and McQuillan 2016).

Create Collaboration and Innovation

Transformation of fashion industry requires system-level change with an unprecedented degree of commitment, collaboration, and innovation. There is a need to adopt comprehensive, broad, and integrated approach to design and production in order to create sustainable design and innovation across the complete industry (Rosen and Kishawy 2012). Innovation offers the means to build a new reality. Fourth Industrial Revolution (4IR) technologies will permit track ability and traceability of garments beyond the point of sale, thereby allowing authentication, resale, and material recovery (Lacy et al. 2018).

Establishing new or closer collaboration with stakeholders within or beyond the traditional supply chain is key in creating sustainable fashion economy (Neergaard et al. 2009). For example, Sustainable Apparel Coalition, with members such as

Nike, Marks & Spencer, Patagonia, Levi Strauss, Walmart, etc., can exert pressure on suppliers by requesting more ethically produced items (Rydberg 2016) like organically grown cotton and new materials with lower environmental impact.

Collaboration provides various benefits to partners through resource sharing, product or service innovation, or access to new markets (Googins and Rochlin 2002). Collaboration plays a vital role in current reuse and recycling initiatives of fashion (Hvass 2016). For example, fashion companies partner with charity organizations to facilitate the collection of used clothes or defective collections. Similarly, fashion brands implement in-store take-back schemes in collaboration with third-party partners. For example, Danish fashion brand, Jack and Jones, partners with the global textile and shoe collection company I:CO to collect used clothing and shoes and give them new life through reuse or recycling (Hvass 2016).

Raise Consumer Awareness

Research suggests that customers are unaware of the need for garment recycling and there is general lack of knowledge among consumers regarding varied textile reuse and recycling possibilities (Joung 2014; Ekström and Salomonson 2014). Fashion companies need to get closer to their consumers in order to understand, influence, and satisfy them. A new approach is required to build customer relationship that aims to make customers involved and responsible partners in the value chain processes (Lüdeke-Freund 2009). For example, Patagonia's Common Threads Initiative requests customers to take a pledge and become partners in the initiative to reduce consumption and keep the products out of landfills while providing various services that equip consumers to reuse and recycle their products (Hvass 2016).

Ethical consumption requires customers to purchase less, use products for longer duration, and generate less waste (Jain 2019; Tilikidou and Delistavrou 2004). This calls for radical transformation in present consumption practices, such as from consumption of resource-intensive products to high-value services (Sustainable Consumption Roundtable 2006). Several product-service system (PSS) business models have recently emerged (Reim et al. 2015), for example, the Albright Fashion Library in New York provides apparels and accessories on membership basis for reasonable fees. Product swapping and consultancy services have also arisen, such as ClosetDash, where the consumers can pay on an hourly basis, for getting one-on-one styling advice in person or online. These services facilitate consumers to learn to use already available clothes in different ways rather than buy new ones (Armstrong et al. 2015). However, a shift from ownership to use-oriented economy needs significant change in consumer lifestyles, values, and daily routines (Billharz and Cerny 2012).

Today, consumers demand for newness, variation, and style in fashion. Therefore, new business models like rental, recommence, sharing, etc. can cater to these demands in an environment-friendly way. These models require radical shift in consumer behavior to achieve scale. As per Accenture and Fashion for Good Report (2019), retailers can overcome this challenge by incentivizing consumers to act differently (e.g., discount vouchers for resale) and ensuring effortless customer experience through investment in the front-end customer interfaces as well as garment collection and delivery capabilities.

Although, in the past few years, consumers' awareness toward eco-friendly clothing has grown, their buying decisions are still hardly governed by sustainability criteria. There are numerous obstructions in adopting sustainable clothing (McNeill and Moore 2015). Consumers, in general, do not have knowledge of the outcomes of the production in the fashion industry (Bhaduri 2011). Therefore, fashion companies should develop a system to provide information to consumers regarding the materials used in producing garments, the people involved in the manufacturing of the clothes, and the environmental impact of the production. They should work toward making the entire value chain transparent.

The Four Actions Sustainable Fashion Value (FASFV) Framework: An Audit Instrument

The FASFV framework can serve as an audit instrument for fashion companies to ensure development of sustainable business practices throughout design, production, and consumption processes (Table 1).

Table 1 The Four Actions Sustainable Fashion Value (FASFV) Framework: An Audit Instrument

Eliminate (E)	Raise (R)
• eliminate traditional business models with take-make-use-dispose orientation.	• raise consumers' knowledge on
	– buying ethical clothes
	– prolonging product life cycle through repair and reuse
	– proper garment care during washing, drying, etc.
	– recycling possibilities (e.g. donation boxes)
Reduce (R)	Create (C)
• reduce unfair trade practices	• create innovative business models
• reduce use of toxic materials, chemicals and water consumption	• create innovative man-made fibers
• reduce use of resource intensive fibers like cotton	• create recycling infrastructure
• reduce waste by:	• create skill development programs for workers
– shortening production cycles	• create collaboration with suppliers and customers
– designing for longevity	• create technology for value chain transparency
– designing for waste minimization	

Source: Designed by author adapted from Kim and Mauborgne (2005)

Doodlage (Case Company)

Doodlage is a fashion and lifestyle brand championing zero material wastage by upcycling fabric scraps into other products like accessories, bags, and even garment tags. Kriti Tula, founder of Doodlage, believes in "conscious fashion." The brand works on the principle of creating sustainability at all levels from cutting of a garment to using eco-friendly dyes and fabrics to ethical packaging. Doodlage has been featured among top 8 Asian sustainable fashion brands by *Harper's Bazaar*. Based on the information collected from the founder and the designers through semi-structured interview, this study determines processes utilized by Doodlage for its upcycled fashion collection (Fig. 2). The process is divided broadly into six heads:

1. Research: Research is an integral part of Doodlage. It conducts extensive research related to fabric sourcing, market trends, colors, style, and design inspiration. The brand also continuously explores the possibilities of working with new innovative alternative fabrics which have minimum impact on the environment. In the past, they have incorporated organic cotton, corn fabric, eucalyptus fabric, recycled wool, and recycled cotton polyester in their collections as they consume lesser resources for production or recycle existing material to create new fabrics.
2. Material Sourcing: Key source of material for Doodlage is industrial scrap such as misprints, post-cutting waste, etc. from large garment and fabric manufacturers. The brand mainly focuses to collect and use natural fabrics which are 100% cotton waste. Obtaining adequate quantity of source materials and sorting of the

Fig. 2 Upcycled Clothing. (Source: https://www.doodlage.in/)

source materials remains a key challenge for the company. The brand believes in experimenting with new ethical alternative materials. It works to spread awareness about sustainable fabric options to other designers so that the burden can be eventually shifted from cotton which is a high resource-consuming fabric. With the ever-increasing global population, the idea of the brand is to encourage a shift in the material mix toward less land-intensive inputs.

3. Design Process: Each garment is carefully designed to ensure durability. It believes in "zero-waste design" philosophy. The brand turns industrial scrap into stylish, well-made unique items. Since the design is based upon the raw material available at a particular point of time, therefore, each piece produced by the company is one of a kind. That is the reason for naming the company as "Doodlage" – which is based on the word "doodle."

   "Everyone might not be able to create art, but doodling comes naturally. Every individual has a personal style of creating doodles. This concept of individuality is evident in our pieces given the nature of our raw material," said Kriti, explaining the name of the company.

4. Production: Doodlage has its production house in Lado Sarai, New Delhi. Its team comprises of pattern makers and artisans skilled in handwork, stitching, and embroidery. The brand works to streamline each collection to reduce overproduction and mainly focuses on producing upon orders. It emphasizes on creating limited edition collections which are designed based upon availability of raw material. Its range of collection includes everything from classic shirts to contemporary jumpsuits. It produces 200–300 pieces every month by upcycling and reusing 600–900 m of fabric. In addition, upcycling is a labor-intensive and slow process which involves a lot of efforts in fixing fabric issues. Hence, products are priced on the higher side.
5. Retail: Doodlage sells high-quality products with an objective of "design for longevity." It uses both online and offline platforms to market its offerings. It has its own website to showcase its products to the target audience. Its collection is also available in 40 designer stores across the country. Doodlage collaborates with various larger brands to promote its label. It has partnered with Goonj, an NGO that also shares the "zero-waste" philosophy. Goonj has a mission to provide reusable sanitary pads to women in rural India, and Doodlage contributes to it.
6. Consumers: Doodlage aims to bring conscious change among the consumers whose final decision is mainly driven by prices. Most consumers opt to buy lower-priced new items as it offers a false sense of value to buyers. But they are often unethically produced and have short life span. The brand encourages its clients to care for their garments, repair it, and dispose it ethically. It provided awareness about how post-purchase care with respect to washing, drying, and ironing can bring huge difference in keeping clothes in good shape for longer period of time. Each new customer gets a postcard talking more about the impact of fashion. Each customer is facilitated with a repair kit to enable them to extend their product life. The brand thrives on the motto of "Care with love, repair with a purpose." Target customers range between the age of 18 and 45 who care for

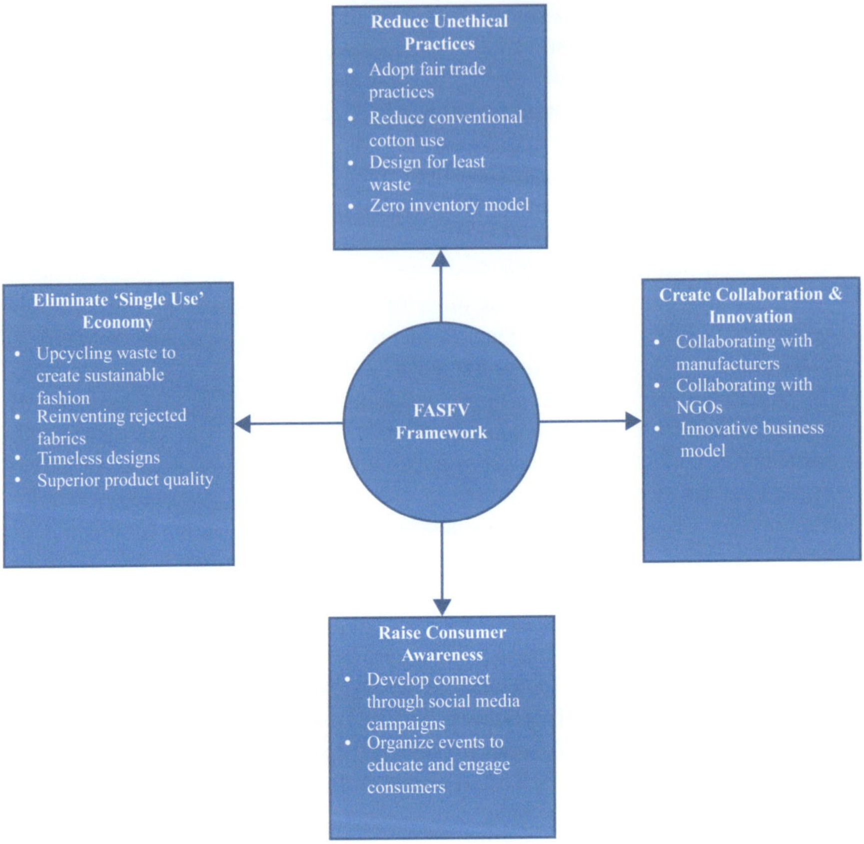

Fig. 3 Analysis of Value Creation Process of Doodlage (Case Company) through FASFV Framework. (Source: Designed by author)

the environment and look for product longevity and product design. Their price range for apparel starts from around $ 60 and goes up to $ 200.

Doodlage works on the principle of building sustainable fashion across the value chain from sourcing and procurement to design and production and from distribution to disposal. The process is analyzed through the FASFV framework (Fig. 3).

Eliminate "Single-Use" Economy

Doodlage believes in creating quality garments which can be worn and re-worn again and again unlike fast fashion. It has built standards and practices for designing clothes that can be effortlessly reused or recycled. It has established mechanisms to make apparel value chain more transparent. Ninety percent of material utilized by

Doodlage is fabric waste, and ten percent is sustainable alternative fabrics. Technological advancements in fabrications allow the brand to look for ethical options. By working with fabrics that are discarded during production process, the brand saves resources that would go into creating virgin fabric, as well as it reduces waste that is generated by the industry.

Kriti explains, "Brands need to understand the potential of new fabrics and learn how to work with them. There is need for movement that is away from consumption of conventional cotton."

Reducing Unethical Practices

Doodlage tries its best to minimize its production waste, and whatever is left is reused in making bags and home furnishings. A major source of fabric for Doodlage is the discarded textiles from large manufacturers, unused post-cut fabrics, end-of-the-line fabric wastages, unsold stock, rejected hand-dyed and hand-woven fabrics, and dead stock. Doodlage believes in ethical design strategy of "design for waste minimization." It develops mood board, sketches, color palette, and designs, with key consideration to utilizing textile waste as source materials in each of their collections. It uses patchwork,[1] paneling, embroideries, and pattern cutting techniques which enables best use of available fabrics (Fig. 4). The brand relies on the philosophy of design for slower consumption; therefore, no surplus is produced. It works on zero inventory model and creates new pieces as per orders only.

In addition, Doodlage believes in fair trade practices. It ensures that each individual in the value chain is paid a fair wage. Further, the brand uses leftover fabric for creating packaging bags, and compostable plastic is used for shipping products.

Creating Collaboration and Innovation

Doodlage continuously looks for creating world-class, sustainable, product range through collaboration with various organizations. While most of their fabrics comes from big manufacturing units, they are always open to collect fabric wastages from small cottage industry as and when the opportunity approaches.

"One such visit to Jaipur we chanced upon a godown full of block printed fabrics that were essentially testers and of no use to the retailers. We bought it and it worked beautifully," says Kriti.

It has partnered with Brahmakarma, a textile and accessory brand creating block printed sarees. Doodlage reuses the rejected sarees stock to create contemporary designs for existing Brahmakarma clients. It also collaborates with Fabindia, which

[1] Technique of creating smaller pattern pieces

Fig. 4 Patched panel overlap jacket. (Source: https://www.doodlage.in/)

focuses on spreading handmade, conventional Indian crafts to consumers in India and abroad. It sources rejected fabrics from various dyers and printers working with Fabindia to develop its new collection called "Re Fab."

Further, the brand is trying to organize garment collection drives and collaborates to create repair center pop-ups as a part of this initiative.

Raising Consumer Awareness

Doodlage sincerely works toward overcoming the greatest hurdle of creating consumer demand for their sustainable fashion items. It actively interacts, involves, and engages with consumers through social media platforms to understand their preferences and accordingly design its new collection. Online presence helps the brand to effectively communicate its ethos to the clients through regular updates about its products, the raw materials used, and processes followed to produce it, the artisans behind the products, etc. Events are organized regularly to help consumers understand the importance of upcycling and encourage them to incorporate sustainable practices toward fashion in their lifestyle.

Kriti says, "Sustainable fashion would not be possible unless consumers are aware."

Figure 5 outlines Doodlage's sustainable fashion business model.

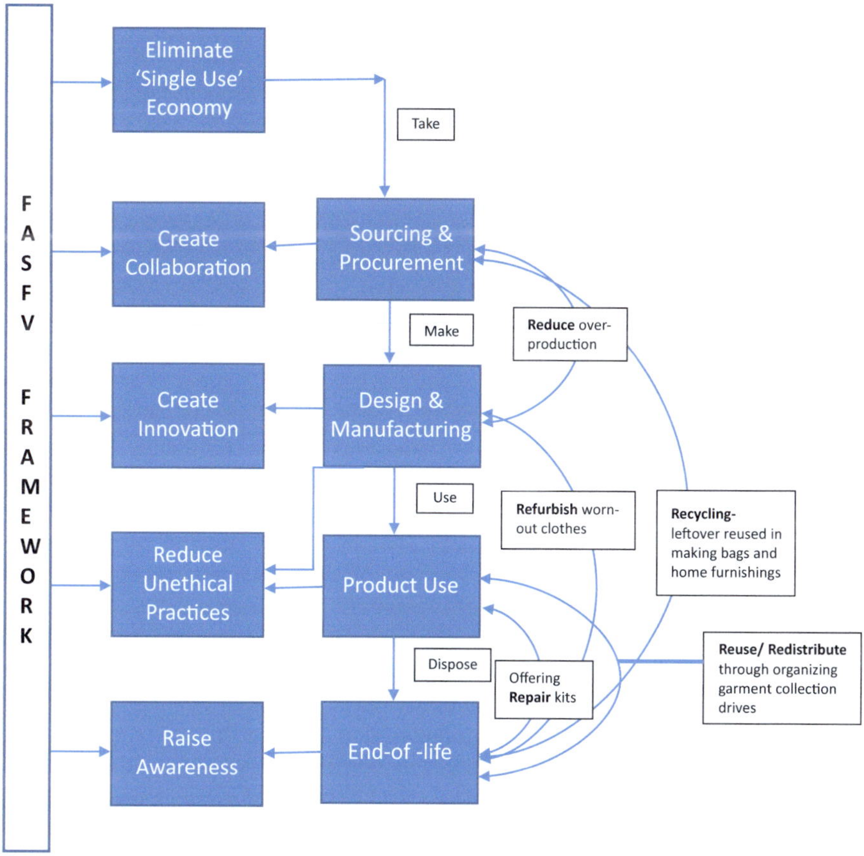

Fig. 5 Doodlage Sustainable Fashion Business Model. (Source: Designed by author)

Doodlage: Road Ahead

Doodlage seeks to transform consumption patterns of consumers who believe in "single-use" way of living. It plans to open repair cafes in the next 5 years which would aim to inculcate a culture of thrift among the millennials. In addition, it aims to tag its garments so that clients can know how individual products were produced along the entire value chain. In future, the brand also seeks to provide "green dry-cleaning" options which do not use harmful chemical solvents. It will continue to reach out to other brands and work together to create demand for responsible dyes.

Further, Doodlage aims to cater to the mass market by making it a more affordable brand. It envisions sustainable consumption to become a "norm" in the society. The brand plans to expand its presence through opening physical stores and develop convenient in-store garment collection system to encourage consumers to drop off unwanted products. The idea is to increase textile collection and recycling rates and decrease waste.

Conclusion

Fashion industry is one of the world's most polluting industries (Shen et al. 2017). The characteristics of fast fashion business model – high volume, rapid lead times, and low prices (Caro and Martínez-de-Albéniz 2015) – generate sustainability issues in relation to society and environment (Securing and Müller 2008; Krause et al. 2009). Therefore, sustainable value creation approach is required where efforts are put on "exploring how to create the value that benefits multiple stakeholders including the environment and society, but not without sacrificing shareholders' benefits" (Yang et al. 2017: 2).

In this study, author developed the *Four Actions Sustainable Fashion Value* (FASFV) framework as an audit instrument for fashion companies to ensure development of sustainable business practices throughout procurement, design, production, and consumption processes. It comprised of the following four value proposition factors for transition toward sustainable fashion economy. *First*, "eliminate single-use economy" – sustainability and closed-loop thinking should be positioned at the heart of each business model. The key concept of circularity lies in the significant reduction of production and consumption of raw materials in blend with a strategy to repair, recycle, and reuse resources from waste (Ellen MacArthur Foundation 2017). *Second*, "reduce unethical practices" – companies must address social and environmental issues that occur along the fashion supply chain, like unfair wages, hazardous working conditions, and unwarranted use of energy, water, and toxic chemicals. *Third*, "create collaboration and innovation" – establishing new or closer collaboration with stakeholders within or beyond the traditional supply chain is key in creating sustainable fashion economy. *Fourth*, "raise consumer awareness" – brands need to develop a new approach to build customer relationship that aims to make customers aware and involved and responsible partners in the value chain processes.

Sustainable practices should be encouraged at each stage of consumer purchase process from buying superior quality textiles to lengthening the product life cycle by carefully using, repairing, refurbishing, reusing, and recycling material that cannot be further used (Dahlbo et al. 2017). There is a need to shift consumer's mindset from quantity to quality (Vehmas et al. 2018). Consumers can be motivated to purchase sustainable fashion through providing them proper education and communication (Goworek et al. 2013) as most of the buyers lack knowledge about the environmental and societal impact of their purchases. For instance, information regarding how much water is consumed in regular T-shirt compared with eco-friendly T-shirt can be provided. Celebrities can also play a significant role in not only promoting the ethical brands but also being part of these brands as designers, entrepreneurs, and advocates (Moorhouse and Moorhouse 2017). Social media platforms can be explored more intensively to reach out target audience (Han et al. 2017). Consumers should also be made aware about the end-of-the-garment life cycle. They could be incentivized for "product take-back system." Advertising campaigns, vouchers, and educating employees who can inform consumers about such schemes are required for taking such system forward.

Through applying a circular business model, Doodlage develops a sustainable fashion value chain to translate the principles of knowledge, purpose, and timeliness into unique ethical products (Ricca and Robins 2012:53; Jain and Mishra 2019). Table 2 illustrates Doodlage's present CE practices and its future goals.

Table 2 Doodlage's present CE practices and its future goals

Business Model Pillars and Building Blocks	Present CE Practices	Future Goals
Product		
Value proposition	• products created from alternative sustainable fabrics (e.g. banana fabric; recycled cotton)	Create 'repair cafes' for consumers to:
	• new collection developed from textile waste	– extend product life
	• timeless designs	– refurbish their worn-out clothes
	• superior product quality	– offer discount coupons on resale
	• prolong product life span through repair and maintenance assistance	– design for disassembly, recycling and re-design
Customer Interface		
Target customer	• consumers between age 18 and 45	• create consumers as suppliers and co-creators of value
	• consumers who care for environment	
	• consumers who look for product longevity and enduring designs	
Distribution channel	• present in 40 fashion stores in India	• expand presence through opening their own physical stores
	• direct reach to consumers through e-store	
Customer relationships	• develop connect through social media campaigns	• incentivize consumers for 'product take-back system'
	• organize events to educate and engage consumers	
Infrastructure management		
Value configuration	• obtaining adequate quantity of source materials	• develop convenient in-store garment collection system
	• sorting of the source materials	• green dry cleaning services
	• no waste philosophy	• create responsible dyes
	• empowering low income craftsmen	

(continued)

Table 2 (continued)

Business Model Pillars and Building Blocks	Present CE Practices	Future Goals
Key capabilities	• entrepreneurial vision • organizational commitment towards environment • promoting traditional handicrafts • innovative business model	• embrace new technologies and innovation in the business model
Partnerships	• partnership with manufacturers (e.g. fabindia; Brandless)for unused post-cut fabrics, dead stock, rejected stock, end-of-the-line fabric wastages • partnership with NGOs (e.g. Goonj)to give its smaller scrap to create items for women in rural India (e.g. reusable sanitary napkins)	• expand partnership for knowledge and resource sharing
Financial aspects		
Cost structure	• costs incurred in scrap collection, sorting, remaking, selling	• create cost efficiencies
Revenue model	• premium pricing strategy for their sustainable garment range • Price range between $ 60 and $ 200	• pricing to cater mass market

Source: Designed by authors (adapted from Osterwalder, 2004)

References

Accenture and Fashion for Good Report (2019) The future of circular fashion- assessing the viability of circular business models. https://d2be5ept72nvlo.cloudfront.net/2019/05/The-Future-of-Circular-Fashion-Report-Fashion-for-Good.pdf.

Accenture (2014) Circular Advantage: Innovative Business Models and Technologies to Create Value in a World without Limits to Growth. https://www.accenture.com/t20150523T053139__w__/us-en/_acnmedia/Accenture/Conversion-Assets/DotCom/Documents/Global/PDF/Strategy_6/Accenture-Circular-Advantage-Innovative-Business-Models-Technologies-Value-Growth.pdf. Accessed 9 Oct 2019

Accenture Strategy Report (2018) To affinity and beyond- From Me to We, The Rise of the purpose-led Brand. https://www.accenture.com/_acnmedia/Thought-Leadership-Assets/PDF/Accenture-CompetitiveAgility-GCPR-POV.pdf.

Allwood J, Laursen S, Russell S, De Rodriguez CM, Bocken N (2008) An approach to scenario analysis of the sustainability of an industrial sector applied to clothing and textiles in the UK. J Clean Prod 16(12):1234–1246

Armstrong CM, Niinimäki K, Kujala S, Karell E, Lang C (2015) Sustainable product–service systems for clothing: exploring consumer perceptions of consumption alternatives in Finland. J Clean Prod 97:30–39

Bhaduri, H. H.-B (2011) Do transparent business practices pay? Exploration of Transparency and Consumer Purchase Intention. 29(3)

Billharz M, Cerny L (2012) big points of sustainable consumption and lifestyle orientations: how do they fit together? In Proceedings of beyond consumption. Pathways to responsible living. 2nd PERL international conference, 2012, technical university Berlin. Hedmark University and

the Partnership for Education and Research about responsible living (PERL) project: Hedmark, Norway; 54–64
Boston Consulting Group (BCG) and Global Fashion Agenda (GFA) report (2017) Pulse-of-the-Fashion-Industry. http://globalfashionagenda.com/wp-content/uploads/2017/05/Pulse-of-the-Fashion-Industry_2017.pdf. Accessed 9 Oct 2019
Caro F, Martínez-de-Albéniz V (2015) Fast fashion: business model overview and research opportunities. In Agrawal N, Smith AS (eds) Retail supply chain management: quantitative models and empirical studies. Springer: Boston. pp. 237–264. ISBN 978-1-4899-7562-1
Christopher M, Lowson R, Peck H (2004) Creating agile supply chains in the fashion industry. Int J Retail Distrib Mgmt 32(8):367–376
Dahlbo H, Aalto K, Eskelinen H, Salmenperä H (2017) Increasing textile circulation – consequences and requirements. Sustain Prod Consump 9:44–57
Deloitte (2013) Fashioning Sustainability. Deloitte, Copenhagen
Ekström KM, Salomonson N (2014) Reuse and recycling of clothing and textiles—a network approach. J Macromark 34(3):383–399
Ellen MacArthur Foundation (2013a) Towards a Circular Economy - Economic and Business Rationale for an Accelerated Transition. https://www.ellenmacarthurfoundation.org/assets/downloads/publications/Ellen-MacArthur-Foundation-Towards-the-Circular-Economy-vol.1.pdf. Accessed 10 Oct 2019
Ellen MacArthur Foundation (2013b) Towards the Circular Economy. Opportunities for the Consumer Goods Sector. https://www.ellenmacarthurfoundation.org/assets/downloads/publications/TCE_Report-2013.pdf. Accessed 10 Oct 2019
Ellen MacArthur Foundation (2017) A new textiles economy: Redesigning fashion's future. http://www.ellenmacarthurfoundation.org/publications. Accessed 10 Oct 2019
Evans R (2011) Case study method in sustainability research. In Franklin A, Blyton P. (eds) Researching sustainability: a guide to social science methods, practice, and engagement, 54–70. Earthscan, Oxon
Gardetti MA, Muthu SS (2015) Sustainable apparel? Is the innovation in the business model? – the case of IOC project. Text Cloth Sustain 1:2
Googins BK, Rochlin SA (2002) Creating the partnership society: understanding the rhetoric and reality of cross-Sectoral partnerships. Bus Soc Rev 105(1):127–144
Goworek H, Hiller A, Fisher T, Cooper T, Woodward S (2013) Consumers' attitudes towards sustainable fashion: clothing usage and disposal. In Gardetti, M.A. and Torres, A.L. (Eds), Sustainability in Fashion and Textiles: Values, Design, Production and Consumption, Greenleaf, Sheffield, p 376–392. https://doi.org/10.9774/GLEAF.978-1-909493-61-2_25. Accessed 10 Oct 2019
Gjerdum Pedersen ER, Reitan Andersen K (2013) The SocioLog.dx experience: a global expert study on sustainable fashion, Mistra Future Fashion; Copenhagen Business School Centre for Corporate Social Responsibility (cbsCSR), Copenhagen
Han SL-C, Henninger CE, Apeagyei P, Tyler D (2017) Determining effective sustainable fashion communication strategies. In: Henninger CE, Alevizou PJ, Goworek H, Ryding D (eds) Sustainability in fashion: a cradle to Upcycle approach. Springer, Heidelberg, pp 127–150
Hvass K. (2016) Weaving a path from waste to value: exploring fashion industry business models and the circular economy. PhD thesis. Copenhagen Business School
Hobbes M (2015) The myth of the ethical shopper. The Huffington Post. http://highline.huffingtonpost.com/articles/en/the-myth-of-theethical-shopper. Accessed 10 Oct 2019
Hethorn, J, Ulasewicz C (2008) Sustainable fashion: why now?: A conversation about issues, practices, and possibilities, Fairchild books
Ho HPY, Choi TM (2012) A five-R analysis for sustainable fashion supply chain management in Hong Kong: a case analysis. J Fash Mark Manag 16(2):161–175
Huber J (2000) Towards industrial ecology: sustainable development as a concept of ecological modernization. J Environ Policy Plan 2(4):269–285
Jain S (2019) Factors affecting sustainable luxury purchase behavior: a conceptual framework. J Int Consum Mark 31(2):130–146

Jain S, Mishra S (2019) Sadhu—on the pathway of luxury sustainable circular value model. In Gardetti M, Muthu S (eds) Sustainable luxury. Environmental footprints and eco-design of products and processes. Springer, Singapore, pp 55–82.
Jain S, Mishra S (2020) Luxury fashion consumption in sharing economy: a study of Indian Millennials. J Glob Fash Market (In press)
Joung H (2014) Fast-fashion consumers' post-purchase behaviors. Int J Retail Distrib Mgmt 42(8):688–697
Kim C, Mauborgne R (2005) Value innovation: a leap into the blue ocean. J Bus Strateg 26(4):22–28
Kim YK (2010) Standards and certifications for sustainable textile & fashion industries. Fash Inform Technol 7:83–90
Krause DR, Vachon S, Klassen RD (2009) Special topic forum on sustainable supply chain management: introduction and reflections on the role of purchasing management. J Supply Chain Manag 45:18–25
Krüger H, Dahl EH, Hjort T, Planthinn D (2012) Guidelines II: a handbook on sustainability in fashion, Sustainable Solution Design Association, SSDA, Svendborg
Lacy P, Long J, Spindler W (2018) The circular advantage- Moving from insight to action https://thecirculars.org/content/resources/The_Circular_Advantage.pdf. Accessed 10 Oct 2019
Lüdeke-Freund F (2009) Business model concepts in corporate sustainability contexts. From Rhetoric to a Generic Template for 'Business Models for Sustainability. Lüneburg: Centre for Sustainability Management
Lüdeke-Freund F (2010) Towards a conceptual framework of business models for sustainability. ERSCP-EMSU Conference 2010 - Knowledge Collaboration and Learning for Sustainable Innovation. Centre for Sustainability Management (CSM), Leuphana University of Lüneburg
McDonough W, Braungart M (2013) The Upcycle: Beyond Sustainability-- designing for Abundance, Macmillan
McKinsey & Company (2016) Style that's sustainable: A new fast fashion formula https://www.mckinsey.com/business-functions/sustainability/our-insights/style-thats-sustainable-a-new-fast-fashion-formula. Accessed 10 Oct 2019
McNeill L, Moore R (2015) Sustainable fashion consumption and the fast fashion conundrum: fashionable consumers and attitudes to sustainability in clothing choice. Int J Consum Stud 39:212–222
Michelini G, Moraes RN, Cunha RN, Costa, JM, Ometto AR (2017) From linear to circular economy: PSS conducting the transition. https://www.sciencedirect.com/science/article/pii/S2212827117301567. Accessed 10 Oct 2019
Moorhouse D, Moorhouse D (2017) Sustainable design: circular economy in fashion and textiles. Des J 20(sup1):S1948–S1959. https://doi.org/10.1080/14606925.2017.1352713
Neergaard, P, Jensen, EC, Pedersen, JT (2009) Partnerskaber mellem virksomheder og frivillige organisationer: En analyse af omfang, typer, muligheder og faldgruber i partnerskaber. Erhvervs- og Selskabsstyrelsen
Nguyen H, Stuchtey M, Zils M (2014) Remaking the industrial economy. McKinsey Quart 1:46–63
Osterwalder A, Pigneur Y (2010) Business model generation: a handbook for visionaries, game changers, and challengers, 1st edn. Wiley, New Jersey
Palomo-Lovinski N, Hahn K (2014) Fashion design industry impressions of current sustainable practices. Fash Pract 6(1):87–106
Preston, F (2012) A global redesign? Shaping the circular economy. Briefing paper, energy, Environment and resource governance, March edition, Chatham House
Remy N, Speelman, E, Swartz S (2016) Style that's sustainable: A new fast-fashion formula. https://www.mckinsey.com/business-functions/sustainability-and-resource-productivity/our-insights/style-thats-sustainable-a-new-fast-fashion-formula. Accessed 10 Oct 2019
Reim W, Parida V, Örtqvist D (2015) Product–service systems (PSS) business models and tactics – a systematic literature review. J Clean Prod 97:61–75
Ricca M, Robins R (2012) Meta-luxury: brands and the culture of excellence. Palgrave Macmillan, London

Rissanen T, McQuillan H (2016) Zero waste fashion design. Bloomsbury
Rosen MA, Kishawy A (2012) Sustainable manufacturing and design: concepts, practices and needs. Sustainability 4:154–174
Rydberg SA (2016) Circular economy business models in the clothing industry Master thesis in Sustainable Development at Uppsala University, 37: 30
Schaltegger S, Ludeke-Freund F, Hansen, EG (2011) Business cases for sustainability and the role of business model innovation: developing a conceptual framework. Centre for Sustainability Management, Leuphana University of Lueneburg, Germany
Securing S, Müller M (2008) From a literature review to a conceptual framework for sustainable supply chain management. J Clean Prod 16:1699–1710
Segran (2019) Nile is launching a guide to sustainability for brands-get it here. https://www.fastcompany.com/90350603/nike-is-launching-a-guide-to-sustainability-for-brands-get-it-here.
Shen B, Li Q, Dong C, Perry P (2017) Sustainability issues in textile and apparel supply chains. Sustainability 9
Shikaleska A V, Gjozinska B, Stojanovikj M (2017) The circular economy- A pathway to sustainable future. http://www.fbe.edu.mk/JSDv17.pdf. Accessed 10 Oct 2019
Stahel WR (1994) The utilization focused service economy: Resource Efficiency. In Allenby BR, Deanna J. Richards (eds) The Greening of Industrial Ecosystems. National Academy Press, Washington, DC, pp. 178–190
Stotz L, Kane G (2015) Facts on the global garment industry. Clean Clothes Campaign. http://www.cleanclothes.org/resources/publications/factsheets/generalfactsheet-garment-industry-february-2015.pdf. Accessed 10 Oct 2019
Stubbs W, Cocklin C (2008) Conceptualizing a "sustainability business model". Organ Environ 21(2):103–127
Sustainable Consumption Roundtable (2006) I will if you will: towards sustainable consumption. Research report by Sustainable Consumption Roundtable, SDC and NCC. http://hdl.handle.net/10023/2312.
Tilikidou I, Delistavrou A. (2004) The influence of the materialistic values on consumers' pro-environmental post-purchase behavior. In Cron WL, Low GS (eds) Marketing theory and applications proceedings of the 2004 American marketing association winter educators' conference. AMA: Chicago IL; 15: 42–49
Tukker A, Tischner U (2006) Product-services as a research field: past, present and future. Reflections from a decade of research. J Clean Prod 14(17):1552–1556
Vehmas K, Raudaskoski A, Heikkilä P et al (2018) Consumer attitudes and communication in circular fashion. J Fash Mark Manag 22(3):286–300
World Economic Forum (2018) A global tipping point: half the world is now middle class or wealthier. https://www.weforum.org/agenda/2018/10/a-global-tipping-point-half-the-world-is-now-middle-class-or-wealthier/. Accessed 10 Oct 2019
Yang Y, Han H, Lee KP (2017) An exploratory study of the mechanism of sustainable value creation in the luxury fashion industry. Sustainability 9
Zhu D, Qui S (2007) Analytical tool for urban circular economy planning and its preliminary application: a case of Shanghai. Urban Ecological Planning 31(3):64–70

Sustainability in Textile Design with Laser Technology

Feristah Unal, Arzu Yavas, and Ozan Avinc

Abstract The textile and fashion industry is one of the oldest and largest industries in the world, and sustainability, which is important in all sectors, is also very important for textile and fashion sector. Nowadays, unfortunately, the textile industry has become one of the sectors that may cause noteworthy environmental damages. The culture of rapid consumption causes depletion of the limited resources in nature and has negative effects on ecosystem. Due to the current world order, resources are rapidly depleting, nature is being destroyed, human health is gradually deteriorating, and therefore many undesirable economic, ecological, and social problem occurrences are on the rise. One of the challenges faced by designers in the textile sector is to provide ecological production solutions while meeting the consumer demands. Finishing processes are generally applied to textile materials to provide the desired appearance and comfort features. However, the high amount of wastewater generation during the application of these finishing techniques and the excessive amount of water and energy consumption lead to environmentally unfriendly processing types. In order to solve this increasing problem in the sector, sustainable materials and production methods should be used and preferred. Therefore, replacement of harmful chemicals with environmentally friendly substances, wastewater reduction, and creation of new design and production processes are of great importance. In particular, alternative methods for the textile finishing processes applied for the creation of new and different designs are being developed. Laser technologies, which are one of these methods, provide both economic and ecological production types. Cutting, engraving, fading, marking, modification, pre-modification, and design are some of the examples of laser technology applications in the textile industry. In this chapter, the contribution of laser technology, as a dry and clean method, to sustainable textile production and design is reviewed in detail.

Keywords Sustainable design · Laser technology · Laser · Sustainability · Textile · Design

---

F. Unal · A. Yavas · O. Avinc (✉)
Textile Engineering Department, Engineering Faculty, Pamukkale University, Denizli, Turkey
e-mail: oavinc@pau.edu.tr

S. S. Muthu, M. A. Gardetti (eds.), *Sustainability in the Textile and Apparel Industries*, Sustainable Textiles: Production, Processing, Manufacturing & Chemistry, https://doi.org/10.1007/978-3-030-37929-2_11

Introduction

Evolving industrialization, population growth, and consumption of natural resources in irreversible amounts cause the natural environment to deteriorate and possible increasing pollution from day to day. This ecological problems cause global warming and climate changes. As in all modern industries, the substances left to the environment by textile manufacturers may also generally not be harmless. For this reason, the usage of less freshwater in textile applications and recycling and re-use of chemicals and wastewater have become extremely important (Öztürk 2018; Türkmen 2009; Can and Ayvaz 2017). Fashion and design is basically a phenomenon that displays rapid and continuous change. It aims for rapid and continuous change and consumption (Mangır 2016; Morgan et al. 2014). Sustainability aims to produce products without damaging vital resources and to have a long product life (Mangır 2016; Morgan et al. 2014). Therefore, fashion and sustainability are known as contrasting concepts (Mangır 2016; Morgan et al. 2014). In the textile industry, the design works to support the purchasing behavior of the customers and meet their aesthetic tastes include dyeing, printing, and finishing processes (Morgan et al. 2014). Different physical processes such as sandpapering, sandblasting, brushing, and embroidery are applied especially in denim production in order to obtain fashionable modern looks. In addition, pre-washing, rinsing, stone washing, sand washing, snow washing, stone washing with enzymes, and chemical processes such as bleaching can also be applied. The energy and water consumption used in these processes may create serious environmental problems. Therefore, these processes are thought to be not environmentally friendly processing options (Morgan et al. 2014; Nourbakhsh et al. 2011; Štěpánková et al. 2010; Chow et al. 2012; Dascalu et al. 2000; Ortiz-Morales et al. 2003; Anderson 2006; Tarhan and Sarıışık 2009; Drago et al. 2008; Khalil 2015). In terms of the production stage only, one jean, according to Levis life cycle analysis data, causes 33.2 kg of carbon dioxide release during its production, therefore consuming 3480 lt of water and 400.1 megajoule of energy. These amounts are equivalent to driving 78 miles with a car, taking a shower 53 times, and watching a plasma screen television for 318 h (Türkmen 2009; Can and Ayvaz 2017). Massive amount of water is generally used to produce 1 kg of cotton (enough for making five T-shirts on average) (Can and Ayvaz 2017). Therefore, sustainability in the fashion and textile sector is very important (Mangır 2016). In order to minimize these problems, new technologies are investigated as an alternative to traditional finishing processes for design purposes. For instance, laser applications, one of these new technologies, significantly reduce the use of chemical, energy, and wastewater in textiles and provide sustainability (Morgan et al. 2014; Nourbakhsh et al. 2011; Štěpánková et al. 2010; Chow et al. 2012; Dascalu et al. 2000; Ortiz-Morales et al. 2003; Anderson 2006). It also provides process flexibility and ease of design (Štěpánková et al. 2010). The laser technology transfers the obtained designs to the textile material more quickly and easily and allows creating the same designs over and over again. Thus, creating high added-value, rich-looking textile products is very easy (Ortiz-Morales et al. 2003; Ondogan et al. 2005; Hassan 2016). Laser technology, which is a fast technology and provides sensitivity in the process, is considered as a clean process which offers low cost and low environmental impact due to its water-free method (Kan 2014a). The most suitable textile materials for laser applications are to be known as flax, cotton, silk, felt, and blends (Hassan 2016; Payne 2010).

Laser technology is one of the cleanest industrial methods (Nourbakhsh et al. 2011). The most common application areas of the laser technology in the textile industry include the removal of indigo dyes from denim, denim fading, pattern formation on textile materials, surface roughness production, cutting, and surface defect determination and its removal (Nourbakhsh et al. 2011). Laser technologies are used previously only for cutting and marking; today fading, laser engraving, controlled cutting, laser welding, and laser bonding can also be carried out with laser. In addition, the laser systems can be utilized in fashion and design for folding, cutting, and modification of the fabric surface to obtain special surfaces (Nourbakhsh et al. 2011).

Laser applications (CO_2 lasers), especially in denim production, have yielded very good results in creating a faded effect on denim. Therefore, its use in textile industry is increasing (Morgan et al. 2014; Dascalu et al. 2000; Ortiz-Morales et al. 2003). The difference of the laser application, especially in the fading process, can be expressed as attitude and touch. Laser technology creates a different tactile effect (Hassan 2016). It is known that laser application (radiation) increases the dye adsorption of synthetic polymer substrates and thus improves dye performance (Morgan et al. 2014). Laser technology is also used to create a visual impact which is close to the devore effect on cotton fabrics (Payne 2010). In addition to this, laser technologies are used to provide durability to the antibacterial properties applied to the fabric for consumer use and comfort (Schindler and Hauser 2004; Ilić et al. 2009). Laser types used in laser technology may vary. They are divided into high- and low-power laser according to their energy density. But the most widely used laser type in textile sector is CO_2 laser which provides high efficiency (Dascalu et al. 2000; Ortiz-Morales et al. 2003; Kan 2014a). These lasers can emit hundreds of kilowatt radiation at 9.6 and 10.6 μm (Nourbakhsh et al. 2011).

The main purpose of the laser application as an alternative to the traditionally applied methods is to provide low cost and low environmental impact way of textile design. This is a very important concept for efficient use of the resources in terms of both producer and consumer leading to more sustainable textile design and therefore world. As aforementioned, laser marking to the surface of the fabric, fading, laser engraving, surface pretreatment, controlled cutting, laser welding, laser bonding, etc. are some of the examples of laser technology applications in the textile industry. In addition, the laser systems are also used in the fashion design and confection for folding, cutting, and modification of the fabric surface to obtain a special surface. The application of CO_2 laser processes instead of not-so-eco-friendly conventional process types increases the effect quality, reduces hairiness relatively, and increases design consistency. In this chapter, the contribution of laser technology to sustainable textile production and design is reviewed in detail.

It is easy to get 2D and 3D design products with laser fabric cutting application. The studies carried out in this sense are explained in detail in the following sections. Laser application is also widely used in denim fabrics (Ozguney 2007). In addition to creating differences in color with fading processes, the laser method studies were also compared with printing processes. The same pattern design was tested both by laser technology and by pigment printing method. Laser technology gives the same effect as pigment printing on denim fabrics; also it was easier, cleaner, and chemical-free. In this sense, laser technology proves itself as a sustainable technology (Ozguney 2007).

In this chapter, firstly laser beams and laser technology are briefly explained. Then, the application areas of laser technology in textile are introduced. Finally, laser application as a modification process on different fibers (cotton, wool, polyester, polyamide, glass), application of fabric cutting and engraving with laser, fading applications on cotton fabric with laser, and denim fabric design applications with laser technology are comprehensively reviewed.

Laser

Optical devices that are developed to obtain very strong, coherent, and single color light are called laser. Laser was formed from the initials of the words "light amplification by stimulated emission of radiation" by Theodore Maiman (Wikipedia Özgür Ansiklopedi 2018; Beyazova and Kutsal 2016; Boyraz and Yıldız 2017). Firstly by Albert Einstein in 1917, the presence of stimulated radiation was demonstrated. In 1960, Theodore Maiman performed laser motion at optical frequency and proved the presence of a ruby laser (Wikipedia Özgür Ansiklopedi 2018; Beyazova and Kutsal 2016; Boyraz and Yıldız 2017). In 1968, Mester revealed that low-energy laser is a stimulating effect on the cells and high-energy laser was an inhibitor effect (Beyazova and Kutsal 2016; Boyraz and Yıldız 2017). In order to obtain laser beams, an active environment providing radiation emission (solid, liquid, gas), energy source, resonance mirror system that accelerates electron movements, and a fiber-optic conductor are required (Boyraz and Yıldız 2017; Webb and Jones 2004; Naeser et al. 2002; Snyder-Mackler and Bork 1988). The basic principle of laser devices is explained as follows. First, photon energy from a light source is passed through a medium. Thus, the rotational speed of the electrons in the atoms of this medium increases. Accelerating electrons form a new beam moving at very different wavelengths and in a single direction (Boyraz and Yıldız 2017; Webb and Jones 2004; Naeser et al. 2002; Drago et al. 2008). The laser beams are located in the electromagnetic spectrum, in visible light and in the infrared region. It is also monochromatic because it consists of rays of a single wavelength (Beyazova and Kutsal 2016; Boyraz and Yıldız 2017; Drago et al. 2008). Due to this feature, it allows the selection of certain wavelengths for certain textures and specific applications (Beyazova and Kutsal 2016; Boyraz and Yıldız 2017). The light from sources such as solar light or electric light bulbs spreads around in a messy manner. Since the waves forming the light are not present in the same phase at the same time, they are propagated in this way.

In contrast to normal light, the laser beam is not dispersed; it consists of parallel waves that center the same direction and the same phase. The reason for the regularity of the laser waves is the stimulated diffusion (Boyraz and Yıldız 2017; Snyder-Mackler and Bork 1988). Normal light is spread in a very short time and distance. The laser beam is spread on a similar thickness to the hair and far distances can be reached the same fineness. This is because the laser beam has a small divergence. For this reason, rectified beam expression is used for laser (Boyraz and Yıldız 2017; Snyder-Mackler and Bork 1988). Laser beams have a large electromagnetic field power. It transfers intensive energy to small surfaces. They can be absorbed, reflected, and transmitted in a similar way to other types of radiant energy (Boyraz and Yıldız 2017; Webb and Jones 2004). Laser technology is an energy source that can be easily applied to the desired material, and its power and density can be easily controlled (Ondogan et al. 2005; Ozguney 2007). It is classified as high- and low-power lasers according to energy densities. High-power lasers are called hot lasers because of the thermal effect they produce. Low-power laser is known as soft laser (Boyraz and Yıldız 2017; Drago et al. 2008). The laser types used vary according to the laser source as gas lasers, liquid lasers, solid-state lasers, semiconductor lasers, and fiber lasers (Boyraz and Yıldız 2017; Ready 1997; Drago et al. 2008). The detailed classification is shown in Fig. 1.

The characteristics of the most commonly used laser types are shown in Table 1 (Ready 1997; Murat 2019; Kan 2014b; Drago et al. 2008).

Laser applications have laser parameters that affect the application. Resolution is measured in points per inch. High resolution means that the dots forming the mold are denser. Thus, a better laser effect is created. Pixel duration refers to the time that the laser pixel point remains on the material surface. In this case, as the pixel duration increases, the amount of energy applied to the specific point increases. Thus, the effect of laser processing is also increased (Kan 2014a). Nowadays, lasers which are widely used in many sectors have a wide usage area in textile sector. In particular, its use for textile design purposes attracts attention. Laser technology is a system that enables the transfer of the shapes and patterns drawn in the computer environment by laser beams to the fabric surface quickly and easily (Webb and Jones 2004; Karagöz 2009). In the infrared spectrum, carbon dioxide lasers with wavelengths of 10.6 μm are commonly used for this purpose (Morgan et al. 2014; Webb and Jones 2004; Ready 1997), since CO_2 lasers have good beam quality and high efficiency. Therefore, it is known as the most widely used laser type in many sectors. Thanks to CO_2 laser, the process is carried out without damaging the volumetric properties of the materials used. With its high sensitivity feature, desired effects, design, and patterns can be easily applied (Boyraz and Yıldız 2017). It is more economical and ecological than the traditional methods used. Since it is an automatic system connected to the computer, it prevents the occurrence of personal errors. In addition, the reproducibility and efficiency of a pattern or effect applied by laser technology is also quite high (Webb and Jones 2004; Karagöz 2009).

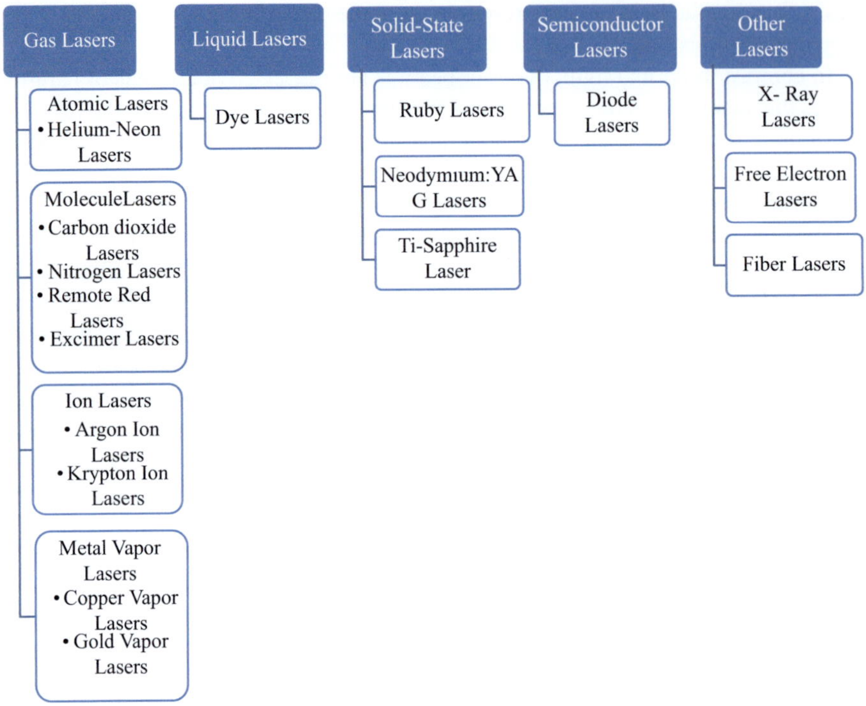

Fig. 1 Classification of lasers (Ready 1997; Murat 2019; Drago et al. 2008)

Table 1 Characteristic of laser types (Ready 1997; Murat 2019; Kan 2014b; Drago et al. 2008)

Lasers	Pulsed or continuous	Configurability	Emission wavelengths
HeNe	Continuous	No	632.8 nm (1.15 μm)
Ar-ion	Continuous	No	351, 455, 458, 466, 477, 488, 497, 502, 515, 529 nm
N_2	Pulsed	No	337 nm
Excimer	Pulsed	No	190–350 nm
CO_2	Both of them	Yes	10.6, 9.6 μm
Dye lasers	Both of them	Yes	365–930 nm
Nd:YAG	Both of them	No	1064 nm (532 nm,355 nm and 266 nm)
Ti-sapphire	Continuous	No	670–1100 nm
Diode Lasers	Both of them	Yes	UV to medium IR

Laser Application Types in Textiles

The combination of traditional and known applications in textile with creative ideas and new technologies enables different designs to be formed (Erbıyıklı 2012). The advantages of innovative technologies in the production process also affect the design process. This makes it possible to create creative products quickly and effectively. Taking advantage of new technologies during the creation of designs is also important in terms of ecology (Bulat and Gürcüm 2016). One of the most important objectives of sustainable design is to integrate environmental awareness and sensitivity at every stage of new product production. Sustainable design increases the market advantage and minimizes environmental impacts through the sustainable product design. Therefore, in design applications, clean and dry physical processes offer more advantages than conventional chemical wet processes. Examples of clean and dry physical processes include laser application. The use of clean and dry technologies such as laser technology in finishing applications significantly reduces water, chemical, and energy consumption (Chow et al. 2012). In the past, laser applications used in the fashion and apparel industry were limited to marking and cutting the textile surfaces. With the technological developments in the recent years, it is possible to apply the desired size and density graphics and designs on any kind of textile surface with laser technology. It also eliminates the limitation of diversity (Hung et al. 2011). The usage areas of laser technology in textiles are shown in Fig. 2.

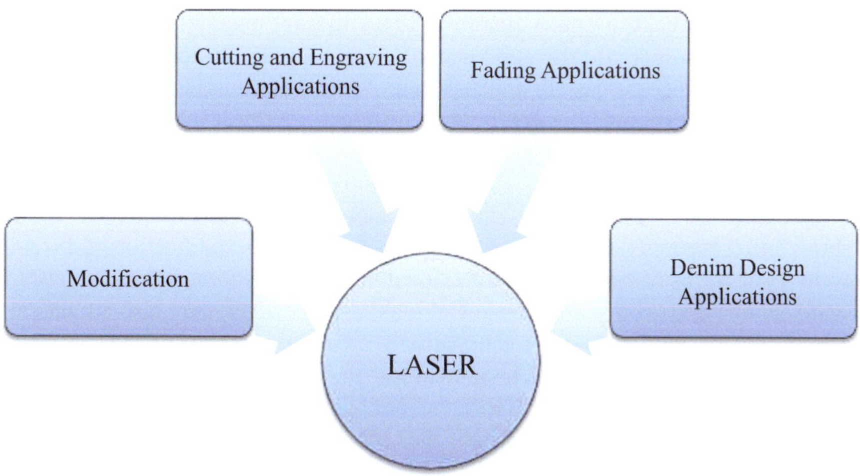

Fig. 2 Major application types of laser technology in textiles

Laser Application as a Modification Process

Textile fibers are typically subjected to a variety of pretreatments to make dyeing and printing processes more effective and feasible. However, these processes can create undesirable process conditions which can lead to an increased waste generation, disturbing working conditions, and higher energy consumption. Hence, pollution reduction in textile production is of great importance to manufacturers worldwide. Moreover, many different finishing processes could be applied to the fabrics after the coloration processes such as dyeing and printing. The textile industry is developing environmentally safer methods to implement sustainable technologies and remain competitive for textile processing. Today, laser technology is considered one of the best methods in modern textile finishing processes to improve the surface functionality of different fibers for various applications, especially for the dyeability of fabrics (Shahidi et al. 2013). Laser applications which are applied as modification processes are given on fiber basis.

Cotton Fabrics

The growth of microorganisms in textile causes problems (Schindler and Hauser 2004; Nourbakhsh and Ashjaran 2012). In order to prevent microorganisms formed on textile products, operations are carried out with antimicrobial agents (Schindler and Hauser 2004). A commonly used substance for antibacterial finishing is silver metals. The durability of the finishing process, such as antibacteriality applied to the fabric for consumer use, is quite important (Schindler and Hauser 2004; Ilić et al. 2009; Nourbakhsh and Ashjaran 2012). Different methods are investigated and developed to increase the antibacterial effects on the fabric. One of them is laser applications (Nourbakhsh and Ashjaran 2012). In cotton fabric experiments, cotton fabric is exposed to laser light at different energy levels. It is aimed to increase the permanent antibacterial effects on the cotton fabric with laser application. The silver nanoparticles are then coated on laser-treated fabrics (Nourbakhsh and Ashjaran 2012). It is coated with silver nanoparticles on laser-untreated cotton fabrics and compared with laser-treated fabric. The surface modification of the cotton by laser forms carboxylic acid functional groups, and those groups can attract positively charged groups. Positive metal ions, such as silver, may also attract carboxylic acid groups. Thus, silver nanoparticles are more absorbed on the surface of the laser-treated cotton fabric. The increase of the absorption of silver nanoparticles on the fabric surface increases the antibacterial properties of the fabric. Therefore, the antibacterial property of the cotton fabric coated with laser-treated silver nanoparticle resists repeated washings for a longer time (Nourbakhsh and Ashjaran 2012).

Wool Fabrics

Another application made for design purposes is the pretreatment and dyeing process of wool fabrics with laser technologies. Laser-assisted dyeing technology has been developed as an alternative to traditional methods using intensive water and chemicals. Thus, the problem of environmental pollution caused by chemical and wastewater can be solved relatively. In addition, it causes reduction in energy consumption. Thus, environmental awareness and sustainability can be provided (Morgan et al. 2014). The most common type of laser used in this process is CO_2 lasers (Morgan et al. 2014; Kan 2014b). Laser technology is used in the pre-dyeing process of the wool fabric which then will be dyed with reactive dyes. This allows the dye to penetrate the fibers at a higher rate. There is a difference on wool fabric in color depth after dyeing process between the areas where the laser beam is applied and the areas where the laser beam is not applied. This is used to obtain different dyeing effects during design. After dyeing, it is proved that the laser beam-applied wool fabric has a darker color and the laser beam-unapplied wool displays lighter color. This tonal separation is used for design purposes to create special surface pattern effects (Morgan et al. 2014). Thus, CO_2 laser applications are an effective surface design tool for wool-based products. In some studies, it has been displayed that the laser irradiation process could decrease the felting and the contraction of the wool fabric. In addition, it is supported by some studies that laser and plasma combination increases the hydrophilicity of woolen textile materials (Morgan et al. 2014). As applied laser power density enhances, the dye performance also increases. However, it should be noted that the laser beam applied at very high power causes the fibers on the wool fabric surface to burn. This means that the color of the wool fabric is deteriorated (Morgan et al. 2014).

The scales on the wool fiber after the laser beam irradiation are almost destroyed. Removing the scales of the wool fiber allows the dye to penetrate the fibers more easily. Thus, dye absorption of wool is improved by laser pre-modification process. Therefore, darker shades were also obtained (Morgan et al. 2014).

The color difference in the treated and untreated areas of the fabric surface is important in terms of design (Morgan et al. 2014). This increases the applicability as a CAD-controlled laser processing technique (Morgan et al. 2014). Visual textures were obtained in different signs which can provide line quality, tone difference, and aesthetics that can be obtained by using laser-assisted dyeing technique (Morgan et al. 2014).

It is possible to obtain different dye intake on the surface of the wool fabric using CO_2 laser (Morgan et al. 2014). Thus, a pattern can be marked on the surface of the wool. It has been proven in the studies that laser-treated areas on wool surface provide better dye absorption than untreated areas. Generally, long dyeing periods are required at high temperatures to obtain maximum dye retention on the wool fiber. However, it is known that laser-pretreated wool provides high dye intake at low temperatures. The laser-reinforced dyeing technique used to create different surface designs contributes to the reduction of energy and wastewater with low dyeing temperatures leading to improved dye performance (Morgan et al. 2014).

In addition to this, it is possible to obtain anti-felting effect with laser technology. Laser technology can be used instead of conventional methods such as chlorination usage for this purpose. Thus, the environmental damage of harmful chlorine components is significantly reduced (Nourbakhsh et al. 2011). One of the biggest problems of woolen fabrics is the felting shrinkage caused by washing. This problem, which adversely affects the use of wool, is caused by mechanical agitation, moisture, and heat-induced scale movement. The most common chlorination method used to reduce the problem is by using dilute hypochlorite solution under acidic conditions. While this method reduces the problem, it creates another bigger environmental problem. Therefore, the use of laser technology, which is an environmentally friendly technological solution for the wool felting problem, becomes very important (Nourbakhsh et al. 2011). If the laser technique could be applied to wool felting-proof finishing process on a commercial scale, this would notably help the environment by completely eliminating the utilization of hazardous chlorine and its derivatives (Nourbakhsh et al. 2011).

Polyester Fabrics

Laser technology is a widely utilized technology in surface modification of polymers. Characteristic modifications of the surface morphology of polymers such as polyamide and polyester could be manufactured by laser radiation (Montazer et al. 2012; Esteves and Alonso 2007). Especially, it is possible to find many studies that have been carried out on polyester. It is known that laser application changes many properties positively or negatively such as fabric weight, fiber diameter, yarn abrasion, tensile strength, tensile elongation, twist, surface gloss, wetting, air permeability, and crystallinity (Kan 2008a, b). The characteristics such as wettability and air permeability are positively affected by laser radiation, while fabric weight, fiber diameter, tensile strength, yarn wear, and bending properties are adversely affected by laser radiation (Kan 2008b). Laser radiation generally has no effect on the mass and structural properties of the polymers. This is explained by the low penetration depth of the laser radiation (Kan 2008a, b). Wong et al. modified poly(ethylene terephthalate) using an excimer laser (248 nm). Thus, it was found that the polyester surface developed a periodic roughness and fluctuation. In addition, it has been shown that polyester wettability after the suitable laser treatment can be reduced (Wong et al. 2001, 2003). In the literature, in order to examine the effect of laser treatment on dyeing properties of polyester fabrics, CO_2 laser treatment is applied before and after disperse dyeing of polyester fabric. After the process, properties of polyester fabric such as color fastness, bending stiffness, wettability, and crystal size were investigated. The outcomes confirm that laser treatment has a positive effect on color fastness. However it was also stated that while no significant color fastness development is monitored in the low-intensity laser process, it enhances the rub fastness property with the high-intensity laser process. Its properties such as wettability and bending stiffness are negatively affected by the effect of laser intensity (Montazer et al. 2012).

It has been reported that the laser process used by Majid et al. after polyester dyeing using three different disperse dyestuffs has a strong effect on color difference. The increase in laser power has led to a great increase in color difference and dye adsorption in all three dye colors. After dyeing, the color difference of the laser-treated samples is higher in blue and red dyes and lower in yellow dye (Montazer et al. 2012).

Laser irradiation prior to dyeing causes surface morphological modification. Laser irradiation after dyeing may result in some changes in the molecular structure of the dyes. That is, anthraquinone dyes are more tagged than laser radiation than azo dyes. In this study, it was pointed out that the color increase observed was caused by the wavy structure of the modified surface (Montazer et al. 2012). The laser heat causes the polyester fiber to melt and adhere to each other, which causes the porosity of the fabric surface to change (Montazer et al. 2012).

Polyamide Fabrics

There are also studies about the modification of polyamide fabrics by laser radiation. Some pretreatment techniques were applied to polyamide fabrics before their dyeing. In addition to conventional pretreatment techniques, the effects of enzymatic treatments and plasma modification processes were also investigated on polyamide fabrics. It was also proved that the modification of polyamide fabrics is possible by using laser technology (Bahtiyari 2011). The effect of laser modification (excimer laser) on dyeing properties of polyamide materials was investigated. It was concluded that the dyeability of the polyamide fabric increased after laser modification. The increase in dyeability of polyamide fabrics is explained by the decrease in the crystallinity of the polyamide. Furthermore, the low intensity of laser modification increases the concentration of amino groups required during dyeing with acid and reactive dyes. Laser treatment results in increased dyeability of polyamide fabrics. This increase is more pronounced in the coloration process with reactive dyes (Bahtiyari 2011). As a result, laser modification using low laser densities ensures the dyeability of polyamide fabrics with minimal fiber damage (Esteves and Alonso 2007; Bahtiyari 2011). After laser treatment, the dyeing characteristics using the disperse and reactive dyes are improved. Thus, the laser process (CO_2 laser) is considered to be a powerful tool that can increase dye absorption (Esteves and Alonso 2007). This provides an alternative boosting option for polyamide fiber dyeing. The laser modification process has a high industrial potential because it is an environmentally friendly dry process that does not contain the chemicals and/or solvents which could be required for wet processing (Shahidi et al. 2013). In addition, with this method, the designers could obtain different designs and patterns. The laser treatment set can allow the production of desired designs in different shades without the utilization of harmful chemicals and water (Bahtiyari 2011).

Glass Fabrics

A different example of pre-modification with laser beam is the application to glass fibers. In a study carried out in 2012, TiO_2 (titanium dioxide) nanoparticles were applied evenly to the surface of a glass fiber mat by laser irradiation. 100 μs laser beam was used in the application. The results show that laser beams can enhance the air permeability of fabric samples. Since the laser irradiation induced ripple structures on the surface of the fibers leading to more air space creation between fibers and fabric. Therefore it is possible that more air could pass through the fabric resulting in better air-permeability (Wiener et al. 2013). Recently, the application of nano-sized TiO_2 as a finishing process in textile has been investigated to provide effective protection (such as UV protection, bacterial protection, and self-cleaning performance) because of its photocatalytic activity. Laser modification on the surface of the material is one of the most studied technologies (Kan 2008b; Wiener et al. 2013). The physical and chemical properties of the material are also affected after laser radiation. It is therefore believed that such surface modifications on the polymer may have significant effects on the textile properties (Wiener et al. 2013).

Application of Fabric Cutting and Engraving with Laser

Textile designers make use of innovative technologies to build their designs on material innovation. For instance, laser cutting technique is widely used by designers for designing the clothes and in patterning processes. The evaluation of fabrics designed by laser cutting in the clothing industry can be given as an example of the use of laser technology in the product innovation. Laser can easily cut the appropriate format drawing created in computer environment. Thus, it automates the manufacturing process. The effects that can be achieved by wrapping around the area that has been cut on the fabric surface in the past are now achieved by laser cutting in less time and with more creative and aesthetic features. During laser cutting, a high-intensity laser beam focuses on a small spot on the surface. The high beam density at that point melts and evaporates the material. Cutting process is carried out in a simple and clean way without the need of fixing the material. In the face of intense laser energy, many materials sublime in a quarter of a second. Cut emissions are in the form of smoke and completely removed, filtered, and discharged. There are no burr, yarn particles, etc. in the processed material. Thus, the laser-treated product is ready for further processing without the need for intermediate processing. With laser cutting systems, fabrics can be cut faster and more accurately. With a 500 watt laser beam, 40–70 m of fabric can be cut per minute. Laser cutting technology could be utilized in the cutting of medical textiles, in the cutting of interior architectural products, in military textiles,and in the cutting of automotive textiles (Gürcüm and Bulat 2016). Laser cutting applications in the early 1990s, the work of many contemporary designers, who used laser technology together with textile materials, pioneered today's design aesthetics. Designer Lauren Moriarty has designed curved and bent surfaces with laser cutting process using heat-sensitive fabrics such as neoprene (Payne 2010). The laser cutting process started to be applied as an alternative to the devore process, which is more commonly known as incineration. A very close effect of the effect obtained by the use of chemicals in the devore process can also be obtained by laser cutting. The woven cotton plush structure can be rastered by laser (Payne 2010). In this process, speed and power adjustment is made according to velvet depth. The laser is moving forward and backward and burns the fiber surface. Thus, a view (an effect) similar to that obtained in devore printing can be obtained. The best devore effect is obtained from woven velvet fabrics made from natural fibers. Knitted plush fabrics can be distributed under laser light (Payne 2010).

It can perform laser printing as well as cutting with laser beam. This operation is performed by reducing the intensity of the laser light. The laser printing process is known as laser engraving and laser marking. The fabric or other tissues are burned or scraped with laser beams, and the desired pattern is transferred to the surface of the material (Gürcüm and Bulat 2016).

Cotton, plush, or velvet fabrics are burned and can be given hollows and embossed effects (Anderson 2006; Gürcüm and Bulat 2016). Laser printing can be applied to different materials including rubber, silk, polyester and other synthetic, wood raw materials and leather, artificial leather, latex, woven fabric, glass, natural felt, artificial felt, and paper. The engraving method that creates embossing effect on textile materials such as leather and felt also enables the textile designer to perform artistic works (Gürcüm and Bulat 2016). Laser cutting method has been applied to industrial felt and silk fabric (Payne 2010; Drago et al. 2008). The felt was cut at very slow speed and high strength. Silk was cut at high speed and slow strength.

Laser cutting technology is a technology that allows designer intervention during the cutting process. During cutting, the laser cutter can be stopped, and the material can be moved on the laser bed. This allows the designer to rethink the part of the design and to instantly apply the different designs that come to mind. The laser cutting process has a good effect on textile materials woven from different types of yarns. Cotton woven fabric (the weft yarn of the fabric is thin metal thread) can be given as an example. The hidden metal weft thread used in the fabric reveals that the cotton fiber is burned (Payne 2010).

In the past, the lace appearance was generally obtained with the intensive handwork, and nowadays knitting industry could generate lace fabrics in a significantly shorter time and in a very economical way with the help of laser cutting. Garments with 2D and 3D designs matching the body shape can be obtained by laser cutting. The same pattern can be applied symmetrically or asymmetrically to the garment, but it is considered extremely important that the visual integrity of the garment is not impaired when patterning. The application of different patterns on different layers increases the visual richness of the garment. With the laser cutting, the garments with aesthetic properties gained by the draped and 3D fabrics can be obtained, where the feeling of weight created by multilayer fabrics is eliminated. By applying the drape technique, by folding the cut pieces on top of each other, or by combining the geometrically cut pieces in various ways, three-dimensional or volumetric effects can be created on the surface and structure of the garment (Gürcüm and Bulat 2016).

Fading Applications on Cotton Fabric with Laser

One of the most important fabric features that consumers should take into consideration when choosing, buying, or wearing a textile product (clothes, etc.) is color (Hung and Kan 2017). It is possible for designers to create design simply by changing the tones of different colors. Especially in denim fabrics, various fabric coloring techniques can be applied (Hung and Kan 2017). However, the incorporation of innovative technologies in production overshadows the traditional fabric coloring methods. Laser technology, which has been used in different design processes in textile, is one of the most popular technologies to achieve different color and shade effects (Hung and Kan 2017). Laser technology can change the physical and chemical properties of different material surfaces such as polymers, metals, and semiconductors. Laser treatment applied to a cotton material causes the color fading effect in cotton (Chow et al. 2012; Hung et al. 2011). This effect is called as the fading effect. The effect of laser radiation process on the performance characteristics of cotton fabrics such as tear strength and the attitude and appearance characteristics such as roughness, hardness, and drape are also very important (Chow et al. 2012). Color fading is also gaining popularity for design purposes, especially in the jean industry. In the recent years, color fading in jeans has started to offer different visual effects in terms of design and texture (Chow et al. 2012; Hung et al. 2011). The excess water consumption of conventional fading techniques is widely known. In addition to environmental pollution, the reproducibility of stoning, bleaching, and enzymatic processes is also very low. It is also difficult to apply design processes to different fabrics with these conventional methods. These time-consuming methods are an important barrier for mass production, and it causes cost increases. Laser (CO_2 laser) applications, which are becoming more and more common in textile fading process, create repeatable models. It also eliminates the problems of traditional methods (Chow et al. 2012; Hung et al. 2011; Drago et al. 2008; Khalil 2015). The laser can deliver graphs of desired diversity, size, and density on textile surfaces, including less water consumption, process flexibility, precision, and reproducibility of designs on both knitted and woven fabrics (Chow et al. 2012; Hung et al. 2011).

In 2017, On-na Hung and Chi-wai Kan investigated the effect of laser technology on the color properties of cotton-based fabrics. They treated cotton fabrics with two different approaches. One of these is laser treatment and then dyeing the cotton fabric. In other approach, cotton fabric is first dyed and then laser treated. CO_2 laser was used in the study, and different combinations of laser processing parameters (the resolution 52, 60, and 68 dpi and the pixel time 110, 120, 130, and 140 s) were applied. Laser-treated fabrics displayed a lighter shade than control samples. This confirms that both approaches can provide color fading effect. Besides, it was observed that the cotton fabrics which were first treated with laser and then dyed had a lighter color tone than the cotton fabrics which were first dyed and then laser treated. For this reason, the method of laser treatment and subsequent dyeing provides a stronger color fading effect (Hung and Kan 2017).

SEM appearances of untreated cotton fiber and cotton fiber after fading process with laser technology were examined. Untreated cotton fiber had a flat fiber surface. However, it is observed that the laser beam opens pores of various sizes on the cotton fiber. It exhibits a sponge-like appearance (Chow et al. 2012; Montazer et al. 2014).

The weight loss is occurring on cotton fabric because of the thermal degradation of cotton fibers during laser treatment and thermal energy is absorbed by the fiber during the laser process. At the same time, water and gases such as carbon dioxide cause the internal volume of the fiber to increase. Consequently, the swelling and expansion effects of the fiber cause a sponge-like structure (Chow et al. 2012).

By controlling the resolution and pixel duration of laser light, it is possible to obtain the desired laser effect on cotton fibers (Chow et al. 2012). Laser processing of cotton fabric and then dyeing allow us to obtain faded tones. Therefore, the color fading effect could be easily determined by controlling the laser process parameters. However, in the conventional methods such as enzyme washing,etc., the color fading effect is largely dependent on experience and trial and error (Chow et al. 2012).

Denim Fabric Design Applications with Laser

Denim fabrics are usually obtained by weaving the blue indigo-dyed warp yarn and the undyed raw weft yarn, and denim fabrics are usually woven in twill weave (3/1 or 2/1) type. Denim fabrics can be produced from yarns such as cotton, polyester, nylon, viscose, and lycra (Anderson 2006; Karagöz 2009; Solaiman 2015; Kan 2014b; Tarhan and Sarıışık 2009; Khalil 2015). Often warp yarns of denim fabrics are dyed with indigo. It is possible to dye in black and other fashion tones or colors (Anderson 2006; Hassan 2016; Khalil 2015). Indigo, which is widely used in classical denim, is one of the oldest textile dyes obtained from plants. The dye is bonded to the fabric by mechanical forces rather than chemically (Anderson 2006). Denim fabrics vary with changing fashion such as denim with fantasy yarn, denim with metal yarn, denim with jacquard pattern, printed denim and washed denim with stone, etc. (Hassan 2016). In the recent years, in order to fulfill different customer needs and desires, different designs can be made on denim fabrics, and different techniques can be applied to support customer purchasing behaviors and address different pleasures. These techniques used for these purposes can be printing, embroidery, and different kinds of final product washing (Ondogan et al. 2005; Ozguney 2007). The indigo dyestuffs used in dyeing the warp yarns of denim fabric have very low rub fastness properties. Therefore, it is easy to fade the fabric color with the different finishing and washing processes (Ondogan et al. 2005; Kan

2014a, b). Sanding, stone washing, sand washing, potassium permanganate spraying, snow washing, acid washing, and stone washing with enzyme are generally used to create different visual effects on denim fabrics. This process, which is applied to create the design by wiping the colors of certain areas of the denim fabric (fading), causes different problems (Ondogan et al. 2005; Kan 2014a, b; Ozguney 2007; Solaiman 2015; Kan et al. 2010; Tarhan and Sarıışık 2009; Khalil 2015; Gürcüm and Bulat 2016). Some of these are problems with workflow due to difficulty in application and time loss. In addition to this, there may be some other problems such as not being able to produce tonal differences at desired levels, damaging the product, decreasing the product quality, nonapplicability of the original designs and writings, and low repeatability.

In addition, it is one of the most important problems that the traditional applied methods generally may not be environmentally friendly, ecological, and sustainable. New technologies are being explored to provide faster, better-quality design effects and display more advantages to solve the problems of these previous conventional design techniques (Ondogan et al. 2005; Kan 2014a, b; Ozguney 2007; Solaiman 2015; Khalil 2015; Gürcüm and Bulat 2016). The studies on improving the visual aspects of denim fabrics and the new techniques developed include laser technology (Ondogan et al. 2005). The use of laser is particularly common in the fading of denim fabrics and in the design studies (Fig. 3). Laser technology can be carried out very precisely and quickly for the transfer of the desired shape, pattern, or graphic to the denim fabric with the desired power, wavelength, and density (Ondogan et al. 2005; Ozguney 2007; Kan 2014b; Kan et al. 2010; Tarhan and Sarıışık 2009; Khalil 2015). In addition, by laser technology, it is possible to obtain

Fig. 3 Different denim fabric fading examples with laser irradiation

the desired designs without causing any wear and deformity in the fabric texture (Ozguney 2007). Laser technology provides process flexibility and ease of operation. It is also a technology that allows the creation of new and unique designs (Dascalu et al. 2000). A pair of high-speed computer-assisted mirrors helps the laser beam to move over the denim material. Different fading levels can be obtained by controlling computer and laser beam parameters or by using different parameters (Dascalu et al. 2000). High laser power allows us to obtain pale tones. However, the removal of the color from denim fabrics is not only relevant to the laser power. Laser processing parameters such as resolution and pixel time are highly effective on the process (Kan 2014a, b). In the laser application, firstly, it is necessary to adjust the power and intensity of laser light according to the material to be applied, and the design to be transferred is prepared on the computer. The laser beam of the selected power and intensity is directed to the textile surface, and the coloring agents on the textile surface are faded. The type of laser used in the case of denim studies is usually carbon dioxide lasers (Ondogan et al. 2005; Ozguney 2007).

Laser beam is moved over the textile material by two computer-controlled scanning mirrors (Dascalu et al. 2000; Ortiz-Morales et al. 2003). The laser beam interacts with the colored textile material, and the color fading occurs in the irradiated areas. Colorant residues are sucked out by means of air filtration system. Laser fading process is carried out in three stages (Niemz 2013). Firstly, the desired patterns and designs are transferred to the computer system. Then the laser system, the screening system, and the conveyor system are controlled, and finally, laser fading operation is performed (Dascalu et al. 2000; Ortiz-Morales et al. 2003).

The laser beam fading process is dependent on two parameters (Dascalu et al. 2000; Ortiz-Morales et al. 2003). One of these is the laser parameters, and the other is the material parameters. Laser beam parameters are expressed as wavelength, power, repetition rate, and pulse duration. The material parameters are thermal variability, reshaping, and reflection coefficient (Dascalu et al. 2000; Ortiz-Morales et al. 2003). All these parameters ensure an efficient operation without damaging the fabric. The sufficient intensity of the laser beam causes photo-dissociation of the dye on the material. Thus, the desired pattern fading process is realized (Kan 2014a). The photon energy emitted by the laser beams during the process is absorbed by the covalent bond in an organic polymer. The lowest dissociation energy of the indigo bonds is higher than 2.7 eV (Dascalu et al. 2000; Ortiz-Morales et al. 2003; Niemz 2013). The photon energy emitted by the Nd:YAG laser is between 1.2 and 2.4 eV. Therefore, Nd:YAG lasers cannot be used for laser fading. Excimer lasers are not preferred for laser fading operations in terms of cost, maintenance, and reliability. The most common laser type used during this process is CO_2 lasers (Dascalu et al. 2000; Ortiz-Morales et al. 2003). Ortiz-Morales, Martin, and their friends have compared different laser types used in denim fading process in their studies in 2003 (Ortiz-Morales et al. 2003). They worked with three laser types in their research. One of them is a 10.6 μm CW CO_2 laser and laser power ranges from 4 to 25 W. Focus and linear scanning speed is changed between 5 and 10 in/s, and different power densities and fluency are obtained. The other laser type used is 1.06 μm at the basic wavelength and at the first harmonic is 532 nm pulsed Nd:YAG laser. The pulse energy of this laser operating with Q-key scanning as electro-optic is between 1.5 and 8 mJ. In order to perform a good fading operation in this type of laser, the dot and beam must be focused well. Another laser type used in their study is the pulsed CTH:YAG laser at 2.09 μm. The pulse energy is between 0.65 and 0.81 J. In such lasers, the laser beam is transmitted by fiber to the focusing head that conducts the line scan. The beam must be focused for a good fading process. After the process, standard tests such as tear strength and shrinkage were applied to denim fabrics, and each test result was quite good. But, only CO_2 laser is a good alternative because of its interaction properties. Investment and energy costs of other laser types are very high. CO_2 laser has high efficiency. It also has a lower energy waste, less heat generation, less cooling systems, and less water consumption.

In a study conducted by Ondogan in 2005, laser technology was applied to half of 40 pairs of jeans, and manual fading procedure was applied to the other half. Time studies were carried out for laser beam design and manual design, and the differences of the two methods were compared. As a result, it has been proven that a company using manual design should run at least six people to be equal to the production capacity of the laser design machine. This means more dependency on the workforce. In addition, the non-reproducibility, the decrease in quality, and the loss of competitiveness reveal the problems experienced in the manual design process (Ondogan 2005).

Arif Taner Ozguney compared pigment printing with laser technology in his study (Ozguney 2007). The same pattern design was applied to denim fabrics of the same characteristics with two different processes. At the end of the application, the effect obtained by the laser design method on the denim fabric is similar to the effect created by the pigment printing method. However, laser design method is faster than pigment printing method. The preparation of pigment printing is more time-consuming. There is no difference between the tensile strengths of the samples obtained by both methods, and it shows that the laser design method does not damage the fabric. It is also very advantageous since the laser design process does not include any steps that increase the production time and costs, such as drying and fixing (Ozguney 2007).

The fabric properties used in denim fading processes using CO_2-type laser may vary. Laser processing of textile materials consisting of synthetic fibers such as polymethyl methacrylate (PMMA), poly(ethylene terephthalate) (PET), polyamide (nylon), and polypropylene (PP) has been developed for many years (Kan 2014a).

In 2009, Tarhan and his colleagues compared the sandblasting method, one of the traditional fading methods, with the laser method. In their processing, they applied laser and sandblasting operations on the same type of denim fabric (100% cotton) at different densities and pressures. At the end of the process, they investigated the effect of the applied methods on the performance characteristics of denim fabrics. As a result of the fading trials, the strength and weight values of the fabrics decreased in the blasting process depending on the pressure increase, the time, and the number of washes. They observed significant color loss in the fabric colors in parallel with the increase in the same parameters. Significant losses were observed in the strength and weight of the fabric as an outcome of laser and sandblasting process. They noted that in order to reduce these losses during production, physical wear should be applied to the lowest level and that the necessary color loss effect should be obtained by regional bleaching which would be directed to the physical wear area (Tarhan and Sarıışık 2009).

In the same year, the effect of laser application on the color and mechanical properties of denim fabrics was investigated by Arif Taner Özgüney and Gonca Özçelik (Özgüney et al. 2009). In this study, they used CO_2 (10.6 μm) laser source for the fading of denim fabrics. They investigated the effect of laser treatment on the properties of denim fabrics such as tensile strength, tear strength, friction strength, and maximum and kinematic friction coefficient and determined the optimum working conditions. The desired color fading effect is achieved with a pulse duration of 100–150 μm. Because of the removal of the carbon particles, they observed that the jaundice index changed significantly after washing. The minimum reduction in tensile strength is achieved with a pulse duration of 250 μm, while tear strength values are optimal at a pulse duration of 500 μm. For friction coefficient and wear resistance values, the pulse time of 100 μm is optimum. Generally, in terms of both color and mechanical properties, it was decided that the pulse duration of 100–150 μm is the most suitable for laser fading (Özgüney et al. 2009).

Kan et al. (2010) studied the fading process applications with CO_2 laser, also called as surface engraving (gravure) (Kan et al. 2010). They investigated the effect of the surface scraping with laser process on the color properties of the denim fabric. They have realized the treatment with 100% cotton fabric that woven with 2/1 twill weave type. The resolution of the laser beam was set at 35, 45, 55, 65, and 75 dpi, and the pixel time was set at 120, 130, 140, and 150 μs. As a result, they obtained a paler surface appearance due to the effect of high laser power. They also observed green-yellow color in fabrics after denim scraping (Kan et al. 2010).

In a symposium held in 2014, laser engraving and coating methods were applied to fashionable denim fabric. It has been emphasized that using the modified water-based polyurethane (PU) resin glue color or the multicolored coating, it softens the elasticity of the fabric and also solves the problem of cotton/polyester woven denim fabric having the problem of tearing during the laser process (Jiang et al. 2014).

The same year, the combinations of different laser process parameters to the denim fabric by Kan, i.e., the resolution and pixel duration, were applied in different irradiation directions, i.e., weft and warp directions. In practice, 100% cotton denim fabric with fabric weight of 378 g/m^2 and pulsed CO_2 laser with 10.6 μm wavelength were used. The laser beam resolution was set at 30, 60, and 80 dots per inch (dpi) and the pixel duration was set at 110 μm, 160 μm, 220 μm, and 300 μm, respectively. The laser power enhanced with increasing resolution and pixel time. The color fading effect had increased by the increase of laser power, but an optimum fading effect had been obtained. Nonetheless, it has been noted that the direction of the laser radiation does not affect the color fading effect and the color level (Kan 2014b).

The effects of laser technological parameters on denim fabric (denim cloth with 3/1 twill weave type with 98% cotton/2% elastane) were studied (Jucienė et al. 2014). In fact, considering the morphological fabric modifications, more detailed studies were carried out to predict the effects of several main laser parameters on color. In their study, researchers took into account the parameters such as speed, step, and beam power in laser application. It was reported that the color difference of the laser-treated specimens was different in the warp and weft directions (Jucienė et al. 2014). The highest color difference was obtained while altering the beam power. The largest effect on the change of color hue and color saturation among all studied laser technological parameters was found for laser power (Jucienė et al. 2014).

A carbon dioxide (CO2) laser was used for the color fading treatment of two types of denim fabrics (the first denim type was manufactured with low-twist yarn spun by torque-free ring-spinning technology, and the other denim type was produced from conventional ring-spun yarn) (Kan 2014a). The warp yarn of both denim fabric samples was dyed with the same indigo dyeing process (Vat Blue 1). In this case, the biggest difference between the two fabrics is the amount of twist and volume in the warp yarn. In this process, two laser parameters were used, namely, resolution and pixel duration. The laser type used in this process is 10.6 μm, pulsed CO_2 laser. Its power is between 60 and 230 watts. The pulse energy is between 5 and 230 mJ, and the pulse activation time is set to less than 45 μs. Laser deformations applied to fabrics cause a certain dimensional change. However, in this study, no difference in the dimensional stability was observed between the torque-free ring-spun yarn and traditional ring-spun yarn. Spinning methods were compared. According to this, spun yarn by the torque-free ring-spinning technology exhibited lower reflectance value. Laser-treated fabrics were observed in green-yellow colors. The denim fabric woven with the torque-free ring-spun yarn performed better than the denim fabric woven with the conventional ring-spun yarn in terms of colour fading after laser treatment (Kan 2014a). In addition, the traditional process method of stone washing and CO_2 laser method were applied to denim fabrics of the same characteristics. Conventional stone washing method has more rinsing steps. The conventional stone washing process takes 45 min, while the laser process is carried out in 3 min. The resulting fading effect is more clearly seen in laser application. Accordingly, it is seen that laser technology saves energy, time, water, and chemicals leading to sustainable process when compared with the traditional methods (Kan 2014a).

Solaiman and Joy Krishna compared traditional fading methods with laser technology the following year (2015). Indigo-dyed 100% cotton 2/1 twill denim fabric was used in this study. Sand brushing and potassium permanganate spray methods were compared with pulsed CO_2 (10.6 μm) laser method. Laser power ranges between 150 and 250 w. As a result, the maximum loss in tensile strength and strength values was observed in the fading process by sandblasting method. From the laser fading method point of view, it was stated that there is no significant loss in strength and tensile strength (Solaiman 2015). Many different kinds of effects and designs can be obtained with the aid of laser technology (Fig. 4) (Anderson 2006).

The effects of seam types applied to denim fabric were studied on fading, patterning, and design effect utilizing laser technology (Hassan 2016). There are designs which are applied to create new and unique effects on the pockets and stitches of denim jeans. Three different seam types [seam type 1 is superimposed

Fig. 4 Different designs created with laser technology

seam type (SSa), seam type 2 is lapped seam type (LSb), and seam type 3 is lapped seam type (LSc) by using two different stitch types (stitch type 516 and stitch type 301)] were utilized to sew denim fabric. In this study, flatbed laser machine was utilized to engrave shapes on denim fabric before and after sewing by utilizing two different speeds (speed 1180 m/s, speed 2400 m/s). Sewing threads were 100% spun polyester with the yarn number of Ne 22/3. The size of the needle was 16. Post-processing durability, efficiency, and appearance properties were evaluated according to the references. According to the evaluation, lapped seam type (LSb) (overedge stitch 516 and lockstitch 301) should be used to obtain more durable stitches, and laser machine operating at 400 m/s speed should be preferred. For more aesthetic seams, the superimposed seam type (SSa) (overedge stitch 516) should be used, and the laser machine operating at a speed of 180 m/s should be preferred (Hassan 2016).

In a study conducted by Dudeja Jai Paul in 2016, the effect of laser denim washing on the wear of the stitched denim was investigated. In this work, Dudeja used 100% cotton fabric in three different weights and applied industrial CO_2 laser to the sewn areas of fabrics. According to the test results, it is advised to apply laser fading on the denim fabrics that have maximum weight for industrial applications. It is also advisable to use superimposed seam instead of stitched lap-felled seam (Dudeja 2018).

In a recent study conducted in 2019, Venkatraman examined the color changes, fiber morphology, and distortions in 100% cotton indigo-dyed denim fabrics. Carbon dioxide laser was used at different power and density levels in order to give faded effect to 100% cotton denim fabrics. It was reported that the laser power affected the color change, and as the grayscale enhanced, the color fading was higher and affected the fabric performance across all fabric weights. As a result, the method using a pulsed CO_2 laser with a wavelength of 10.65 μm to give a fading effect to 100% cotton denim fabric had a lower risk when compared with conventional methods (Venkatraman and Liauw 2019).

Thanks to laser technology, the forms and designs suitable for the user's requirements are transferred to the fabric surface, thus giving the fabric various visual features (Kan 2014a; Petkov et al. 2008). With the help of computer, it is possible to draw technical and geometrical shapes on the material to be applied to the design (Ondogan et al. 2005; Kan 2014a). The wavelength, power, and density of the laser beam to be applied can be adjusted according to the surface texture and structure of the material of designs to be formed. Laser fading process is a dry and clean process and there is no use of chemicals and water leading to less threat creation to the environment. The process flexibility allows for the reproduction of existing stone washing designs and the creation of new patterns (Kan 2014a, b). Ease of application, even on ready-made clothing, provides applicability. Automatic control of the laser technology allows the addition and removal of the designs without repeating the design. It does not create any deformation in the areas where the laser irradiation is not applied. In addition, the applied designs are not limited to two dimensions. In order to enrich the visual features of the product, it is possible to obtain three-dimensional (3D) designs with deliberate deformations (Kan 2014a). A flexible application model and rapidly responding to various demands in the international market, laser technology provides full-time production. Thus, high value-added products can be created. In addition, laser technology is compatible with programs such as AutoCAD, Photoshop or CorelDRAW where 2D designs could be created (Kan 2014a). The most important feature of laser technology is its clean production type. It provides less chemicals and water consumption than traditional processes (Kan 2014a). Thus laser technology generates less environmental pollution and enables a more sustainable production (Dascalu et al. 2000).

Conclusions

The combination of innovation, creativity, and production is the first step of creating patterns and designs in textile and fashion industry. Original, new, innovative designs and fast production attract customers' attention, support their purchasing behavior, and address their aesthetic tastes. Laser technology is very suitable for producing ecological, fast, and original designs. Eliminating and/or minimizing the environmental, economic, and humanitarian damages arising from the production of textile and fashion sectors is very important and vital for the world and humanity. It is very important to use dry and clean production systems and technologies that do not cause waste formation while achieving the intended design successfully. New and different design works performed in textiles have become an important factor in the marketing of textile products. With laser technology, the design strengthens the visual characteristics of the finished product, enriches its aesthetic properties, and facilitates the creation of high value-added products. It is one of the new and exciting technologies that enable innovative design work. It is possible to obtain surprisingly aesthetic results with laser technology. Therefore, laser technology results in questions about the use of traditional non-ecological methods for design purposes. The high added value and quality of a product is one of the factors that affect the purchasing behavior of consumers. Laser technology is a technology that serves the desired purpose and gives successful results.

Laser technology is applied commonly in the modification process of cotton, wool, polyester, and polyamide and glass fibers. Laser technology affects the dyeability of fabrics and improves surface functionality. In this way, it is possible to obtain different designs from fabrics obtained from different fibers. In this manner, the usage areas of the fabrics are actually increased. Thanks to the laser fabric cutting application, it is possible to obtain patterns that are very similar to devore printing. Thus, aesthetic designs that have never been imagined until now are provided with laser technology. In addition, laser treatment of cotton fabrics and subsequent dyeing of cotton fabrics result in faded tones. This makes it easier to create different designs in light and dark tones or with a faded effect. One of the areas where laser technology is used most is denim fabric applications. In general, the studies on color change, fiber morphology, and fiber deterioration of dyed denim fabrics are common. Laser technologies are more advantageous than conventional methods due to their easier and faster creation of the desired effects, even on very difficult designs with respecting the environment. During all these studies, water is saved, and the release of harmful chemicals to the nature is minimized. Thus, it becomes possible to create a sustainable living space with the aid of laser technology.

In the future, it can be expected that non-ecological techniques used for design purposes can be replaced by laser technology. Textile companies use more and more dry and clean technologies such as lasers in their production, leading to more sustainable manufacturing types. In the coming years, textile companies will need to develop environmentally safer methods for their designs in order to remain competitive in the textile and fashion industries. Supporting sustainable fashion and design and realizing the production of many companies with this awareness are on the way to become a rising trend in the fashion world.

References

Anderson K (2006) There's a New Technology in Town: Laser Engraving http://t2jeans.free.fr/page3.html. or www.techexchange.com.

Bahtiyari MI (2011) Laser modification of polyamide fabrics. Opt Laser Technol 43(1):114–118. https://doi.org/10.1016/j.optlastec.2010.05.014

Beyazova M, Kutsal YG (eds) (2016) Fiziksel tıp ve rehabilitasyon. Güneş Tıp Kitabevleri, Ankara

Boyraz İ, Yıldız A (2017) Lazer Çeşitleri ve yüksek yoğunluklu lazer kullanımı. Journal of Contemporary Medicine 6(1-Ek (Olgu Sunumları)): 104–109. doi: https://doi.org/10.16899/ctd.55797

Bulat F, Gürcüm BH (2016) Tekstil Tasarımında Inovatif Bir Yaratıcılık Aracı Olarak Lazer Kesim İdil Sanat ve Dil Dergisi Winter II(28):107–130

Can Ö, Ayvaz KM (2017) Tekstil ve Modada Sürdürülebilirlik. Acad J Soc Sci 1(3):110–119

Chow YLF, Chan A, Kan CW (2012) Effect of CO_2 laser irradiation on the properties of cotton fabric. Text Res J 82(12):1220–1234. https://doi.org/10.1177/0040517511429609

Dascalu T et al (2000) Removal of the indigo color by laser beam–denim interaction. Opt Lasers Eng 34(3):179–189. https://doi.org/10.1016/S0143-8166(00)00087-7

Drago N, Villoresi P, Bertogli G (2008) Laser technologies: A step forward for small and medium sized enterprises. http://institute.unido.org/wp-content/uploads/2014/11/10.-Laser-technologies-a-step-forward-for-small-and-medium-enterprises.pdf. Accessed 12 June 2019

Dudeja JP (2018) Laser application in apparel industry. Int J Adv Sci Technol Eng Mgmt Sci. doi: https://doi.org/10.13140/RG.2.2.15620.58243

Erbıyıklı N (2012) Tekstil ve Moda Tasarımı Açısından Sanat ve Bilim. Akdeniz Sanat Dergisi 4(7):0–0

Esteves F, Alonso H (2007) Effect of CO_2 laser radiation on surface and dyeing properties of synthetic fibres. Res J Text Appar 11(3):42–47. https://doi.org/10.1108/RJTA-11-03-2007-B006

Gürcüm BH, Bulat F (2016) Tekstil Tasarımında Inovatif Bir Yaratıcılık Aracı Olarak Lazer Kesim. İdil Dergisi 6(28):107–130. https://doi.org/10.7816/idil-06-28-08

Hassan NNE (2016) The effect of using Laser engraving on seam properties of weaving denim products. Int Dgn J 6(1):225–231. https://doi.org/10.12816/0036448

Hung ON, Kan CW (2017) A study of CO_2 Laser treatment on colour properties of cotton-based fabrics. Coatings 7(8):131. https://doi.org/10.3390/coatings7080131

Hung ON, Song J et al (2011) Using artificial neural network to predict colour properties of laser-treated 100% cotton fabric. Fibers and polymers 12(8):1069–1076

Ilić V, Šaponjić Z, Vodnik V, Potkonjak B et al (2009) The influence of silver content on antimicrobial activity and color of cotton fabrics functionalized with ag nanoparticles. Carbohydr Polym 78(3):564–569

Jiang H, Yu RZ et al (2014) Innovative denim fabric design by color coating using laser etching and hollowing technologies. Paper presented at the Proceedings for the 9th International Shibori Symposium at the China National Silk Museum, The Hong Kong Polytechnic University, China, 31 October – 4 November 2014

Jucienė M, Urbelis V et al (2014) The effect of laser technological parameters on the color and structure of denim fabric. Text Res J 84(6):662–670

Kan CW (2008a) Effects of laser irradiation on polyester textile properties. J Appl Polym Sci 107(3):1584–1589

Kan CW (2008b) Impact on textile properties of polyester with laser. Opt Laser Technol 40(1):113–119

Kan CW (2014a) CO_2 laser treatment as a clean process for treating denim fabric. J Clean Prod 66:624–631

Kan CW (2014b) Colour fading effect of indigo-dyed cotton denim fabric by CO_2 laser. Fibers Polym 15(2):426–429

Kan CW, Yuen CWM, Cheng CW (2010) Technical study of the effect of CO_2 laser surface engraving on the colour properties of denim fabric. Color Technol 126(6):365–371. https://doi.org/10.1111/j.1478-4408.2010.00270.x

Karagöz G (2009) Denim Yıkama İşlemlerinde Ortaya Çıkan Zararlar Nedenleri ve Çözüm Olanakları. Yüksek Lisans Tezi, Ege Üniversitesi

Khalil E (2015) Sustainable and ecological finishing technology for denim jeans. AASCIT Commun 2(5):159–163

Mangır AF (2016) Sürdürülebilir Kalkınma İçin Yavaş ve Hızlı Moda. Selçuk Üniversitesi Sosyal Bilimler Meslek Yüksekokulu Dergisi 19(41. Yıl Özel Sayısı):143–154

Montazer M, Taheri SJ, Harifi T (2012) Effect of laser CO_2 irradiation on various properties of polyester fabric: focus on dyeing. J Appl Polym Sci 124(1):342–348

Montazer M, Chizarifard G, Harifi T (2014) CO_2 laser irradiation of raw and bleached cotton fabrics, with focus on water and dye absorbency. Color Technol 130(1):13–20

Morgan L et al (2014) Laser enhanced dyeing of wool for textile design. In: re-thinking textiles and surfaces, University of Huddersfield, UK, 26–27 November 2014

Murat A Lazer Çeşitleri. https://staff.emu.edu.tr/alimurat/Documents/BMET%20301/LAZER%20Cesitleri.ppt. Accessed 11 June 2019

Naeser MA, Hahn KAK et al (2002) Carpal tunnel syndrome pain treated with low-level laser and microamperes transcutaneous electric nerve stimulation: a controlled study. Arch Phys Med Rehabil 83(7):978–988

Niemz MH (2013) Laser-tissue interactions: fundamentals and applications. Springer Science & Business Media, Berlin

Nourbakhsh S, Ashjaran A (2012) Laser treatment of cotton fabric for durable antibacterial properties of silver nanoparticles. Materials 5(7):1247–1257. https://doi.org/10.3390/ma5071247
Nourbakhsh S, Ebrahimi I, Valipour P (2011) Laser treatment of the wool fabric for felting shrinkage control. Fibers and Polymers 12(4):521–527
Ondogan Z (2005) A laser surface design machine to improve the productivity of textile manufacture. Lasers Eng 15(5):375–385
Ondogan Z, Pamuk O, Ondogan EC et al (2005) Improving the appearance of all textile products from clothing to home textile using laser technology. Opt Laser Technol 37(8):631–637
Ortiz-Morales M, Poterasu M, Acosta-Ortiz SE et al (2003) A comparison between characteristics of various laser-based denim fading processes. Opt Lasers Eng 39(1):15–24. https://doi.org/10.1016/S0143-8166(02)00073-8
Ozguney AT (2007) The comparison of laser surface designing and pigment printing methods for the product quality. Opt Laser Technol 39(5):1054–1058
Özgüney AT, Özçelik G, Özkaya K (2009) A study on specifying the effect of laser fading process on the colour and mechanical properties of the denim fabrics. Tekstil ve Konfeksiyon 19(2):133–138
Öztürk Ç (2018) Tekstil sektöründe temiz üretim tebliği uygulamaları. Master's thesis, Namık Kemal University, Tekirdağ. http://acikerisim.nku.edu.tr:8080/xmlui/bitstream/handle/20.500.11776/2659/0053281.pdf?sequence=1&isAllowed=y. Accessed 10 June 2019
Payne J (2010) cutting through the surface: the use of laser cutting technology with traditional textile process. In: Textile Society of America Symposium Proceedings - 12th biennial symposium, University of Nebraska, Lincoln, 6–9 October 2010
Petkov PV, Dimov SS, Minev RM, Pham DT (2008) Laser milling: pulse duration effects on surface integrity. Proc Inst Mech Eng Part B: J Eng Manuf 222(1):35–45
Ready JF (1997) Industrial applications of lasers. Elsevier, New York
Schindler WD, Hauser PJ (2004) Chemical finishing of textiles. Elsevier, London
Shahidi S, Wiener J, Ghoranneviss M (2013) Surface modification methods for improving the dyeability of textile fabrics. In: Günay M(ed) eco-friendly textile dyeing and finishing, IntechOpen, Croatia, p 34–50
Snyder-Mackler L, Bork CE (1988) Effect of helium-neon laser irradiation on peripheral sensory nerve latency. Phys Ther 68(2):223–225. https://doi.org/10.1093/ptj/68.2.223
Solaiman SJ (2015) Comparative analysis of manual fading and laser fading process on denim fabric. Sci Discov 3(6):44–49. https://doi.org/10.11648/j.sd.20150306.11
Štěpánková M, Wiener J, Dembický J (2010) Impact of laser thermal stress on cotton fabric. Fibres Textiles East Eur 18(3):70–73
Tarhan M, Sarıışık M (2009) A comparison among performance characteristics of various denim fading processes. Text Res J 79(4):301–309
Türkmen N (2009) Tekstil ve Moda Tasarımı Açısından Sürdürülebilirlik ve Dönüşüm. Yayınlanmamış Sanatta Yeterlik Tezi, Mimar Sinan Güzel Sanatlar Üniversitesi
Venkatraman PD, Liauw CM (2019) Use of a carbon dioxide laser for environmentally beneficial generation of distressed/faded effects on indigo dyed denim fabric: evaluation of colour change, fibre morphology, degradation and textile properties. Opt Laser Technol 111:701–713
Webb C, Jones J (eds) (2004) Handbook of Laser technology and applications: Laser design and laser systems, vol 2. CRC Press, United Kingdom
Wiener J, Shahidi S, Goba MM (2013) Laser deposition of TiO_2 nanoparticles on glass fabric. Opt Laser Technol 45:147–153. https://doi.org/10.1016/j.optlastec.2012.07.012
Wikipedia Özgür Ansiklopedi (2018) Laser. https://tr.wikipedia.org/wiki/Lazer. Accessed 10 June 2019
Wong W, Chan K, Yeung KW, Lau KS (2001) Chemical surface modification of poly (ethylene terephthalate) by excimer irradiation of high and low intensities. Mater Res Innov 4(5–6):344–349. https://doi.org/10.1007/s100190000116
Wong W, Chan K, Yeung KW, Lau KS (2003) Surface structuring of poly (ethylene terephthalate) by UV excimer laser. J Mater Process Technol 132(1–3):114–118

University Intervention in Inculcating Design Practices for Sustainable Fashions

Veena Rao, Rajesh Kumar, Aysha Shaima, and A. Venkatachalam

Abstract Sustainable fashion is clothing that we get substantial use of by never buying that's less than contextually fabulous. Changing consumer consumption habits and technological intervention has created a world of fashion with short fashion cycles and planned obsolescence. The fast fashions and their ecological impact have drawn the attention of the researchers, with greater emphasis given to the intervention of the higher education institutions in sustainable development. Aligning to changing global requirements and readjusting to the region specific human resource needs arising out of ever-changing technology and industrial processes are some of the crucial contemporary issues that the higher education institutions have to learn to handle. Higher education institutions could no more remain as isolated centers of excellence, completely insulated from the socioeconomic mainstream. Being an epitome of knowledge creation and knowledge dissemination for the benefit of the societies, the institutions have to play a proactive role in identifying the growing impact of technological changes and unsustainable consumption patterns in different industries leading to environmental degradation. The institutes of higher education should aim at constant efforts to design and revitalize design courses with the aims of skill and competence inculcation along with the sensitivity for economic, social, and ecological sustainability. The collaborative efforts with different stakeholders play a prominent role in implementing sustainable practices. This paper as an attempt to address ecological sustainability emphasizes on the model involving the stakeholders and the institution of higher education in inculcating sustainable design practices among the budding designers.

Keywords Sustainability · Fast fashion · Ecological impact · Design practices · Design education

V. Rao (✉) · R. Kumar · A. Shaima · A. Venkatachalam
Department of Design, Manipal School of Architecture and Planning, Manipal Academy of Higher Education, Manipal, Karnataka, India
e-mail: veena.rao@manipal.edu

S. S. Muthu, M. A. Gardetti (eds.), *Sustainability in the Textile and Apparel Industries*, Sustainable Textiles: Production, Processing, Manufacturing & Chemistry, https://doi.org/10.1007/978-3-030-37929-2_12

Introduction

This study arose from unbridled consumption of clothing by the customers, which has created the world of fashion with short fashion cycles and planned obsolescence, which in turn has led to increased disposability, thus highly contributing to ecological degradation. The globalization has changed many industrial sectors wherein fashion and related industries stand out (Anguelov 2015). It is evident through the profound transformation witnessed by the industry due to various changes in business environment (Bhardwaj and Fairhurst 2010), the most significant being shortened life cycle of a fashion being adopted and disposed theorized as fast fashion business model. The model satisfies the consumer demand of quick fashion with multiple styles developed in small quantity and priced low, making the new fashion trends affordable. The fast fashion model has brought in the concept of clothing being considered as disposable, which in turn requires the companies to produce clothing in lead time (Yao and Zhang 2018; Bick et al. 2018). Firms adopting fast fashion business model rely on orchestrating network of third-party manufacturers which is generally located in developing countries that have to respond very quickly to the continuous new requests (Lorenzoni 2016). The globalization has resulted in meeting the requirements of the fast fashions with the manufacturing sectors using cheap materials and cheap labor from anywhere in the world in the production of fashion trends (Joung 2014). Augmenting the same is the increased purchasing power and affluent lifestyle of the consumers which have aggravated the phenomenon of fashion styles becoming obsolete with shorter fashion cycles. The increased demand from the consumers and financial pressure to produce fashion products at an increasingly fast rate have associated fast fashion with the concepts of "disposable clothing" and "throwaway culture" among the consumers (Yee et al. 2016). Fashion sector is widely seen as indivisible from consumer capitalism and the capitalist logic of perpetual growth based on increasing throughput of materials (Fletcher and Williams 2013). Fast fashion is responsible for the tendency of overconsumption practices throughout the society, which has led to negative societal and environmental consequences (D'Souza 2015).

The societal consequences include the health and safety of the professionals working in global clothing industries owing to condensation of the manufacturing base for mass production in low- and middle-income countries (Bick et al. 2018). The reported incidents of garment industries in these middle and less developed countries include poor working conditions, use of child labor, exploitation of women workers, and exposure to occupational hazards such as solvent and adhesives used in the manufacturing process (Hobson 2013).

The environmental consequences of fast fashion are the result of pollution footprint generated from each step of the clothing life cycle (Claudio 2007).

(a) Fiber stage: depletion of nonrenewable resources (fibers from petrochemicals, which are widely used for fast fashion).
(b) Production stage: water and air pollution as a result of production of many items of clothing.
(c) Discarding stage: landfills due to frequent disposal of fashion items (Cuc and Vidovic 2014; Gupta et al. 2016).

The fast fashion industry with serious social and environmental costs involved in textile manufacturing has not yet made suitable efforts to achieve sustainability. Sustainably produced clothing is seen as a potential solution to lessen the current strain of the fast fashion on social and environmental well-being globally. In this context, the importance of educational institutions as a decisive factor for achieving a culture of sustainability through training for future professionals has garnered importance with UN declaring 2005–2014 to be the decade of education for sustainable development (ESD) (Kozar and Connell 2013). The role of institutes in creating new generation of environmentally skilled designers is discussed as a means of combating the environmental degradation caused by fashion industry. Through this paper, the designer attempts to understand the impact of fast fashion industry on the environment and sustainability measures adopted, the role of educational institutions in training professionals toward sustainable society, and the design strategies that address the sustainability. Emphasis in this paper is on the effort of the university in engaging stakeholders and designing resourceful solutions through classroom activities to address the need for sustainable fashion industry. It is intended that graduates who participate in these activities are better equipped with sustainable solutions and are expected to assume responsibility for creating sustainable future.

Fast Fashion Industry and Sustainability

Fashion industry contributes to approximately 20% of global waste and 10% of global carbon emissions, making it the biggest polluters in the world only next to oil industry (UNEP 2018). Fast fashion, which rose to prominence at the turn of the twenty-first century, had added to the industry's enormous greenhouse gas emissions and devastating environmental impact (Boggon 2019). With increasing production levels, it is predicted that the water consumption by the industry will grow by 50 percent (31.17 trillion gallons), the carbon footprint by 2791 million tons, and the waste generation by 148 million tons (Sumner 2019).

Fast fashion referred to as disposable or throwaway fashion is linked with clothing brands such as Zara or H&M, which rose to fame through production of popular fashion trends from runways and luxury clothing brands. The industry is based on quick response (QR) manufacturing methods that combine quick response production capabilities with enhanced product design capabilities. The business model incorporates quick time response to agile, lean retailing, and leagile process (Buzzo and Abreu 2018) aimed at providing latest styles to all classes of consumers. It largely depends on the production chain and supplies to meet the current market demands and reflect the same effectively in constantly changing merchandise assortments (Peterson et al. 2010; Buzzo and Abreu 2018). In other words, it is a clothing range meant to sell quickly so that the retailers can constantly change the items they offer (Drašković et al. 2018). The fast fashion centered at cheap, easy, and short lead times of production and supply requires the business to make use of low-cost materials, low-cost labor, short lead times of design and production, and effective volume of production (Fletcher 2011). This shorter lead time has allowed the clothing manufacturers to introduce new lines more frequently. For instance, Zara offers 24 new clothing collections each year, whereas H&M offers 12–16 lines in a year and refreshes them weekly (Remy et al. 2016). The successful business models of the fast fashion brands have inspired retailers to explore fast fashion product lines. Currently, even the traditional luxury brands such as Versace offer fast fashion options to reach a larger market of demographic fashion consumers (Anguelov 2015). This dissemination of the quick response model to the luxury brands helps to satisfy the deeply held desires among young consumers for luxury fashion, even as it demonstrates no sustainability (Kaikobad et al. 2015).

From financial rewards to ever-changing product lines, fast fashion continues to present multiple benefits for retailers and their consumers. The idea of fast fashion firm offering large quantities of clothing at cheap prices leads to domino effect causing garment consumption to skyrocket (Lorenzoni 2016), (Bick et al. 2018). Although this transition indicates "democratization" of fashion, making it available for wide range of customers, the environmental and sociological risks (Binet et al. 2019) associated with production of inexpensive clothing, cannot be neglected. The risks are hidden throughout the life cycle of the garment (Bick et al. 2018). It includes the vast amount of resources it consumes in the production of fiber-yarn-fabric, the water pollution related to the use of toxic chemicals in processing, and increasing levels of textile waste. These issues require to be addressed with innovative ways of how clothes are produced in pace with the demand and supply of shorter fashion styles (Remy et al. 2016). Though many retail brands have moved toward sustainable initiatives to reduce water, chemical, and energy use, new dyeing technologies such as carbon dioxide dyeing to reduce the water consumption, the problem that fashion industry faces, are toward the consumption side.

The fashion industry has not made much progress in the area of sustainability and is yet to make contribution toward universally acceptable definition of sustainable fashions and exact specifications of sustainable fashion goods (Caro and Martìnez-De-Albèniz 2015). Sustainability endorses social and environmental protection, out of which the social issues pertaining to exploitation of human rights in the textile and apparel industry are sufficiently addressed. It is the environmental protection that has drawn attention in a subtle way and not recognized under the umbrella of fashion brands (D'Souza 2015). The need of the hour is to drastically alter the patterns of consumption and production to ensure preservation of the limited available natural resources as a means of addressing the environmental sustainability and hence the survival of the planet (Vehmas et al. 2018; UNEP 2018).

Role of Educational Institutions

Educational institutions play a pivotal role in inculcating the culture of sustainability through training of future professionals (Kozar and Connell 2013). Today's millennial students and professionals have ingenuity to bring change in social, cultural, environmental, economic, and political practices worldwide (Pasricha and Kadolph 2009). The complex web of sustainable problems instigated by the fast fashion industry needs the attention of fashion designers who have been educated at the university to demonstrate concern for the environment, society, and economy (Obregón 2012). Also, the budding designers need to be made aware of the problems associated with the industry and work toward sustainable goals. The paradigm of sustainable education required for the hour is to be based on holistic approach in contrast to the conventional educational process which is focused on "taking things apart" (Fletcher and Williams 2013).

Sustainability in fashion industry can be achieved through consumer-based approach or design-based approach. While the consumer-based approach heavily relies on the informed and the responsible decisions made by individuals, the design-based approach requires the designer to evolve design strategies with focus on making environmental improvements emphasizing on eco or green design (Gwilt and Rissanen 2012). Universities of higher education have to play a prominent role in educating the customer about the impact of fast fashions on the society and environment and hence advocating ethical fashion consumptions. Also, the universities of higher education as a base of knowledge development and knowledge dissemination can focus on inculcating design-based approach for sustainable fashion design.

The industry of fashion with fast fashions is in a state of unsustainability with both social and ecological impacts. To achieve sustainability, it is imperative that the classrooms move to the real world or the real world is made a part of the classrooms. It is prudent that the industries make a commitment of creating a demand for professionals equipped with knowledge, skills, and ethics toward sustainability. The demand created should be met by the universities which need to infuse the sustainability into the apparel and textile curricula, hence equipping the graduates with competencies to meet the current and future sustainability needs (Kozar and Connell 2013).

This paper focuses on educating the design students with strategies for sustainable development. The intention here is on training of professionals toward sustainability who in turn make efficient use of the resources and design the sustainable options from which choices are made.

Sustainable Design Strategies

The sustainable design strategies referred in this section are associated with the strategies that help in reducing the negative impacts associated with the life cycle of a product designed.

Tischner and Charter (2013) identify four Rs' approach to the development of sustainable design. It includes:

- Repair: It emphasizes on repair approach/modifications to the existing products.
- Refine: It addresses the eco-efficiency of the existing products.

- Redesign: It aims at the use of technologies and materials to meet the product requirements with reduced environmental impact.
- Rethink: It is the most radical shift, which requires a designer to rethink of strategies that can meet the consumer requirements for new lifestyles, in a sustainable manner.

Ceschin and Gaziulusoy (2016) design for sustainability follows design approach categorized under four innovation levels, which include product design innovation level, product-service system innovation level, spatio-social innovation level, and socio-technical system innovation level. Different levels of innovation are detailed with supporting approaches reviewed for sustainable design strategies applied in fashion design.

Level 1: Product design innovation level—design approaches focusing on improving existing or developing completely new products. The different strategies suggested include the following:

- Green design and eco-design: Green design addresses lowering of environmental impact by redesigning of individual qualities of individual products. Eco-design focuses on the whole life cycle of the product (extraction of raw materials to final disposal).
- Emotionally durable design addresses the psychological obsolescence of the customers in discarding the products. Emotionally durable design can be addressed by a designer who can create deeper bond between the user and the product by understanding how the wearers use and engage with clothing. For instance, a garment designed for longevity can be designed with slow fashion approach that embodies specific functions encouraging the wearer to use a garment in a particular way. Different strategies can be used in designing a garment which is multifunctional or transformable, or highly durable, or may purposefully change with age becoming different through use (Gwilt 2014).
- Design for sustainable behavior states the role of the designer in using design strategies to reduce the environmental impact of the product throughout its life cycle.
- Nature-inspired design talks about imitating the nature's materials and processes in achieving sustainability in production-consumption system. Two strategies that are suggested include cradle-to-cradle design and biomimicry design. Cradle-to-cradle design approach in fashion design aims at closed-loop system where the goal is to extend the use time of the garments and that the material is recycled in several rounds. The system requires the designer to appreciate waste as a valuable resource for design (Niinimäki 2017). Biomimicry requires a designer to study the models and processes of the nature that can be adapted to solve the human problems. It also requires the designer to use ecological standards to judge the "rightness" of the innovation. Biomimicry in fashion design is seen mostly in the form of mimicking the texture, color, and shape through different surface ornamentation techniques such as embroidery and creation of fashion design inspired from nature or using discarded pieces such as plywood in developing a pattern (Chen and Peng 2013; Gwilt 2014).

- Design for the base of the pyramid emphasizes on the need for designing for the poorest portion of the global population.

Level 2: Product-service system innovation level—here the focus is beyond individual products toward integrated combinations of products and services (e.g., development of new business models).

Level 3: Spatio-social innovation level—here the context of innovation is on human settlements and the spatio-social conditions of their communities. This can be addressed on different scales ranging from neighborhoods to cities. The different levels of innovation are given below:

- Design for social innovation which requires complementing the technological innovations with that of social innovations.
- Systemic design identifies mimicking natural ecosystems for sustainability wherein the waste from one productive process becomes input to the other processes.

Level 4: Socio-technical system innovation level—here design approaches focus on promoting radical changes on how societal needs, such as nutrition and transport/mobility, are fulfilled and thus on supporting transitions to new socio-technical systems (Ceschin and Gaziulusoy 2016).

Sustainable Product and Service Development (SPSD) conceptualizes the approach of making products and/or services in a more sustainable way throughout their entire life cycle, from conception to end of life. SPSD is aimed at achieving an optimum balance between triple bottom line of three Ps: people, planet, and profit. In other words, the goal of SPSD is to produce products that address the environmental protection, social equity, and economic prosperity and still meet the customer requirements with cost-effectiveness (Maxwell and Van der Vorst 2003). The summarized SPSD concept is given in Fig. 1.

Gwilt (2014) focuses on the sustainable design strategy (Fig. 2) with focus on a specific approach in making improvements in design, production, use, and/or disposal phases of a product life cycle.

Design strategy suggested by Gwilt (2014) is given in Table 1.

Design Strategies Implemented at Manipal Academy of Higher Education

The requirement of fashion designers with sustainable design skills is explored through the design strategies reviewed and conceptualized by the authors as given in Fig. 3. The conceptualized model incorporates the product design innovation level proposed by Ceschin and Gaziulusoy (2016) integrated with the design strategy suggested by Gwilt (2014). Engagement of stakeholders in sustainability represents a powerful drive for value creation. Hence, the model proposes collaborative efforts with different stakeholders in implementing sustainable practices. The paper

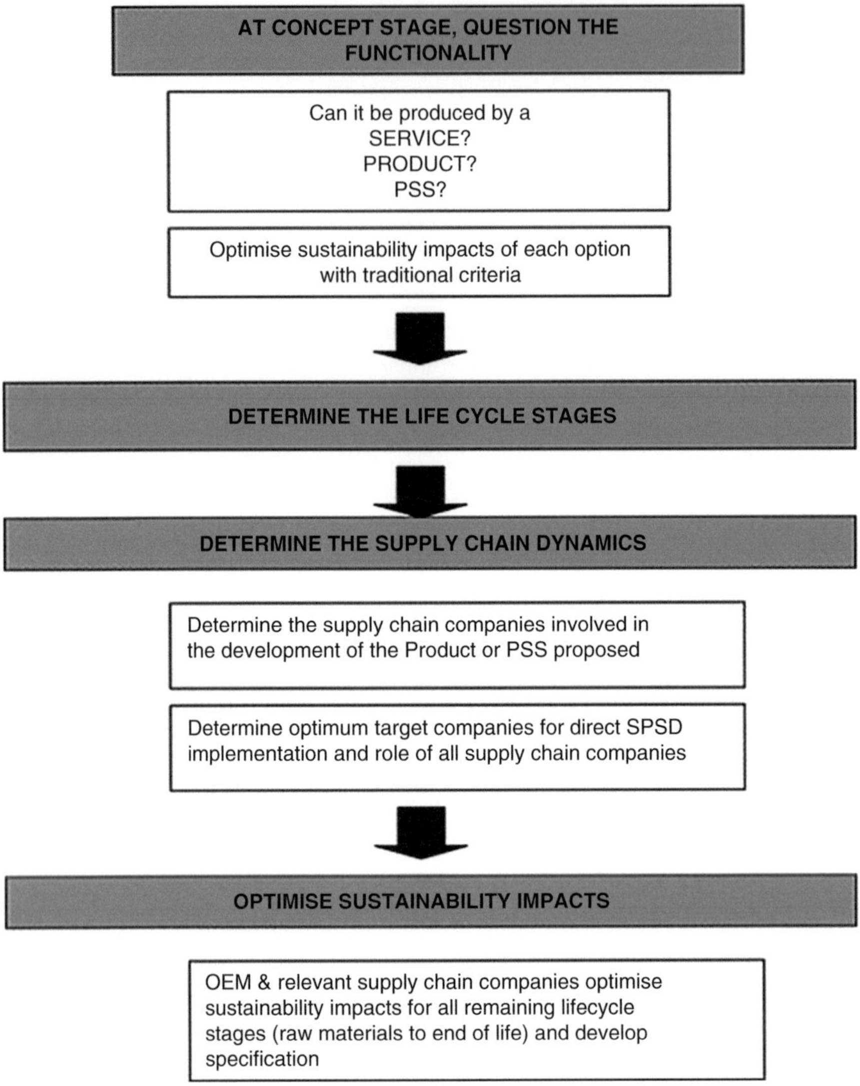

Fig. 1 Sustainable Product and Service Development
Source: (Maxwell and Van der Vorst 2003)

is supported with the design projects aimed at inculcating sustainable design practices among the fashion design students of the Manipal Academy of Higher Education (MAHE), Manipal.

Design Stage: Vezzoli and Manzini (2008) argues that an environmental conscious strategy included from the beginning of design process helps preventing or limiting the associated problems instead of losing time (health and money) involved

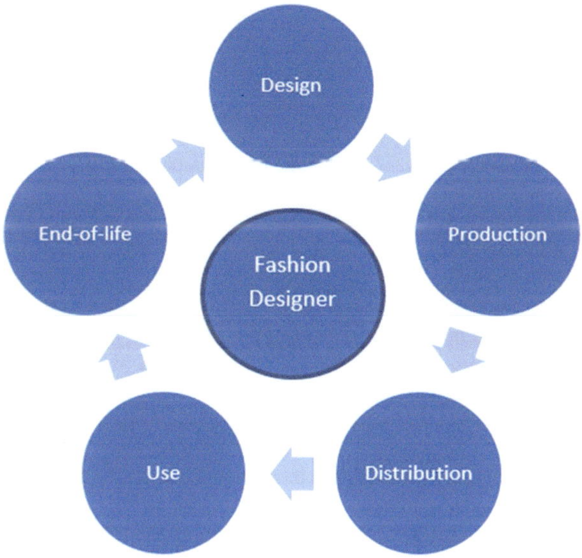

Fig. 2 Design strategy (Gwilt 2014)

Table 1 Design strategy proposed by Gwilt (2014)

Stage of the garment life cycle	Inputs	Outputs	Design strategies
Design	What materials and supplies do you intend to use?	What outputs arise from textile treatments such as dyeing and printing?	Design for empathy
	What are they made from?	What outputs arise from raw material production processes?	Design for well-being
	What textile processes will you use?	Can you see outputs related to the design of your garment?	Design for low-impact materials and processes
	Do these processes require any resources?		Design for mono-materials
Production	Who makes your supplies and garments?	What happens to the textile waste from your production?	Design for zero waste
	Where are the garments assembled?	Will any other waste be created during production?	Design for durability
	What resources are needed for production?	Will there be any impact on the health and lives of people involved in production?	Design for efficient use of materials and resources
			Design for ethical and fair trade production

(continued)

Table 1 (continued)

Stage of the garment life cycle	Inputs	Outputs	Design strategies
Distribution	What distance do your goods (pre- and post-production) have to travel?	What is the energy and resource consumption associated with your transportation?	Design for need
	How are goods transported?	How is packaging waste managed?	Design to minimize transportation
	What are your packaging needs?		Design to reduce/reuse packaging
	Do products need to be stored at any point?		Design to engage with local communities
Use	What services are required during use phase?	What outputs arise from your preferred laundering method?	Design for multifunction
	What is your preferred process of laundry?	Is the energy used for ironing or tumble-drying?	Design for modularity
	How frequently does the garment need laundry?	Are chemicals used in dry-cleaning process?	Design for low-impact care
	What is needed to maintain the garment?	Is there any impact on human health?	Design for customization
			Design for repair
			Design for product/service systems
End of life	What happened to the garment at the end of its life?	What impact arises from the process of managing the waste material?	Design for reuse
	What services are required during the disposal phase?	Can the materials be reused or recycled?	Design for disassembly
		What will happen if the garment is placed in landfill or incinerated?	Design for recycling/upcycling
		Is there any impact on human health?	Design for remanufacturing
			Design for closed-loop system

in redressing the damage already done. In fashion industry also as in other industries, design stage is considered as the starting point for addressing the negative environmental and ethical impacts associated with the phases of the garment's life cycle (Gwilt 2014).

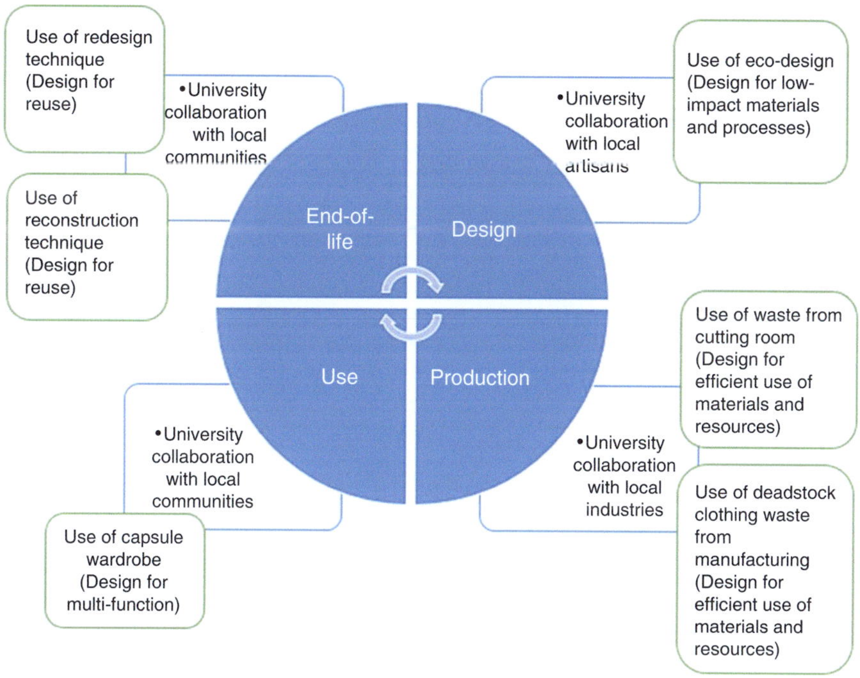

Fig. 3 Conceptual model

Project 1: Use Eco-Design (Design for low-Impact Materials and Processes)

Importance of the use of natural and biodegradable fibers as a solution for sustainability in fashion and textile industry has garnered considerable amount of research mainly in the areas of substituting synthetic fibers with biodegradable fibers. The project (Fig. 4) in collaboration with the local weavers is an upshot, owing to the reason that students are motivated to explore the materials available in the local area and make a choice-decision of selecting the material based on the social impact and the environmental impact of the material. The design students of the project explored the local region to understand the possible use of close by available sustainable fiber along with the processing of the fiber to yarn and to fabric. The extract of the interview is tabulated and presented in Table 2.

The fiber used for the project involves the use of pseudo-stems of banana plant, which are generally thrown away by the local farmers after harvesting. The literature hints that banana farming generates several tons of biomass, most of which goes as waste due to non-availability of suitable technology for its commercial utilization (Sharma and Sunita 2014). The fiber extracted from banana pseudo-stem is blended with cotton (25:75) and woven into fabric with the involvement of local weavers who otherwise are associated with weaving of locally available material called "*neer panche*" commonly used for religious use as dhoti in the local area.

Fig. 4 Eco-design

Table 2 Design stage

Variables	Input (Brainstorming by students)	Output (Response from the exploratory study)	Design strategy followed
Livelihood status of weavers	What are the products woven?	Weavers in the regional area weave "*neer panche*" and handloom sarees	**Design for low- impact materials and process**
	What is the income generated through weaving?	One saree requires a day to be woven. We earn Rs. 300–500/day	
Locally available fibers (sustainable fibers)	What is the fiber generally used?	We receive the yarn used for weaving the products from the cooperative society	
	What are the other sources of fiber that can be used for apparel?	Dakshina Kannada is known for production of areca nut and banana	
		The areca nut fiber, as per the literature reviewed, is generally used for development of non-woven fabric,whereas the fibers extracted from banana pseudo-stem can be explored for apparel	

The design students as a part of the market research explored the production and supply status of the material "*neer panche*," which highlighted on the limited requirement, thus leading to the low-income status of the weavers. The project aided as a means of contributing to the ecological sustainability (use of natural waste resource in developing the fiber) and economical sustainability (involvement of the weavers for economic benefit) in a very subtle way. The aesthetics of the garment is addressed by the design of reversible garment, which adds to the possibility of multiple use of the garment. The value addition to the garment is the addition of the Toda embroidery documented by the students as a part of curricular requirement of the program.

Production Stage: The production stage involves a generic sequence of events that includes sketching of a garment, developing paper patterns and toile, and preparation of marker, the full sample range being cut and made (Gwilt 2014). These different stages of production yield wastage at different levels (Rathinamoorthy 2018):

Production wastes: These include the leftover items in the apparel industry such as trims, leftover fabrics, proofs, etc.

Pre-consumer wastes: These include the materials that are discarded before it is ready for consumer use.

Post-consumer wastes: These include the apparel or home textiles that an individual does not require anymore and chooses to dispose them.

This section is oriented toward the pre-consumer waste synonymously termed as pre-consumer textile wastes (PrCTW), which are those wastes that originate with the manufacturers and never make it to the consumers. The examples include spinning waste, weaving waste, fabric cutting waste, apparel manufacturing waste, etc. (Vadicherla et al. 2017). The projects in this section are an interpretation of pattern-making and cutting to develop garments from pre-consumer waste generated in the local industries. The design students of the project explored the local manufacturing units to understand the different types of pre-consumer waste generated in the production of a garment. Table 3 is the consolidated extract of the interview with the industries.

The design students who conducted interviews with the local manufacturing units concluded that the industry generally produces pre-consumer waste (the waste generated in the cutting room) and dead stock clothing waste (due to overproduction or end-of-roll fabric). The projects developed utilizing these wastes are inspired from the brand "Doodlage" which envisions the upcycling of the industrial waste collected from the large garment and fabric manufacturers.

Project 2: Use of Waste from Cutting Room (Design for Efficient Use of Materials and Resources)

Cutting room in the apparel industry is considered as one of the important departments in terms of waste generation. There are two types of losses in cutting process—marking loss (waste fabric that is present in the un-usable area of the marker)

Table 3 Production stage

Variables	Input (Brainstorming by students)	Output (Response from the exploratory study)	Design strategy followed
Types of waste generated	What are the different types of waste generated in the manufacturing unit?	Waste generated from different segments of the industry:	**Design for efficient use of materials and resources**
		Pre-production: Fabric inspection, trims, proofs, etc.	
		Production: Left over roll (after spreading), cutting loss (marker efficiency of 85%).	
		Post-production: Left over garments (mismatch in print, excess production), sometimes dead stock (fashion unsold).	
		Pre-production: The fabric with major defect is returned to the supplier. Such fabrics are sold to the local vendors for lesser cost than the original price.	
	How does the industry respond to the waste generated?	Production: Left over rolls are laid separately for cutting smaller segments of the garment. The cutting loss is given to the collection centers who use the same for making door mats, stuffing for pillows, mattress, etc.	
		Dead stock is put out at the industry for second sale. A few of them are donated to orphanage.	

and spreading loss (waste fabric outside the marker) (Rathinamoorthy 2018). The same is evident in the institutes teaching fashion design where a lot of material is wasted due to improper lay and cutting. Creative approach at patternmaking and cutting is required to address the sustainability issues at the production stage.

The project (Fig. 5) is an example of upcycling fashion that involves collecting the waste generated, sorting the waste, and creating the new product with added value and life. For the project, the students used the following materials:

(a) Denim waste generated in the cutting room from the apparel manufacturing unit.
(b) Indigo-dyed Kalamkari waste generated in the studio at the university.

The design development of reversible jacket aimed at utilizing the waste to production requires the designers to draft the pattern and arrange the denim and indigo strips accordingly. The attachment of the strips involved use of overlock machines with contrasting thread to aid in reversing of the garment. The denim strips bleached and dyed using tea and coffee in different concentration augment the aesthetics of the garment.

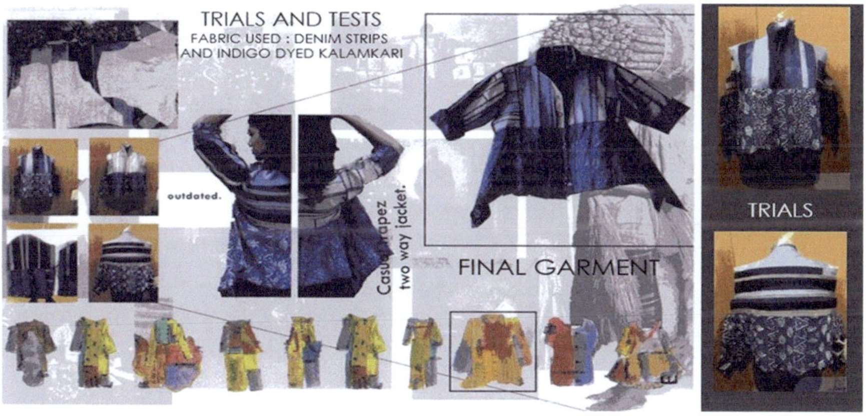

Fig. 5 Design from cutting room waste

Project 3: Use of Dead Stock Clothing Waste from Manufacturing (Design for Efficient Use of Materials and Resources)

Dead stock clothing waste is generated due to improper stocking, mismatch in the prints, excess production, and unsold fashion products, all of which contribute to the generation of waste from the apparel industry. These goods instead of being moved to the landfills could be put to reuse, remanufacture, and recycle so that the value can be re-generated (Choi and Li 2015).

The project (Fig. 6) involving recycling to address the problem comprised of collecting the dead stock from the apparel industry to develop new designs. The design development involved disassembly of the components of the garment to repair, remake, and eventually recycle the material at the end of the cycle. For the project, the XXL sizes that are returned to the industry have been collected and disassembled. The design process followed mix and matching of the garment components aided with patching technique.

Use: Use in design process refers to the care phase of the life cycle of the garment, which includes the knowledge of washing and drying of clothes. However, since the fast fashion is associated with the increased demand from the consumers for new styles and trends, the paper focuses on design strategies that can be used to fulfill the consumer desire for new fashions. The same is justified with lack of social dimension found among the consumers who exhibit continuous desire for fashionable renewal, in order to achieve personal affirmation and to distinguish themselves from others (Cimatti et al. 2017). By designing clothes for multifunction, the social and cultural desire for continuous renewal of costumes can be achieved.

Fig. 6 Design from dead stock clothing waste from manufacturing

Project 4: Use Capsule Wardrobe (Design for Multifunction)

The project (Fig. 7) in collaboration with communities involved the designer sensitizing the communities on the impact of the fast fashions on the society and provides design solutions through multifunctioning garments, without compromising on the style or function. It requires a designer to educate customers on how to build capsule wardrobe based on their preferences. The knitwear collection developed as a studio project is inspired from the works of designers like Tara Marie MacSharry who have designed fashion with multiple purposes.

The project required the designer to understand the client requirements and innovate designs that can be worn in multiple ways by adding or editing the look of the garment. The clients for the project have been adolescents in the age group of 18–22 years who are eager to experiment with new changes in fashion. The design

Fig. 7 Design using concept of capsule collection

collection involved the use of knitwear module in developing the range of garments that could be worn in multiple ways.

End of Life

Fast fashions, which have grabbed the industry, have mainly strived on the desire of the younger fashion customers who desire for the new trends at the shorter duration (Morgan and Birtwistle 2009) that has led to increased production and increased disposal of the garments (Joung 2014). The environmental issues involved with fast fashions not only include the ecological footprint created from the textile industries in the production process but also the waste generated due to disposal of clothing by the consumers. These clothing disposal decisions have long-term ecological effects because the decomposition process of certain materials may take many years. Growing increase in textile and clothing waste associated with fast fashions has led to the development of techniques of reuse and recycling, which is sometimes termed as eco-efficiency approach. However, these techniques have faced critique that they do not provide solutions to real sustainable problems associated with increased consumption and growing waste problems of the textile production and social sustainability problem of textile and clothing industry (Niinimäki and Hassi 2011). The projects in this section, in collaboration with local communities are designed to address the sustainable issues of garments being thrown away due to multiple reasons such as fitting issues, being worn-out, out of fashion, etc. The local communities for the project are encouraged to drop in used garments that included the sarees, kurtis, shirts, denims, etc. The projects focused on using the shirts and denims, which are considered to be classical garments with continuous life cycle.

Project 5: Use of Reconstruction Technique (Design for Reuse)

The project (Fig. 8) involved the local communities to drop in unused shirts, which are disassembled and reconstructed into unique pieces. These techniques help the consumers' desire for variety and change and even accommodate complete reassembly of the old garments into an entirely new product.

Project 6: Use of Redesign Technique (Design for Reuse)

The project (Fig. 9) involved the local communities being encouraged to drop in the worn-out jeans to be designed into new garments using redesign technique. Redesign requires deconstruction and reconstruction of a garment, which is a more involved process than alterations, a tailor might make to improve a garment's fit. Redesign could vary by the extent of the garment's change, from adding minor design detail such as a decorative trim, to the change of the garment's silhouette such as attaching a peplum, and to the complete transformation of the garment's original purpose such as changing from a dress to a top (Janigo and Wu 2015). Figure 8 is the designer interpretation of redesigning the denim trousers to jumpsuit.

Insight on the Sustainable Projects of Students

Fashion design education at MAHE follows a triangular approach for developing core competencies, which involves an integration of cognitive competencies (to think), experiential competencies (to explore), and technical competencies (to build). The projects undertaken by the students helped in exploring the sustainable problems related to the fashion industry, brainstorming different sustainable design strategies, and the use of the same for building the sustainable design solutions. The projects undertaken have posed opportunities and challenges in working within the framework of fashion education.

The sustainability projects involving manufacturing units posed a challenge of creating platform wherein the students interact with the industry workforce. Interaction with the professionals from the industry in understanding the production process and the sustainability-related issues helped the design students connect with

Fig. 8 Design using reconstruction technique

Fig. 9 Design using redesign technique

reality. The students, through design and development of the projects, gained an insight into the real-world problems associated with different segments of the fashion industry. The students reported a sense of purpose and exhibited willingness of engaging themselves in coming up with applicable sustainable solutions to the existing problems. The manufacturing units surveyed had their own methods of disposal of the waste generated that encouraged cradle-to-cradle concept wherein the waste from the cutting room became the raw material for another process. However, the involvement of the students in designing from the waste generated in the cutting room provided a new way of looking at solving the sustainability problems. The involvement of industries often commonly seen in engineering streams with industries sponsoring capstone projects can add value and serious participation of the faculty and students in designing sustainable solutions.

The sustainability projects involving the local weavers requires understanding of the working conditions of the units, willingness of the weavers to engage in sustainable projects, and student's ability to experiment with sustainable fibers and processing techniques. The projects contributed to improvement of the livelihood of the weavers engaged in a small way, and they can be successful when the economical sustainability of the weavers will be addressed through large-scale of projects.

The sustainability projects involving the customer can be achieved only by building awareness about the ecological problems posed by the fast fashion and by engaging the willing audience in designing and using sustainable fashion. Workshops on capsule collection and the ways of upcycling the old garments help engage the customers in sustainable fashion consumption.

The main limitation of the projects is the customer acceptance and feedback on the sustainable designs developed. However, the aim of the paper is to demonstrate that the series of collective small actions from the designers and the stakeholders can bring change within the reach of all.

Conclusions

The role of educational institutions in generating synergies to address ecological issues is irrefutable. Providing platforms in education to explore sustainable design projects prepare the designer with the sensitivity for economic and ecological sustainability. As an attempt towards the same, this paper has set out the projects done by the fashion design students of MAHE, addressing the problems faced by the fashion industry. The inclusion of the projects with practical constraints has helped the design students to master design skills of observing, analyzing the problem, design thinking, making, and doing. It helped the students to stay connected to the industries, local artisans, and customers. The projects strongly emphasized on the involvement of stakeholders and the designers for successful implementation of the design strategy. It is hoped that this document contributes to the generation of ideas and discussion among the educational practitioners involved in inculcating design educational practices in the field of fashion design.

Acknowledgments The authors place on record the support received from the students of Manipal Academy of Higher Education for their contribution in design development and documentation:

Design Courtesy—Ms. Nikhila Anandan and Ms. Aysha Shaima.

Modelling Courtesy—Ms. Anila Alexander.

Photo Courtesy—Mr. Janardan Massar and Mr. Sharvan K Acharya.

The authors also thank the local weavers and manufacturing units for their support in developing the projects.

References

Anguelov N (2015) The dirty side of the garment industry: fast fashion and its negative impact on environment and society. CRC Press, Boca Raton

Bhardwaj V, Fairhurst A (2010) Fast fashion: response to changes in the fashion industry. Int Rev Retail Distrib Consum Res 20:165–173. https://doi.org/10.1080/09593960903498300

Bick R, Halsey E, Ekenga CC (2018) The global environmental injustice of fast fashion. Environ heal a glob access. Sci Sour 17:1–4. https://doi.org/10.1186/s12940-018-0433-7

Binet F, Coste-Manière I, Decombes C, Grasselli Y, Ouedermi D, Ramchandani M (2019) Fast fashion and sustainable consumption. In: Muthu SS (ed) Fast fashion, fashion brands and sustainable consumption. Springer Singapore, Singapore, pp 19–35

Boggon C (2019) How polluting is the fashion industry. https://www.ekoenergy.org/how-polluting-is-the-fashion-industry/.

Buzzo A, Abreu MJ (2018) Fast fashion, fashion brands & sustainable consumption. In: Muthu SS (ed) Fast fashion, fashion brands and sustainable consumption. Springer Singapore, Singapore, pp 1–17

Caro F, Martìnez-De-Albèniz V (2015) Fast fashion: business model overview and research opportunities. Int Ser Oper Res Manag Sci 223:237–264. https://doi.org/10.1007/978-1-4899-7562-1_9

Ceschin F, Gaziulusoy I (2016) Evolution of design for sustainability: from product design to design for system innovations and transitions. Des Stud 47:118–163. https://doi.org/10.1016/j.destud.2016.09.002

Chen T, Peng L (2013) Nature-inspired fashion design through the theory of biomimicry. https://www.academia.edu/5119373/Nature-inspired_Fashion_Design_through_The_Theory_of_Biomimicry. Accessed 30 June 2019

Choi TM, Li Y (2015) Sustainability in fashion business operations. Sustain 7:15400–15406. https://doi.org/10.3390/su71115400

Cimatti B, Campana G, Carluccio L (2017) Eco design and sustainable manufacturing in fashion: a case study in the luxury personal accessories industry. Procedia Manuf 8:393–400. https://doi.org/10.1016/j.promfg.2017.02.050

Claudio L (2007) Waste couture: environmental impact of the clothing industry. Environ Health Perspect 115(9):A449–A454

Cuc S, Vidovic M (2014) Environmental sustainability through clothing recycling. Oper Supply Chain Manag An Int J 4:108. https://doi.org/10.31387/oscm0100064

D'Souza C (2015) Marketing challenges for an eco-fashion brand: a case study. Fash Theory - J Dress Body Cult 19:67–82. https://doi.org/10.2752/175174115X14113933306824

Drasković N, Marković M, Petersen C (2018) The early days of swatch: birth of the fast fashion watch business model. Market-Tržište 30:93–108. https://doi.org/10.22598/mt/2018.30.1.93

Fletcher K (2011) Slow fashion: an invitation for systems change. Fash Pract 2:259–265. https://doi.org/10.2752/175693810x12774625387594

Fletcher K, Williams D (2013) Fashion education in sustainability in practice. Res J Text Appar 17:81–88. https://doi.org/10.1108/RJTA-17-02-2013-B011

Gupta C, Vaid N, Jain A (2016) Recycling pre-consumer textile waste using water soluble film technology for promoting environmental sustainability. https://www.ijsr.net/archive/v5i11/ART20162750.pdf. Accessed May 2019

Gwilt A (2014) A practical guide to sustainable fashion. Fairchild, London

Gwilt A, Rissanen T (2012) Shaping sustainable fashion: changing the way we make and use clothes. Earthscan, London

Hobson J (2013) To die for? The health and safety of fast fashion. Occup Med (Chic Ill) 63:317–319. https://doi.org/10.1093/occmed/kqt079

Janigo KA, Wu J (2015) Collaborative redesign of used clothes as a sustainable fashion solution and potential business opportunity. Fash Pract. https://doi.org/10.2752/175693815X14182200335736

Joung HM (2014) Fast-fashion consumers' post-purchase behaviours. Int J Retail Distrib Manag 42:688–697. https://doi.org/10.1108/IJRDM-03-2013-0055

Kaikobad NK, Bhuiyan ZA, Sultana F, Mahmudur Rahman (2015) Fast fashion: marketing, recycling and environmental issues. Int J Humanit Soc Sci Invent ISSN (Online 2319–7722)

Kozar JM, Connell KYH (2013) Socially and environmentally responsible apparel consumption: knowledge, attitudes, and behaviors. Soc Responsib J 9:315. https://doi.org/10.1108/SRJ-09-2011-0076

Lorenzoni G (2016) The fast fashion business model. In: Krogstie J, Mouratidis H, Su J (eds) Advanced information systems engineering workshops. Springer International Publishing, Cham, pp 65–71

Maxwell D, Van der Vorst R (2003) Developing sustainable products and services. J Clean Prod 11:883–895. https://doi.org/10.1016/S0959-6526(02)00164-6

Morgan LR, Birtwistle G (2009) An investigation of young fashion consumers' disposal habits. Int J Consum Stud 33:190. https://doi.org/10.1111/j.1470-6431.2009.00756.x

Niinimäki K (2017) Fashion in a circular economy. In: Henninger CE, Alevizou PJ, Goworek H, Ryding D (eds) Sustainability in fashion: a cradle to upcycle approach. Springer International Publishing, Cham, pp 151–169

Niinimäki K, Hassi L (2011) Emerging design strategies in sustainable production and consumption of textiles and clothing. J Clean Prod 19:1876–1883. https://doi.org/10.1016/j.jclepro.2011.04.020

Obregón C (2012) Cumulus sustainable fashion education: From trend to paradigm ? Paper presented at Cumulus conference, Cumulus the international association of universities and colleges of art, design and media. Santiago, Chile. https://www.academia.edu/2052716/Sustainable_Fashion_Education_From_Trend_to_Paradigm. Accessed 12 June 2019

Pasricha A, Kadolph SJ (2009) Millennial generation and fashion education: a discussion on agents of change. Int J Fash Des Technol Educ 2:119. https://doi.org/10.1080/17543260903390183

Peterson JH, Chang J, Wong Y-H, Lawrence CA (2010) The emergence of the fast fashion business model and imposed quick response challenges for chinese fabric manufacturers. In: Cheng TCE, Choi T-M (eds) Innovative quick response programs in logistics and supply chain management. Springer, Berlin, pp 387–422

Rathinamoorthy R (2018) Sustainable apparel production from recycled fabric waste. In: Muthu SS (ed) Sustainable innovations in recycled textiles. Springer Singapore, Singapore, pp 19–52

Remy N, Speelman E, Swartz S (2016) Style that's sustainable: A new fast-fashion formula. https://www.mckinsey.com/business-functions/sustainability/our-insights/style-thats-sustainable-a-new-fast-fashion-formula. Accessed 14 June 2019

Sharma AK, Sunita C (2014) Utilization of pectinases for fiber extraction from banana plants waste. Int J Waste Resour 04:4. https://doi.org/10.4172/2252-5211.1000162

Sumner M (2019) Can the fashion industry ever really be sustainable?. http://www.thefashionlaw.com/home/can-the-fashion-industry-ever-really-be-sustainable.

Tischner U, Charter M (2013) Sustainable solutions: developing products and services for the future. Greenleaf, Sheffield

UNEP (2018) Putting the brakes on fast fashion. https://www.unenvironment.org/news-and-stories/story/putting-brakes-fast-fashion.
Vadicherla T, Saravanan D, Muthu Ram M, Suganya K (2017) Fashion renovation via upcycling. In: Muthu SS (ed) Textiles and clothing sustainability: recycled and upcycled textiles and fashion. Springer Singapore, Singapore, pp 1–54
Vehmas K, Raudaskoski A, Heikkilä P, Harlin A, Mensonen A (2018) Consumer attitudes and communication in circular fashion. J Fash Mark Manag 22:286–300. https://doi.org/10.1108/JFMM-08-2017-0079
Vezzoli C, Manzini E (2008) Design for environmental sustainability. Springer, London
Yao L, Zhang J (2018) research on the development status and strategy on sustainable fashion of fast fashion brands. In: Proceedings of the 2018 international conference on energy development and environmental protection (EDEP 2018), Oct 2018

Index

S. S. Muthu, M. A. Gardetti (eds.), *Sustainability in the Textile and Apparel Industries*, Sustainable Textiles: Production, Processing, Manufacturing & Chemistry, https://doi.org/10.1007/978-3-030-37929-2

MIX
Papier aus verantwortungsvollen Quellen
Paper from responsible sources
FSC® C105338
FSC
www.fsc.org

If you have any concerns about our products,
you can contact us on
ProductSafety@springernature.com

In case Publisher is established outside the EU,
the EU authorized representative is:
Springer Nature Customer Service Center GmbH
Europaplatz 3, 69115 Heidelberg, Germany

Printed by Libri Plureos GmbH
in Hamburg, Germany